Electromagnetic Foundations of Electrical Engineering

Electromagnetic Foundations of Electrical Engineering

J. A. Brandão Faria

Instituto Superior Técnico – Technical University of Lisbon, Portugal

WILEY

A John Wiley and Sons, Ltd, Publication

Registered office
John Wiley & Sons Ltd, The Atrium, Southern Gate, Chichester, West Sussex, PO19 8SQ, United Kingdom

For details of our global editorial offices, for customer services and for information about how to apply for permission to reuse the copyright material in this book please see our website at www.wiley.com.

Library of Congress Cataloging in Publication Data

Faria, J. A. Brandão, 1952–
Electromagnetic foundations of electrical engineering / by J. A. Brandão Faria.
p. cm.
Includes bibliographical references and index.
ISBN 978-0-470-72709-6 (cloth)
1. Electric engineering. 2. Electromagnetic fields. 3. Electromagnetism.
I. Title.
TK146.F37 2008
621.3—dc22

2008016867

British Library Cataloguing in Publication Data

A catalogue record for this book is available from the British Library

ISBN 978-0-470-72709-6 (H/B)

Set in 10/12pt Times by Integra Software Services Pvt. Ltd, Pondicherry, India
Printed and bound in Great Britain by CPI Antony Rowe, Chippenham, Wiltshire

To the memory of my wife,
Fernanda Faria

'Let there be light', and there was light
Genesis 1:3

$$\begin{cases} \operatorname{curl} \mathbf{E} = -\dfrac{\partial \mathbf{B}}{\partial t} \\ \operatorname{div} \mathbf{B} = 0 \\ \operatorname{curl} \mathbf{H} = \mathbf{J} + \dfrac{\partial \mathbf{D}}{\partial t} \\ \operatorname{div} \mathbf{D} = \rho \end{cases}$$

Contents

About the Author

J. A. Brandão Faria received his PhD degree and aggregate title in electrical engineering from the Instituto Superior Técnico of the Technical University of Lisbon, where, since 1994, he has been a Full Professor of Electrical Engineering teaching undergraduate and graduate courses in electromagnetics. His teaching activities also include lecturing courses at the Portuguese Air Force Academy.

Professor Brandão Faria was a senior scientist with the former Centro de Electrotecnia Teórica e Medidas Eléctricas where he served as President from 1994 until 2000. Since 2008, he has been at the recently created CIEEE (Center for Innovation in Electrical and Energy Engineering).

Professor Brandão Faria was the recipient of two Honorable Mentions awarded in 1994 and 2007 by the Portuguese State Department of Science and Technology, and by the Technical University of Lisbon, respectively. His areas of interest include electromagnetic field problems, power lines, and wave propagation phenomena in multiconductor transmission lines. He is the author of two books on electrical engineering subjects, namely *Optica* and *Multiconductor Transmission-Line Structures*, and has published over 100 technical papers.

Professor Brandão Faria is a member of the Editorial Board of the *European Transactions on Electrical Power* (John Wiley & Sons, Ltd) and a senior member of the IEEE.

For more details follow the author's website link at:

https://fenix.ist.utl.pt:443/homepage/ist11545.

Preface

This book has been written bearing in mind not only my own students but also electrical engineering students in general, including European students now facing the challenges of the Bologna Reform.

The primary goal of this textbook, *Electromagnetic Foundations of Electrical Engineering*, is to provide undergraduate students taking courses in electrical engineering with a scientifically founded and unified basis of fundamental knowledge on electromagnetic field phenomena, which will enable them to grasp advanced topics and specialized applications that will be dealt with later in their courses, or that they will come across in their professional lives as engineers.

Several distinguishing features make this new textbook unique in its area. It is primarily a balanced foundations book with a broad scope. The emphasis is on basic principles, concepts and governing laws that can be used precisely by electrical engineering students pursuing studies in areas as diverse as power and energy systems, telecommunications, electronic circuits, control systems, bioengineering, etc. In order to reach and serve as large a readership as possible, bias towards specific areas has been deliberately avoided. Electrical engineering professionals (practitioners) with a need for a refresher course in electromagnetic foundations will also find the book a valuable asset.

A project-solving oriented posture is adopted to capture more easily the reader's interest. However, it is not my intention to provide ready-made recipes or rote procedures for students; my approach emphasizes problem solving as a thought process based on concepts and on concept linking. Right at the beginning of the book, a project portfolio is proposed and offered to students in order to capture their attention and trigger their curiosity (project solutions will be available separately). These projects tie together a diversity of knowledge components whose roots lie in different chapters in the text; this salient feature, it is hoped, will help readers understand the big picture, avoiding segmented perspectives. The key idea is to enable students' knowledge integration skills so that, after completing the book, they can solve the various problems and questions included in the proposed project portfolio. When they do, both the students and the book will have accomplished their goals.

In addition, in all chapters, several fully worked-out application examples are presented to illustrate the theory and concepts that have just been introduced and developed. End-of-chapter homework problems, intended to help guide students in their learning process, are also included; these problems are of practical interest and focus on engineering applications.

The material covered in the book is assumed to be taught in the fourth or fifth semester of the first cycle of studies leading to a Master's degree. Its content, its smooth build-up, as well as its presentation and style, make it suitable for adoption in any top-tier university in the world.

The topics addressed in the textbook are confined to the teaching/learning cycle of a single semester, before which students are supposed to have already acquired the necessary basic skills in and knowledge of both mathematics and physics. Therefore, given the allotted time limitations, a very judicious choice of not only the subject matter, but also its methodological presentation, becomes an imperative and difficult task. In addition to these time limitations (typical lecture times do not globally exceed 40 hours per semester), another challenge faced by this book concerns the average preparation background of students. Although the panorama may change from country to country, my own teaching experience indicates that a great deal of caution may be needed. Taking it for granted that students have already mastered key concepts in electromagnetism can be the first step to failure. Even worse, I have seen in many cases that students have been exposed to some misleading ideas, meaning that additional efforts aimed at deconstructing some preconceived or pre-acquired concepts cannot be avoided for the sake of a sound lifelong preparation.

The book is organized into four parts containing several chapters. The starting point is Maxwell's equations. From them, the fundamental laws and principles governing static and time-varying electric and magnetic fields are derived. Results are subsequently particularized for slow time-varying electromagnetic field problems (steady-state sinusoidal circuit analysis and transient phenomena) and for rapid time-varying electromagnetic field problems (electromagnetic waves and transmission-line theory).

The presentation of the book's subject matter starts with very simple phenomena and proceeds, chapter by chapter, to consider progressively difficult topics. Although the material is arranged in traditional chapter form, a unique approach with this book is that its topics are not tightly compartmentalized. Matter belonging to advanced chapters is frequently built upon preceding topics, taking advantage of existing similarities among the governing equations and making use of contact points that may exist among different concepts. This approach not only contributes to a unified vision of the book's content, but also allows students to correlate apparently distinct topics, enabling them to develop a correct frame of thought where knowledge integration is a prominent objective.

Students using this book are expected to attain a level of competence that will enable them easily to follow up advanced classes taught subsequently in their courses, namely electromagnetic waves, radiation, antennas, microwaves, optics, instrumentation and measurement, electromagnetic compatibility, electrical machines, power systems analysis, etc.

The subjects dealt with throughout the book obey the school of thought traditional to top-tier universities: rigorous concepts, solid ideas, clear introduction of approximations, use of deductive methodology and rejection of ready-made recipes. This approach is seasoned with a friendly presentation of all topics aimed at drawing the students' attention to the central issues under discussion. In addition, special emphasis is placed on the examination and criticism of a few aspects where wrongly preconceived ideas are suspected to exist.

Formal demonstrations of certain results and theorems are absolutely necessary; however, in some instances they will be avoided or alleviated whenever possible. In fact, some results will be derived by simply invoking duality principles or by making use of existing analogies with previously treated subjects; this methodology not only saves time, but also contributes yet again to the global goal of knowledge integration.

The organization and style of this book reflect my experience as a faculty member of the Instituto Superior Técnico (IST)[1] – the School of Engineering of the Technical University of Lisbon.

[1] The IST is a European higher education establishment belonging to the CLUSTER (Consortium Linking Universities of Science and Technology for Education and Research) network, which includes Universitat Politècnica de Catalunya – Barcelona, Technische Universität Darmstadt, Eindhoven University of Technology, Institut National Polytechnique de Grenoble, Universität Karlsruhe (TH), École Polytechnique Fédérale de Lausanne, Imperial College London, Université Catholique de Louvain, Kungl Tekniska Högskolan (KTH) – Stockholm, Politecnico di Torino, and TKK Teknillinen Korkeakoulo (formerly Helsinki University of Technology), as well as Ecole Polytechnique Montreal (Canada), Georgia Tech (USA), Tomsk Polytechnic University (Russia) and Tsingua University Beijing (China).

For Electrical Engineers/Practitioners

As mentioned above, the main target of this book is the university population (students and faculty staff) concerned with electrical engineering studies. However, it is widely recognized today that any practitioner must keep up to date with new developments in their area of expertise. This means lifelong learning. Despite a university education, the skills and knowledge acquired there by practitioners may be insufficient for a professional career spanning several decades. What is more, corporate policies regarding employment quite often impose very rapid and dramatic changes on the tasks assigned to employees, who in many cases have to retrain for a new line of work.

Electrical engineering technology is evolving at a very rapid pace in almost all of its branches and, therefore, it is virtually impossible for anyone to be able to stay on top of all of its novel developments (which, they too, will soon become obsolete). The only thing that really remains stable and imperishable is the foundations of electrical engineering. These foundations can provide practitioners with a refreshing of the key concepts and theories underlying their professional activities, or even open doors to a new start in a different area.

Such readers will find in the book not only the necessary electromagnetic basics, but also a vast collection of useful illustrative application examples and problems that will help them solidify their knowledge. A key feature of this book, which I believe can attract their attention and interest, is a project portfolio that precedes the presentation of the theory. It includes a series of elaborated projects (focusing on engineering problems) that tie together the multiple topics dealt with throughout the book. If, after reading the project portfolio, practitioners feel unsure about how to solve the proposed projects, then they will have a strong additional reason to acquire this book.

For Students

The material presented in the book is built on a substrate of knowledge already provided by the basic sciences of mathematics and physics. Students are supposed to be acquainted with certain topics, such as linear algebra, differential equations, integral calculus, vector analysis and complex functions. If students still have difficulties with these topics, they may have to recap them in order to refresh their skills.

This book is not a treatise on electricity and magnetism – its scope is far less ambitious. Its content can be delivered in a single-semester course, and is aimed to provide a scientifically founded and unified basis of fundamental knowledge on electromagnetic field phenomena that will help students follow up more advanced subjects covered in their courses. Topics are introduced in a systematic and friendly manner, proceeding from the simpler to more difficult ones, using a slow build-up process. In addition, a series of application examples and homework problems have been prepared to help students through the learning process. The fact that the book is partitioned into chapters does not imply that some of them can be skipped. Because the subject matter is deeply interrelated, students must try to adhere to the normal chapter sequence, otherwise they may be wasting their time or fail to get an integrated comprehensive view of the electromagnetic phenomena.

At the beginning of the book there is a project portfolio which includes examples of problems that students may encounter during their life in electrical engineering. These projects were conceived so as to merge a variety of knowledge components from different chapters in the text. Students may start by skimming through the proposed project themes just to get an idea of what the book is about and, also, to realize what will have to be learned. Once students have finished reading the book they should return to the project portfolio and try to solve the proposed projects. If they succeed, it will mean that their goals have been accomplished, and they should be confident about their newly acquired skills. Project solving is a well-proven methodology in any learning process. But students should be aware that they themselves, not the instructor, are supposed to do that job. Let us use an analogy to remind students that no one can learn to ride a bicycle by seeing others do it – practice is required to develop the skill.

A final word: do not believe people who say that learning is fun. Learning involves a lot of hard work and persistent effort, especially when the subjects being studied are of an abstract kind. Do not be worried when difficulties arise, do not give up, recap and recap again until the source of difficulty is clearly identified, and then debate the issue with fellow students or ask the instructor for assistance. Further, I will gladly help (brandao.faria@ieee.org). Fun comes only at the end of the process, after many hours of struggle. The moment when students realize they have been empowered with valuable new knowledge and become intellectually richer, then, and only then, will they really experience the feeling of fun.

For Instructors

The failure or success of delivering a university course (any course) relies mainly on the pedagogical skills and scientific preparation of the instructor in charge. A good book can help a lot, but, just by itself, it is not a guarantee of success.

The present book on the electromagnetic foundations of electrical engineering has been conceived in order to assure that its subject matter is presented in a coherent and logical arrangement. In addition, application problems and final work projects have been prepared to guide the students through the learning process.

The content of this textbook has been tested and subjected to proof for many years with thousands of students. I bear witness not only that lectures have been well received and enjoyed by those students, but also that their final exam success rate has been high (around 80 %).

From my own experience three recommendations stand out. When teaching a given subject, the scientific preparation level of the instructor must be several notches above the one that would strictly be required for lectures, otherwise instructors may find it difficult to answer unexpected questions raised by more advanced students. A second aspect has to do with the utilization of audiovisual aids: they should be used very sparingly and prudently, otherwise students may become 'disconnected'. Finally, as far as project solving is concerned, students should be provided with orientation guidelines and have their mistakes corrected, not have problems solved for them, otherwise they will hardly be able to assess their own skills correctly.

Acknowledgements

Now a retired Emeritus Professor with the Technical University of Lisbon, Professor Borges da Silva superbly guided my teaching and research activities for two decades. His thorough and very rigorous approach to electrical engineering subjects made him a role model for me. The writing of this book would never have become possible without the long-lasting influence he has had on me.

I am indebted to the Centro de Electrotecnia Teórica e Medidas Eléctricas – an R&D center for applied electromagnetics with the Instituto Superior Técnico – for the unsparing support I received during the preparation of this work. A special tribute is owed to Ms Idalina Rosa for her generous assistance in the production of the electronic files for the artwork.

A word of appreciation goes to the publisher's staff in general, and to Ms Simone Taylor, Ms Nicky Skinner, Ms Erica Peters, and Ms Alexandra King, in particular, for all their cooperation and help provided during the book's production.

Last but not least, I would like to strongly emphasize my endless heartfelt gratitude and love to my deceased wife Fernanda, who, despite her terminal illness, never failed to back me up in this enterprise.

Project Portfolio

This project portfolio – placed here at the beginning of the book – is a teaser aimed at triggering your curiosity about the subjects to be developed throughout the book.

The portfolio is intended to be a key tool for assessing the skills and knowledge that you should acquire after completing the reading, understanding and comprehending of the contents of this book.

The portfolio includes a series of 20 final-year projects which tie together and interlink the many subjects that you are going to learn and hear about during this semester. While some projects are more focused on a specific theme, others (most of them) encompass and mix several topics from several chapters. The main objective is to provide you with an integrated global view of the many aspects dealt with in the book, aiming to empower you with a sound and lasting grasp of the electromagnetic foundations of electrical engineering.

It is absolutely amazing how a set of only four equations – Maxwell's equations – can be so fruitful as far as their range of applications is concerned. In fact, the same basic tool will allow you to analyze a huge variety of electromagnetic phenomena and electrical engineering devices and systems.

In this portfolio you will be faced with real-life engineering subjects such as rotating induction machines, sound columns, microwave power splitters, and much more.

A list of the proposed projects follows. Their sequence is purely alphabetical in order to avoid any unsubstantiated clues not only on the importance of the proposed themes, but also about their intrinsic difficulties.

Project 1	Analysis of a power delivery system
Project 2	Cylindrical type transmission lines
Project 3	DC transducer
Project 4	Determination of the conductivity of a circular conducting disk
Project 5	Directional coupler analysis
Project 6	Ill-defined grounding problems
Project 7	Induction machine analysis
Project 8	Line-matching technique using an exponential transmission-line section

Electromagnetic Foundations of Electrical Engineering J. A. Brandão Faria

Project 9	Linear variable differential transformer
Project 10	Magnetic actuator and sensor device
Project 11	Overhead-line protection by ground wires
Project 12	Power line carrier communication
Project 13	Pseudo-balanced three-phase lines
Project 14	Screened high-voltage three-phase installation
Project 15	Shielded three-phase cable analysis
Project 16	Three-route microwave splitter
Project 17	Transmission-line system with balun transformer for even- to odd-mode conversion
Project 18	Transmission-line system with transformer-stage matching
Project 19	Two-way loudspeaker analysis
Project 20	Variable reluctance transformer

We hope you find this set of projects challenging and interesting. Enjoy solving them.

You may be wondering about the projects' solutions. They have been prepared and are available. Our advice is that you should work out the solutions by yourself and only as last resource read the answers that have been prepared. Solutions are provided separately at www.wiley.com/go/faria_electromagnetic.

Project P1

Analysis of a Power Delivery System

The high-voltage power delivery system in Figure P1.1 includes a 50 Hz generator, a single-phase transmission line and three industrial consumers.

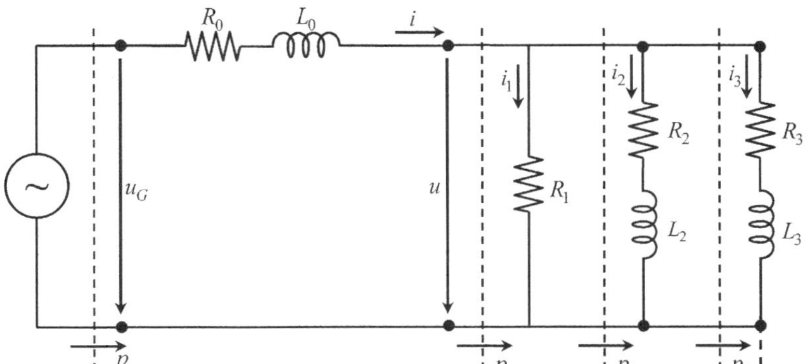

Figure P1.1 Power delivery system comprising a voltage generator, a transmission line and three industrial consumers

The transmission line, represented by its simplified low-frequency equivalent model, is characterized by the following parameters:

length: 200 km; per-unit-length resistance: 0.2 Ω/km; per-unit-length reactance: 1.0 Ω/km

Electromagnetic Foundations of Electrical Engineering J. A. Brandão Faria

Part I Steady-State AC Analysis

The voltage at the far end of the transmission line is given by $u(t) = \sqrt{2}\, U_{rms} \cos(\omega t)$, where $U_{rms} = 110\,\text{kV}$.

Information on the instantaneous powers $p_1(t)$, $p_2(t)$ and $p_3(t)$ is available (see Figure P1.2) where

$$(p_1)_{\max} = 27.207\ \text{MW};\ (p_2)_{\max} = 12.114\ \text{MW};\ (p_3)_{\max} = 8.076\ \text{MW}$$

$$(p_1)_{\min} = -5.323\ \text{MW};\ (p_2)_{\min} = -11.955\ \text{MW};\ (p_3)_{\min} = -7.970\ \text{MW}$$

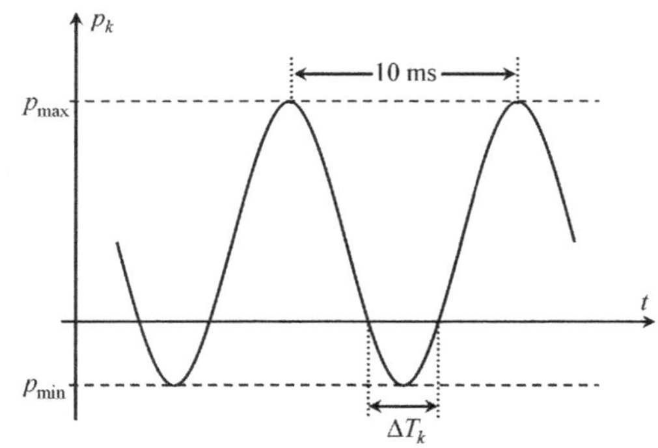

Figure P1.2 Graphical plot of the instantaneous power against time, $p_k(t)$, for $k = 1,\ 2,\ 3$

1. For each power plot $p_k(t)$ determine the corresponding time interval ΔT_k.
2. With the help of the complex Poynting theorem determine the parameters R_k and L_k that characterize each industrial consumer.
3. Obtain the complex amplitudes of u_G, i_1, i_2, i_3 and i.
4. Determine the line power losses.
5. Determine the line voltage drop (as a percentage) defined as

$$\Delta U(\%) = \frac{(U_G)_{rms} - (U)_{rms}}{(U_G)_{rms}} \times 100\%$$

Part II AC Analysis with Power Factor Compensation

In order to minimize the magnitude of the transmission-line current a capacitor of capacitance C is placed at the far end terminals of the line (Figure P1.3).

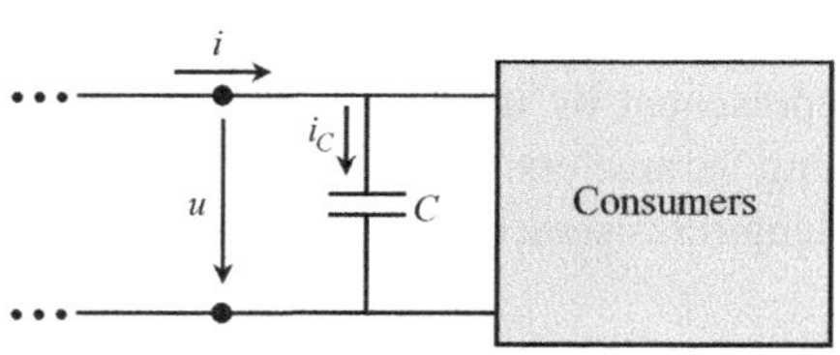

Figure P1.3 A capacitor is placed at the far end of the line for power factor compensation purposes

1. Determine C, and obtain the complex amplitudes of the capacitor current, new line current and new generator voltage.
2. Recompute the line power losses and the line voltage drop, showing that they have decreased appreciably.
3. Consider that the generator voltage is the one that has been determined in 1. Next, assume that consumers 1 and 3 have been disconnected. Determine the overvoltage across the terminals of consumer 2.

Part III Transient Analysis

Take the generator voltage as

$$u_G(t) = \sqrt{2}\ (U_G)_{rms} \cos(\omega t + \pi/2), \text{ with } (U_G)_{rms} = 140\,\text{kV}$$

Consider that the compensation capacitor is absent.

Consumer 1 is permanently out of operation.

Consumers 2 and 3 are initially disconnected (switches S_2 and S_3 are open – Figure P1.4).

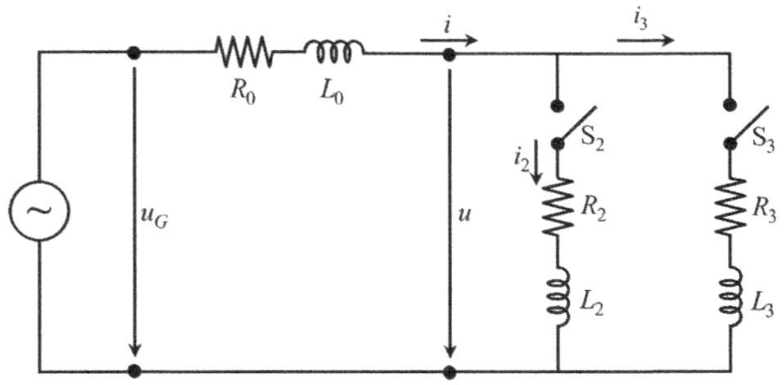

Figure P1.4 Switches S_2 and S_3 are closed at $t = 0$ and $t = t_0$, respectively, giving rise to transient phenomena

1. Consider that consumer 2 is connected to the power line at $t = 0$. Determine the transient regime of $i_2(t)$.
2. Assume that sufficient time has elapsed until $i_2(t)$ can be described by its steady-state regime. Consider that switch S_3 is closed, at $t = t_0$, when i_2 goes through its maximum value. Determine the subsequent transient regimes that describe $i_2(t)$ and $i_3(t)$.
3. For the above situation, comment on the possible event of occurring overcurrents and circuit-breaker activation.

Project P2

Cylindrical Type Transmission Lines

The electrostatic field produced by a symmetrical pair of linear filaments of charge is characterized, in the transverse plane xy, by a family of equipotential lines described by a set of non-coaxial circumferences, and by a family of **E**-field lines described by a set of circumference arcs starting and ending on the positively and negatively charged filaments (which are positioned at $x = \pm a$). The two families of curves are mutually orthogonal. See Figure P2.1.

At a generic point P in space the potential function V and the electric field vector **E** are given by

$$V = \frac{q}{2\pi\varepsilon} \ln \frac{r_2}{r_1}; \quad \mathbf{E} = \frac{q}{2\pi\varepsilon} \left(\frac{\vec{e}_1}{r_1} - \frac{\vec{e}_2}{r_2} \right)$$

where q is the per-unit-length charge density of the positively charged filament, r_1 is the distance between the positive charge and P, r_2 is the distance between the negative charge and P, and $\vec{e}_1$ and $\vec{e}_2$ are radial unit vectors directed to P with origin at the positive and negative charges, respectively. The linear homogeneous dielectric medium is characterized by its permittivity ε and permeability μ.

This basic arrangement of equipotential lines and **E**-field lines can be extremely useful for analyzing a variety of conductor configurations, provided that the conductor surfaces are chosen so as to coincide with existing equipotential surfaces.

Electromagnetic Foundations of Electrical Engineering J. A. Brandão Faria

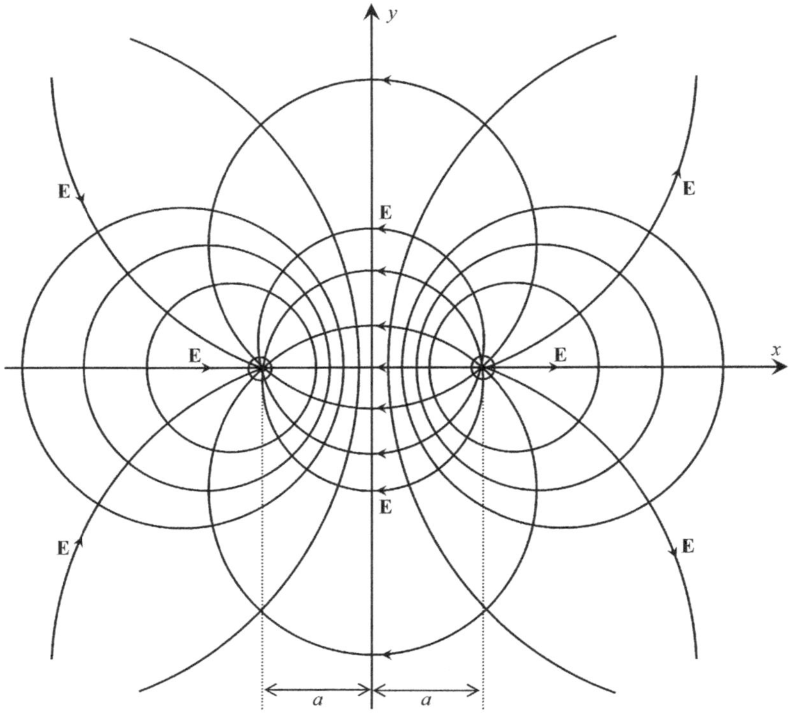

Figure P2.1 Transverse plane view of the electric field lines and equipotential surfaces produced by two parallel symmetrical filaments of charge $\pm q$ positioned at $x = \pm a$

Part I Electrostatic Analysis

1. Consider, as shown in Figure P2.2, that two symmetrical equipotentials are materialized; that is, replaced by conducting surfaces. The configuration obtained describes a symmetrical two-conductor line made of two cylindrical conductors of radius R whose axes are a distance d_{12} apart.

   (a) Given R and d_{12} determine the x coordinates of the conductor axes, as well as the adequate positions for the location of the fictitious filaments of charge, $x = \pm a$.
   (b) Determine the per-unit-length capacitance C of the symmetrical line configuration.
   (c) Determine the per-unit-length external inductance L of the line configuration.

2. Consider, as shown in Figure P2.3, that the left half-plane $x \leq 0$ is filled with a conductor, and that one of the equipotential lines on the right half-plane $x > 0$ is replaced by a cylindrical conductor. The configuration obtained describes a conductor parallel to ground (monopolar line). The radius of the cylindrical conductor is R, and the distance between the conductor axis and the ground is h.

   (a) Given R and h determine the adequate positions for the location of the fictitious filaments of charge, $x = \pm a$.
   (b) Determine the per-unit-length capacitance C of the monopolar line configuration.
   (c) Determine the per-unit-length external inductance L of the line configuration.

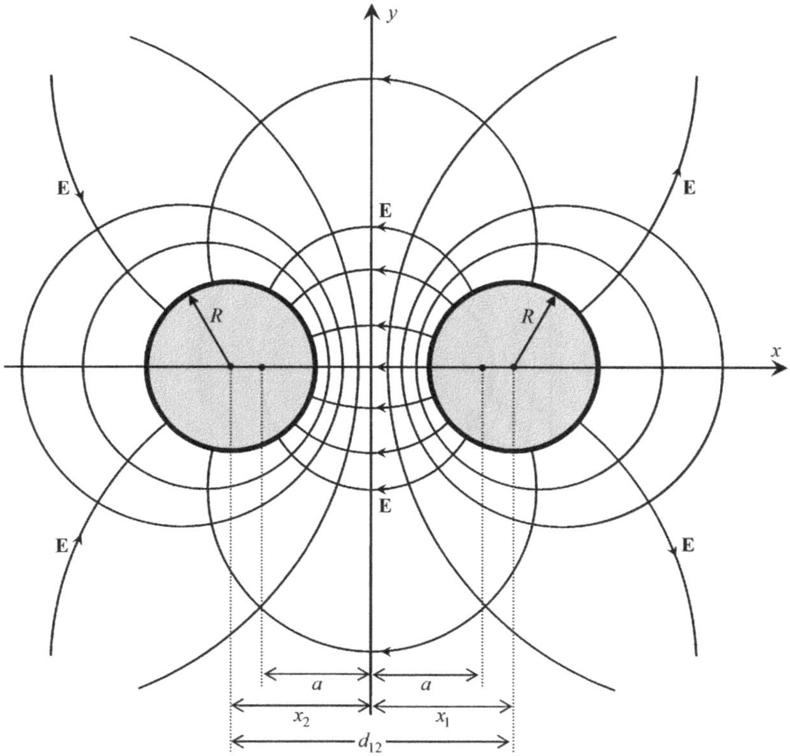

Figure P2.2 Transverse plane view of the electric field lines and equipotential surfaces produced by a symmetrical two-conductor transmission line. The cylindrical conductors, of radius R, have their axes separated by a distance d_{12}

3. Consider, as shown in Figure P2.4, that two asymmetrical equipotentials are replaced by conducting surfaces. The configuration obtained describes an asymmetrical two-conductor line made of two cylindrical conductors of radius R_1 and R_2 whose axes are a distance d_{12} apart.

   (a) Given R_1, R_2 and d_{12}, determine the x coordinates of the conductor axes, as well as the adequate positions for the location of the fictitious filaments of charge, $x = \pm a$.
   (b) Determine the per-unit-length capacitance C of the asymmetrical line configuration.
   (c) Determine the per-unit-length external inductance L of the line configuration.
   (d) Check that the results obtained here are coherent with those obtained in 1 when $R_1 = R_2$.

4. Consider, as shown in Figure P2.5, that two equipotentials, belonging to the right half-plane, are replaced by conducting surfaces. The configuration obtained describes an asymmetrical cable with inner radius R_1 and outer radius R_2. The axes of the two conductors are a distance h apart.

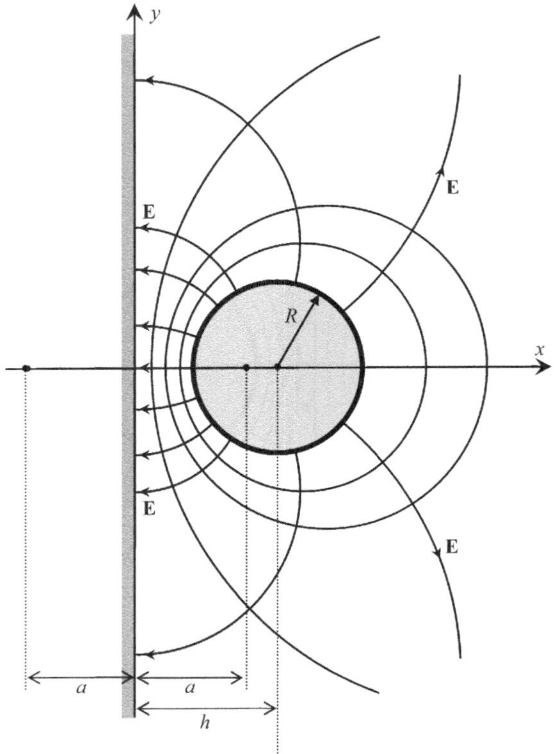

Figure P2.3 Transverse plane view of the electric field lines and equipotential surfaces produced by a cylindrical conductor parallel to a conducting plane. The cylindrical conductor of radius R is a distance h away from the conducting plane

(a) Given R_1, R_2 and h, determine the x coordinates of the conductor axes, as well as the adequate positions for the location of the fictitious filaments of charge, $x = \pm a$.
(b) Determine the per-unit-length capacitance C of the asymmetrical cable configuration.
(c) Determine the per-unit-length external inductance L of the cable configuration.
(d) Check that the results obtained here are coherent with those regarding the (concentric) coaxial cable when $h = 0$.

Part II Transmission-Line Analysis

Assume that all of the transmission-line configurations analyzed in Part I are subjected to a harmonic steady-state regime whose frequency is such that each line is a quarter-wavelength long, $l = \lambda/4$. Assume the lines are lossless, the dielectric medium (air) being characterized by $\varepsilon = \varepsilon_0$ and $\mu = \mu_0$.

All the lines are terminated on a non-matched resistor load $R_L = 300\ \Omega$. Line mismatching is characterized by a load reflection factor $\overline{\Gamma} = 0.5$.

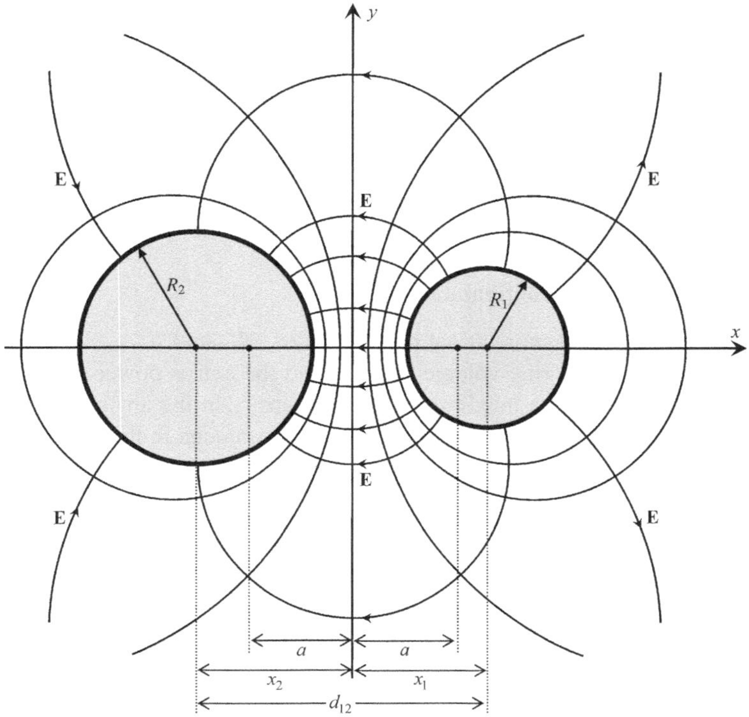

Figure P2.4 Transverse plane view of the electric field lines and equipotential surfaces produced by an asymmetrical two-conductor transmission line. The cylindrical conductors, of radius R_1 and R_2, have their axis separated by a distance d_{12}

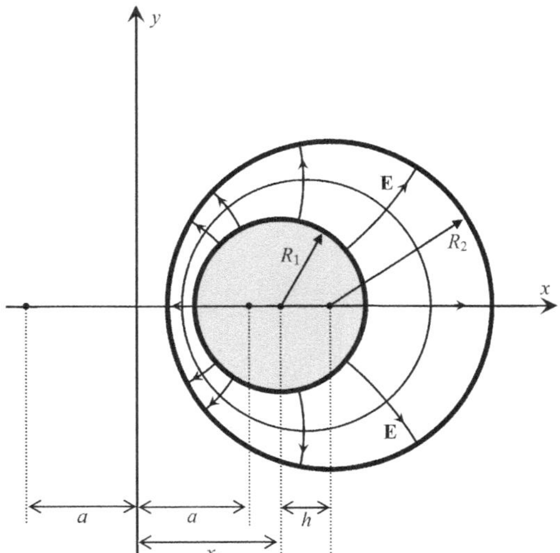

Figure P2.5 Transverse plane view of the electric field lines and equipotential surfaces produced by an asymmetrical cable. The internal and external cylindrical cable conductors, of radius R_1 and R_2, have their axis separated of a distance h

The following parameters are known:

Line 1 (symmetrical two-conductor line): $R = 2\,\text{mm}$.
Line 2 (monopolar line): $R = 2\,\text{mm}$.
Line 3 (asymmetrical two-conductor line): $R_1 = 2\,\text{mm},\ R_2 = 5\,\text{mm}$.
Line 4 (asymmetrical cable): $R_1 = 2\,\text{mm},\ R_2 = 12\,\text{mm}$.

1. Compute the characteristic wave resistance R_w, and the per-unit-length C and L parameters, common to all the transmission lines.
2. For each transmission line configuration:
   (a) Evaluate the unknown geometrical parameters, a, d_{12} and h.
   (b) Evaluate the generator rms voltage $(U_G)_{rms}$ and the active power P delivered to the load, considering that the maximum **E**-field intensity in the air is allowed to reach a threshold of 20 kV/cm (preventing breakdown phenomena in the air from occurring).
3. Given the radii R_1 and R_2 of an asymmetrical cable, together with a specific goal for R_w, can you take for granted that the cable could be designed?
4. Assume that the four line sections are chain connected. Determine the transfer matrix of the chain. Does the result depend on the sequence order of the line sections? Can you permute them?
5. Consider the above chain connection, with the last line section terminated on R_L.
   (a) Taking into account the constraint for the maximum **E**-field intensity referred to in 2(b), find the generator rms voltage $(U_G)_{rms}$.
   (b) Find the active power P delivered to the load.
   (c) Find the maximum **E**-field intensities occurring in each of the line sections.

Project P3

DC Transducer

In this project you will learn how to measure a direct current using saturated transformers.

Figure P3.1 shows the schematic arrangement of a DC transducer, which includes two identical nonlinear supermalloy toroidal transformers, with square cross-section $S = R^2 = 1\,\text{cm}^2$.

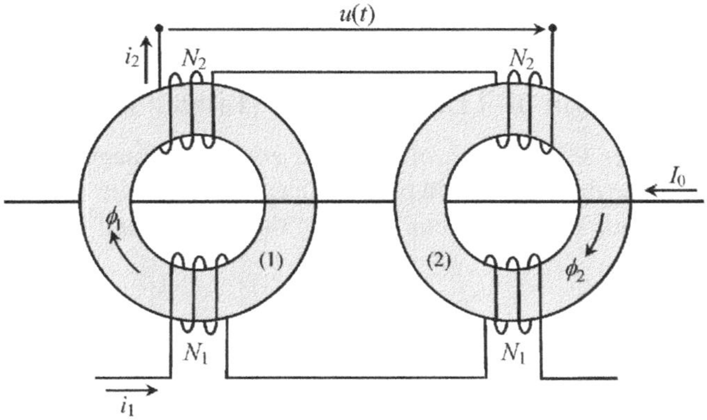

Figure P3.1 Schematic arrangement of a DC transducer made of two nonlinear toroidal cores. The detection voltage $u(t)$ provides information on the current I_0 being measured

I_0 is the direct current to be measured, $i_1(t)$ is an enforced 1 kHz sinusoidal current. The voltage $u(t)$ at the terminals of the detection winding with $2N_2$ turns is monitored, and from it a measure of the unknown direct current is obtained.

Figure P3.2 shows a cross-sectional view of toroidal transformer 1 being utilized, whose size and dimensions are shown. The primary winding with a single layer of N_1 turns is

Electromagnetic Foundations of Electrical Engineering J. A. Brandão Faria

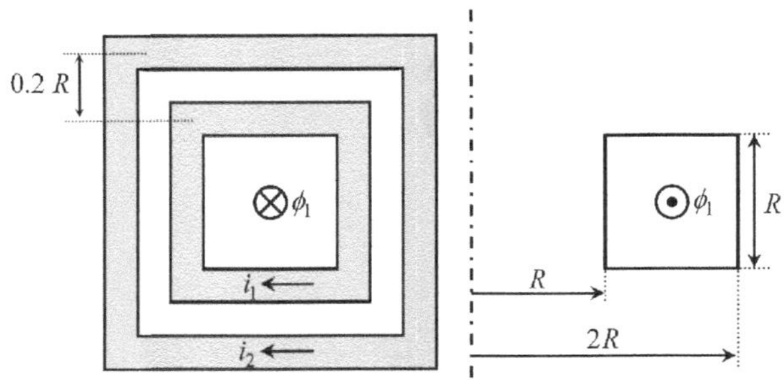

Figure P3.2 Cross sectional view of toroidal core 1

wound directly over the transformer core. The secondary winding with a single layer of N_2 turns is wound around and over the primary winding with an insulation layer left between them. Both windings have a per-unit-length resistance $r_{\text{pul}} = 0.1\ \Omega/\text{m}$ at 1 kHz.

Neglect dispersion field phenomena associated with the primary winding.

The nonlinear magnetic characteristic $B(H)$ of the supermalloy core can be approximated, for small magnetic field intensities ($H < H_{\text{lim}}$, with $H_{\text{lim}} = 7$ A/m), by the following expression: $B = \mu H - \gamma H^3$, where $\mu = 0.25$ H/m and $\gamma = 1.6 \times 10^{-3}$ Hm/A^2. The saturation magnetic induction field is $B_S = 0.6$ T.

Part I Characterization of Transformer 1 (Linear Behavior)

Consider that transformer 1 is working in the linear zone of its magnetic characteristic (for $B \leq B_S$). Assume that the dispersion field phenomena associated with the secondary winding are such that the magnetic coupling factor between the primary and secondary windings is $k = 0.98$.

1. Take $i_2 = 0$.
   (a) Determine the inductance of the primary winding L_{11} knowing that the primary impedance is $\overline{Z}_1 \approx j500\ \Omega$.
   (b) Determine N_1. Find the primary resistance r_1 and check that it is negligibly small compared to Z_1.
   (c) Determine the minimum amplitude $I = (i_1)_{\min}$ of the primary current that leads the core to its saturation point $B_S = 0.6$ T.
2. The number of turns of the secondary winding is $N_2 = 10$. Determine the inductances L_M and L_{22}. Find the resistance of the secondary winding r_2.
3. Consider that the primary current is given by $i_1(t) = 2I\cos(\omega t)$. If the secondary is open ($i_2 = 0$) the core will be saturated. However, if the secondary winding is loaded with a resistor R_L the secondary current will provoke a demagnetization effect whose importance increases with decreasing values of R_L. Determine the maximum value of R_L that prevents the core from going into saturation.
4. Show that the equivalent circuit in Figure P3.3 appropriately simulates the transformer behavior in the linear zone, when the above value for R_L is considered.

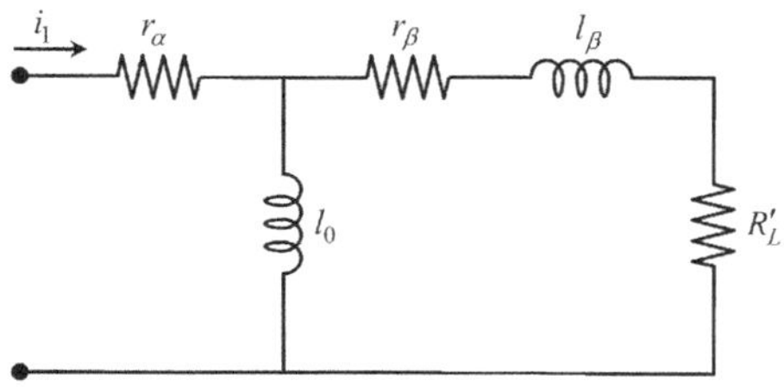

Figure P3.3 Transformer 1 equivalent circuit (for linear operation). The respective parameters are: $r_\alpha = 68.0\,\text{m}\Omega$, $r_\beta = 161.8\,\text{m}\Omega$, $l_0 = 79.58\,\text{mH}$, $l_\beta = 3.28\,\text{mH}$, $R'_L = 299.4\ \Omega$

Part II Transducer Analysis

Consider the configuration shown in Figure P3.1 above, where I_0 is the direct current to be measured. The secondary winding is open ($i_2 = 0$). The excitation current $i_1(t)$ is given by $i_1(t) = I_1 \cos(\omega t)$, with $I_1 = 2I$, therefore assuring that both transformer cores are saturated.

1. Find the magnetic fluxes $\phi_1(t)$ and $\phi_2(t)$.
2. Determine the minimum and maximum direct current intensity I_0 that still allows for the utilization of the approximated expression $B = \mu H - \gamma H^3$. Discuss the influence of the polarity of the direct current.
3. Determine the instantaneous voltage $u(t)$ at the terminals of the detection winding. Show that it is a 2 kHz sinusoidal voltage whose magnitude is proportional to I_0.
4. Determine the sensitivity of the current transducer $S = U_{rms}/I_0$.
5. Knowing that $U_{rms} = 586\,\text{mV}$, find I_0.

Project P4

Determination of the Conductivity of a Circular Conducting Disk

The non-destructive evaluation of the conductivity of materials is an important electrical engineering problem. The technique used to perform such an evaluation is greatly dependent on the geometric shape of the conducting material sample under test. Consider the case of a circular disk of conductivity σ, thickness h and radius r, where four very thin parallel cylindrical electrodes of radius r_0 and height h are bonded onto the disk periphery. The external medium conductivity is $\sigma_{ex} = 0$. The set of four electrodes is symmetrically arranged so that the angle between consecutive electrodes is $\pi/2$.

The current injected into the disk is I.

The voltage between the active electrodes (a) and (a′) in Figure P4.1 is U, whereas the voltage between the passive electrodes (p) and (p′) is U_0.

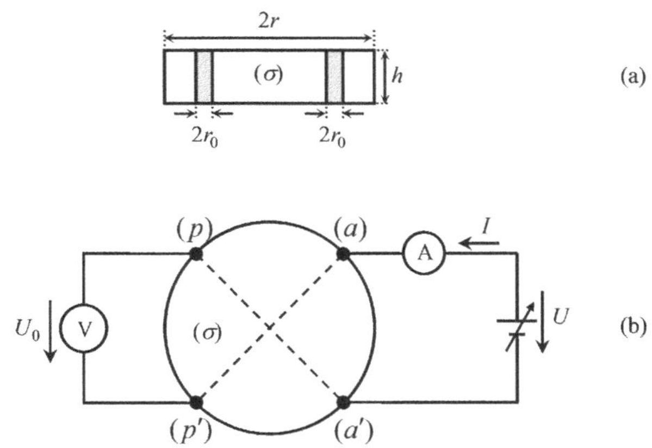

Figure P4.1 Conductivity measurement of a circular disk. (a) Front view of the disk. (b) Top view of the disk and instrumentation apparatus

Electromagnetic Foundations of Electrical Engineering J. A. Brandão Faria

The main goal of this project is to devise a way do determine the disk conductivity σ by measuring I and U_0. The suggested theoretical tool to be utilized is the analogy between electrostatic field equations and stationary current field equations.

Part I Electrostatic Analysis

1. Consider the electrode set immersed in a homogeneous dielectric medium with permittivity ε. Assume that the enforced potential at the active electrodes is $V_a = -V_{a'} = U/2$. Find the electrode charge Q as well as the potential at (p) and (p′).
2. Consider that the homogeneous dielectric medium is partitioned into two areas of different permittivity. The interior of the circle of radius r is characterized by ε_{in} while the region outside is characterized by ε_{ex}. Keep assuming that the enforced potential at the active electrodes is $V_a = -V_{a'} = U/2$. Find the new electrode charge $\hat{Q}$ as well as the potential at (p) and (p′). Particularize the results for the case $\varepsilon_{in} \gg \varepsilon_{ex}$.

Part II Stationary Current Analysis

1. Making use of the existing analogy between electrostatic field equations and stationary current field equations, determine the relationship between I and U_0 (Figure P4.1).
2. Application example. Data: $h = 2\,\text{mm}$, $r = 30\,\text{mm}$, $r_0 = 0.5\,\text{mm}$, $I = 1\,\text{A}$, $U_0 = 1103\,\mu\text{V}$.

   (a) Find σ and identify the material of the disk.
   (b) Determine the resistance R measured between active electrodes.
   (c) Find the voltage between the active electrodes U.
   (d) Evaluate the power corresponding to the energy dissipated in the disk by the Joule effect.

Project P5

Directional Coupler Analysis

With this project you will learn about a device – the so-called directional coupler – whose main function is to measure the rms voltage of both the incident and reflected waves propagating along a given transmission-line structure. In addition, this measurement device also allows for the determination not only of the standing wave ratio, but also of the active power delivered to the load (an antenna usually).

Consider the steady-state sinusoidal regime of an aerial two-conductor symmetrical line, of length $l = 200\,\text{m}$, which connects a 2 kW medium-wave broadcasting station ($f = 1\,\text{MHz}$) to an antenna whose input impedance is $R_A = 125\ \Omega$. The conductors of the transmission line structure are a distance $d = 20\,\text{cm}$ apart and their radius is $r = 1.5\,\text{mm}$. Assume that line losses are negligibly small.

The voltage at the antenna terminals is given by $u_A(t) = \sqrt{2}\,(U_A)_{rms}\cos(\omega t)$.

Part I Transmission-Line Analysis

1. Determine the per-unit-length parameters L and C of the two-conductor line.
2. Determine the characteristic wave resistance R_w of the line. Determine the wavelength λ and the phase constant β for the specified working frequency.
3. Determine the load reflection coefficient $\overline{\Gamma}$ and the corresponding standing wave ratio SWR.
4. Compute the complex amplitudes and rms values of the incident and reflected voltage waves. Determine the complex amplitudes of the line voltage and line current at both ends of the line. Find the evolution of the rms values of the line voltage and line current U_{rms}, I_{rms}, along the line length; specify the maximum and minimum values of those quantities as well as the places where they occur.

Electromagnetic Foundations of Electrical Engineering J. A. Brandão Faria

Part II Directional Coupler Design and Analysis

Figure P5.1 shows the internal constitution of the directional coupler, which includes a voltage arm and a current arm.

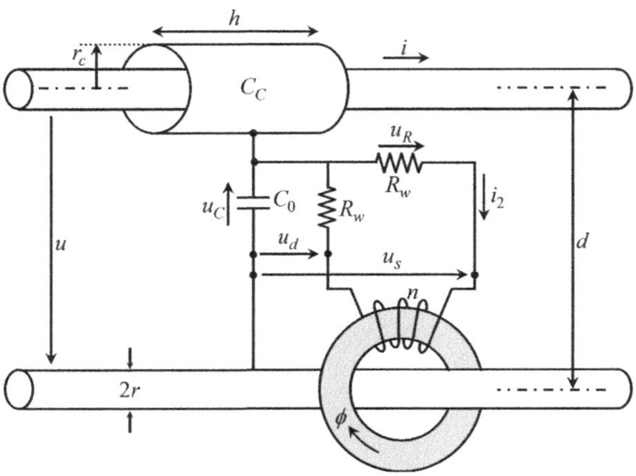

Figure P5.1 Directional coupler connected to a two-conductor transmission line

The voltage arm contains two series-connected capacitors, one a lumped capacitor of capacitance C_0, the other a small coaxial cable section whose internal conductor is the line conductor itself of radius r, and whose external conductor is a cylindrical sheet of length h and radius r_C; the dielectric medium is the air. The capacitance of the small coaxial cable configuration is denoted by C_C.

The current arm is made of a ferrite transformer whose primary circuit is the line conductor itself, and the secondary circuit is an n-turns winding. The secondary winding is loaded by two series-connected resistors, each one with a resistance equal to the characteristic wave resistance R_w of the two-conductor transmission line. For simplification purposes assume that the ferrite transformer is an ideal transformer.

The information conveyed by the pair of voltages u_s and u_d marked in Figure P5.1 is later processed (by software or hardware) in order to provide the user with all the relevant data concerning the transmission-line functioning, namely the active power, the standing wave ratio and the rms values of the incident and reflected voltage waves. The block diagram depicted in Figure P5.2 shows the algorithm used to produce the above data.

1. Determine the voltage ratio $T_u = u_C/u$. Determine the current ratio $T_i = i_2/i$.
2. Make those ratios coincide, $T_u = T_i = -A/2$.

   (a) Determine the relationship between the pair u and i and the pair u_s and u_d.
   (b) By using the transmission-line equations, relate u_{inc} with u_s, and relate u_{refl} with u_d.
   (c) Show that the functional diagram in Figure P5.2 actually meets the goals for which the device was conceived.

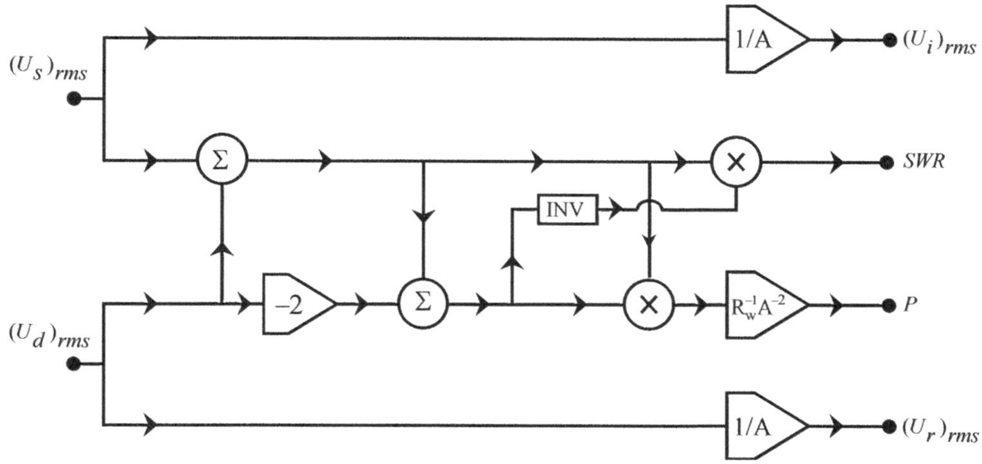

Figure P5.2 Block diagram showing the algorithm used to retrieve information about $(U_i)_{rms}$, $(U_r)_{rms}$, SWR and P, from the input quantities $(U_s)_{rms}$ and $(U_d)_{rms}$. The block INV is an inverter

3. The measuring device should not significantly perturb the normal functioning of the transmission-line system under test. For that reason the impedance of the capacitive divider in the voltage arm should be as large as possible. On the contrary, the transformer input impedance in the current arm should be as small as possible. Use the following criteria:

$$Z_{\text{capacitive divider}} \geq 100R_w; \quad Z_{\text{transformer}} \leq \frac{R_w}{100}$$

   (a) Determine the number of turns n of the secondary winding.
   (b) Determine A, C_0 and C_C.
   (c) Determine the radius r_C of the outer conductor of the coaxial cable section.

4. Consider the internal constitution of the directional coupler shown in Figure P5.1.

   (a) Determine the complex amplitudes of u_C, i_2, u_R and ϕ when:

      (a_1) the directional coupler is placed at the emitter site;
      (a_2) the directional coupler is placed at the antenna site.

   (b) For both cases determine the readings $(U_s)_{rms}$ and $(U_d)_{rms}$.
   (c) Do those readings depend on the place where the directional coupler is installed?

Project P6

Ill-Defined Grounding Problems

The analysis of grounding effects on the performance of transmission systems poses no difficulties as far as low-frequency regimes are concerned. Nonetheless, communication technology trends point towards the use of higher and higher frequencies, where quasi-static approaches fail to provide a credible basis for the analysis of ground connections. In this project you will be alerted to ill-defined grounding problems and learn how to deal with them correctly.

Consider the structure sketched in Figure P6.1 which shows a 50 Ω matched RG-58U coaxial cable of length $l = 207$ mm. The non-magnetic dielectric medium filling the cable is characterized by a relative permittivity $\varepsilon_r = 2.1$. The radius of the external conductor of the cable is $r_2 = 1.48$ mm. Both the high-frequency generator and the cable external conductor are grounded.

Neglect system losses.

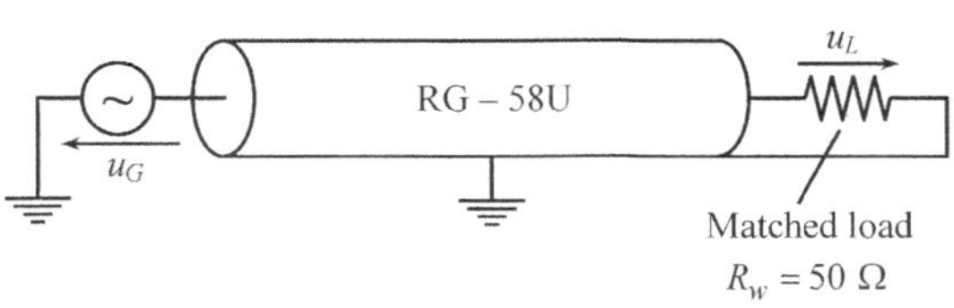

Figure P6.1 A matched coaxial cable with the outer conductor grounded

Part I Cable Analysis

1. Determine the phase velocity v_C of the waves that propagate inside the cable. Determine the inductance L and the capacitance C per unit length of the cable.
2. Determine the radius r_1 of the inner conductor of the cable.
3. What is the relationship between the complex amplitudes of the load and generator voltages? (Note: Think twice!)

Electromagnetic Foundations of Electrical Engineering J. A. Brandão Faria

Part II Grounding Analysis in the Frequency Domain

The structure sketched in Figure P6.1 does not provide you with critical information about the point where the cable is grounded. What is more, it does not tell you where the ground plane is located. You must realize that the problem you are trying to solve is, at this stage, an ill-defined one.

Consider now the detailed situation presented in Figure P6.2, where the cable's grounding point is identified (distance d from the generator), and where the cable distance to the ground is specified, $h = 9\,\text{mm}$.

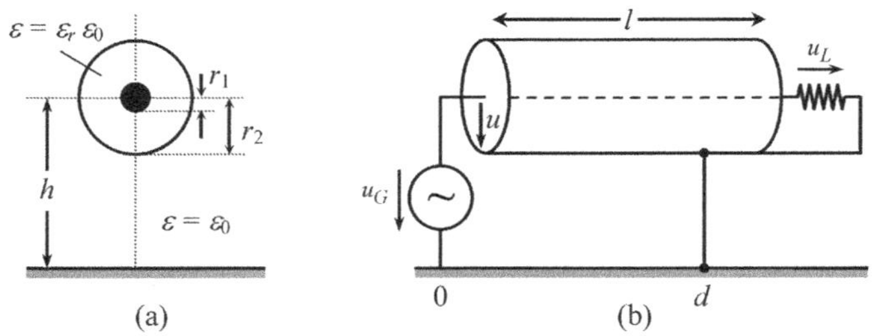

Figure P6.2 Grounded coaxial cable. (a) Cross-section. (b) Longitudinal view explicitly showing where grounding takes place

1. Show that the physical structure in Figure P6.2 can be described by the equivalent scheme depicted in Figure P6.3, which includes the series connection of two different transmission lines: one is the cable itself, and the other is an aerial line (TL) constituted by the cable external conductor and the ground plane.

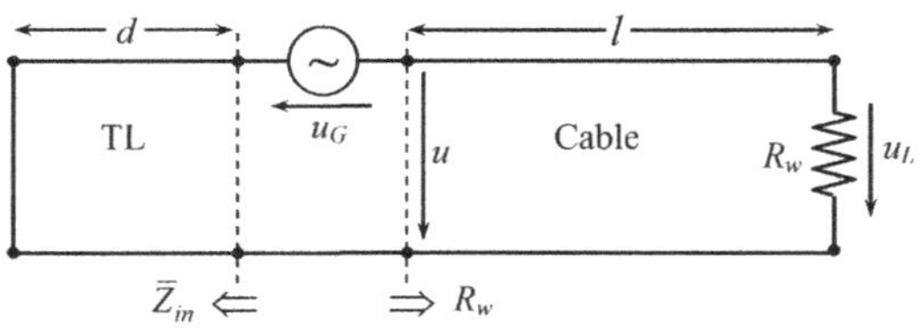

Figure P6.3 Equivalent transmission-line structure corresponding to the cable configuration shown in Figure P6.2

2. Concerning the short-circuited aerial line TL, of length d, obtain:

   (a) The numerical values of the phase velocity v_0 and characteristic wave resistance R_{w_0}.
   (b) The expression of the input impedance $\overline{Z}_{in}$ as a function of d and h.

3. Determine the transfer function $\overline{T}(\omega) = \overline{U}_L/\overline{U}_G$ and plot it graphically for the cases $d = l/2$ and $d = l$.
4. Check that, for certain frequencies, the load voltage turns out to be zero; compute them for the cases $d = l/2$ and $d = l$.
5. Check that, for certain frequencies only, you really do have $\overline{U}_L = \overline{U}_G\, e^{-j\omega l/v_C}$ (your most probable answer to I.3). Compute those frequencies for the cases $d = l/2$ and $d = l$.

Project P7

Induction Machine Analysis

Figure P7.1 shows an induction machine made of two cylindrical iron parts, the rotor and stator, separated by a small air gap of thickness δ, and mean radius R_0.

Neglect the magnetic reluctances of the iron parts.

The axial length of the machine is l.

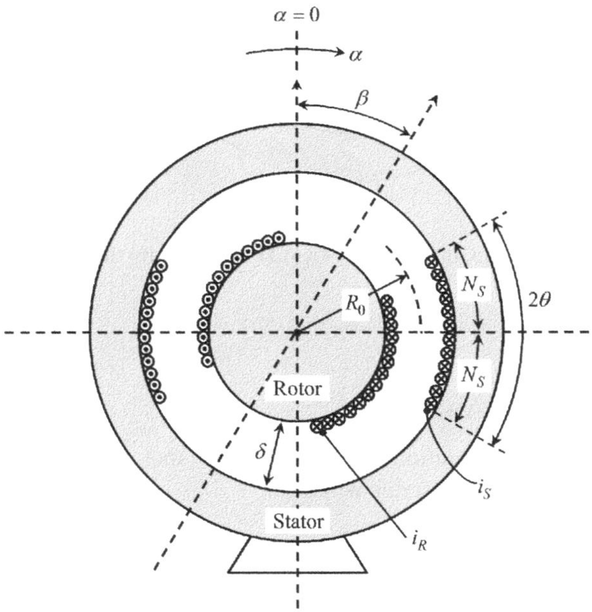

Figure P7.1 Cross-sectional view of the induction machine showing two windings, one on the rotor and the other on the stator. The magnetic axes of both windings make an angle β between them. Note that $\delta \ll R_0$

Along the periphery surface of the rotor, a winding with $2N_R$ turns exists. Likewise, a winding of $2N_S$ turns exists along the stator internal periphery. Each of the windings covers an angular sector of size 2θ. The rotor winding makes an angle β with the machine vertical axis $\alpha = 0$.

Note that the winding turns are placed inside longitudinal slots carved along the iron parts (not shown in the figure).

For illustrative purposes a family of **H**-field lines inside the structure, originated by the stator currents, is shown in Figure P7.2. It should be noted that the magnetic field in the air gap is a radial field, $\mathbf{H} = H(\alpha)\vec{e}_r$.

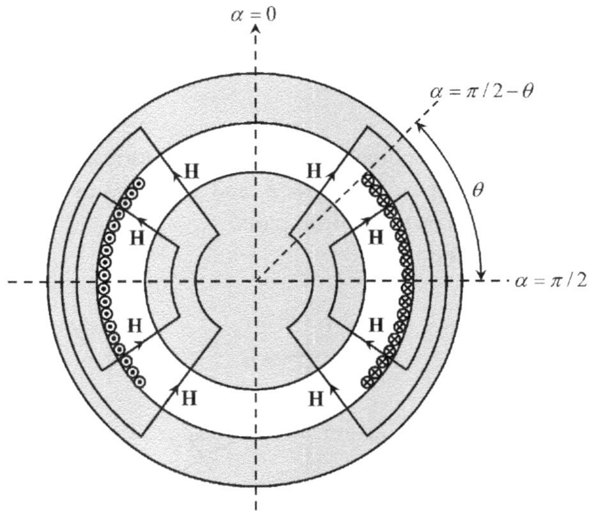

Figure P7.2 In the air gap, the magnetic field **H** produced by the stator currents is radial

Data: $N_S = 60,\ N_R = 20,\ \delta = 2\,\text{mm},\ R_0 = 5\,\text{cm}$.

Part I Magnetic Field Analysis

The rotor is blocked at the position $\alpha = \beta$. Consider that currents $i_R = I_R$ and $i_S = I_S$ respectively flow in the rotor and stator windings.

1. Determine the angular distribution of the stator's magnetic induction field $B_S(\alpha)$ in the air gap. Write the result in the form of a Fourier series expansion. Determine the appropriate coverage angle θ so that $B_S(\alpha)$ can be approximated by a sinusoidal distribution (neglect terms equal to or higher than fifth order).
2. Considering the same coverage angle for the rotor winding, determine the angular distribution of the rotor's magnetic induction field $B_R(\alpha)$.
3. Obtain an expression for the total field in the air gap $B(\alpha) = B_R(I_R) + B_S(I_S)$, as a function of the peripheral angle α.
4. Determine the magnetic energy stored in the machine, and from it find the rotor and stator self inductances L_R, L_S, as well as the mutual inductance $L_{RS}(\beta)$.

5. Determine the axial length l of the machine so that $L_S = 90\,\text{mH}$.
6. Obtain numerically L_R and $(L_{RS})_{\max}$.
7. Determine the torque exerted on the rotor part as a function of the windings' currents, and as a function of the β angle.

Part II Single-Phase Sinusoidal Generator

Figure P7.3 is a schematic symbolic representation of the actual structure depicted in Figure P7.1, where the stator winding can be connected to a resistive load $R_L = 58.77\ \Omega$ (depending on the switch status). Assume that the internal resistance of the stator winding is negligibly small compared to R_L.

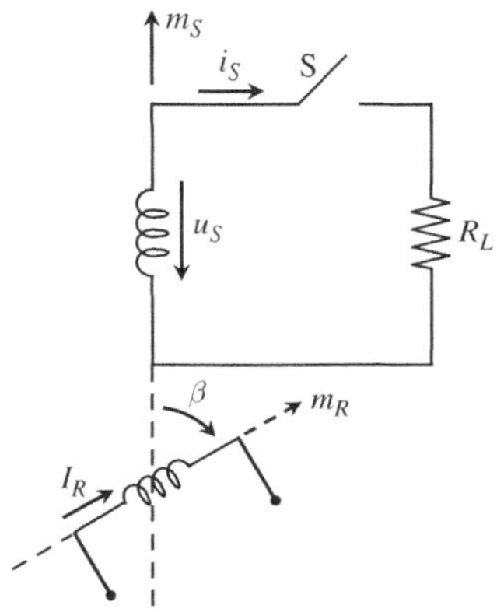

Figure P7.3 Symbolic representation of the machine structure depicted in Figure P7.1. The stator winding can be connected to an external load

Consider that the electric current in the rotor winding is a direct current I_R, and that the rotor rotates with an angular velocity $\omega = 2\pi f$ with $f = 60\,\text{Hz}$ $(\beta = \omega t + \beta_0)$.

1. Consider that the stator winding is left open, $i_S = 0$ (switch S open).
   (a) Determine the voltage at the stator winding terminals, $u_S(t)$. Show that it is a 60 Hz sinusoidal voltage.
   (b) Determine the rotor current I_R such that $(U_S)_{rms} = 110\,\text{V}$.
   (c) Determine the rotating magnetic induction field in the air gap $B_R(\alpha, t)$.
2. Take the current I_R determined above, but now consider that the switch S is closed. For simplification purposes assume that $\beta_0 = \pi/2$.
   (a) Determine the complex amplitude of the stator current i_S as well as the complex amplitude of the new stator voltage u_S.
   (b) Determine the active power at the load terminals.
   (c) Determine the time-averaged external torque T that should be exerted on the rotor to keep it turning at a speed of 60 rotations per second.

(d) Determine the external power necessary for that purpose, and check that it corresponds to the power delivery to the load.
(e) Determine the magnetic induction field in the air gap produced by the stator current $B_S(\alpha, t)$.

Part III Three-Phase Sinusoidal Generator

Consider Figure P7.4 where not one, but three identical windings have been placed around the stator periphery. The three windings make angles of $2\pi/3$ to each other. The rotor characteristics are the same as before.

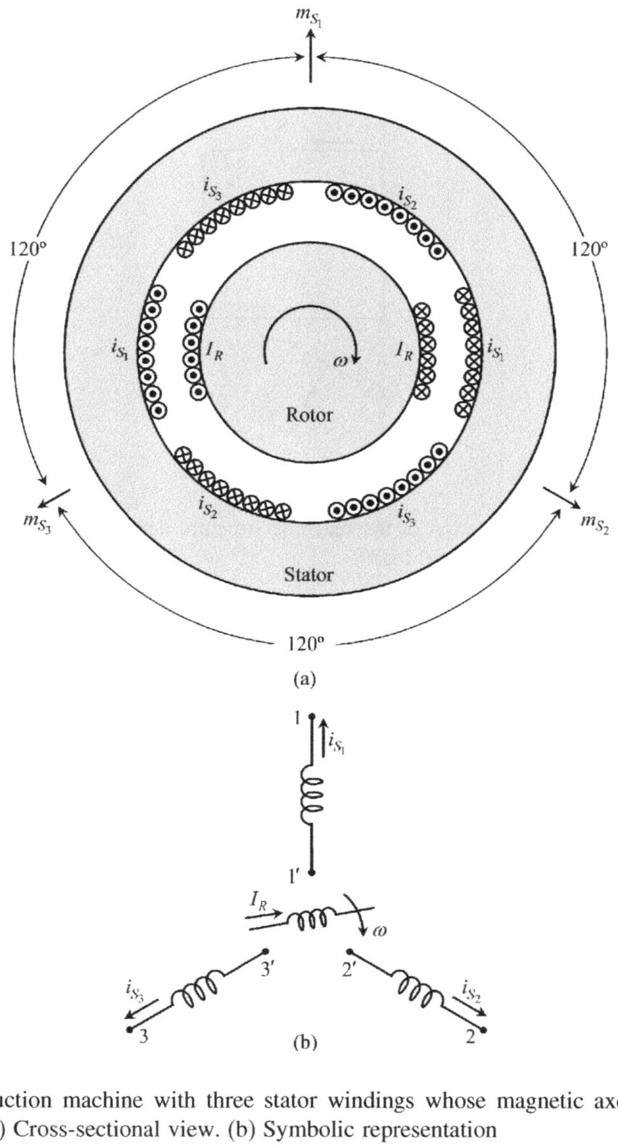

Figure P7.4 Induction machine with three stator windings whose magnetic axes are sequentially shifted by $2\pi/3$. (a) Cross-sectional view. (b) Symbolic representation

1. Assuming that all of the stator windings are left open, $i_{S_1} = i_{S_2} = i_{S_3} = 0$, determine the voltages at the terminals of the three stator windings, u_{S_1}, u_{S_2} and u_{S_3}.
2. Consider that the stator windings are connected to a balanced three-phase load, the stator currents being given by $i_{S_k}(t) = I_S \cos(\omega t - \xi_k)$, where $\xi_k = 2\pi (k-1)/3$. Determine the resulting magnetic induction field in the air gap produced by all of the stator currents $B_S(\alpha, t)$.

Project P8

Line-Matching Technique using an Exponential Transmission-Line Section

Exponential line sections are non-uniform lines whose transverse profile gradually changes with the longitudinal coordinate and whose main purpose is to provide load-matching conditions.

In this project you will analyze and design a matching section made of an exponential line.

Part I Electric Field Analysis

1. Consider two thin parallel cylindrical conductors of radius $r_0 = 1.5\,\text{mm}$, whose axes are a distance *2d* apart. The dielectric medium is the air. When a voltage U is applied between the conductors they become electrically charged with symmetrical per-unit-length charges $\pm q$ (Figure P8.1).

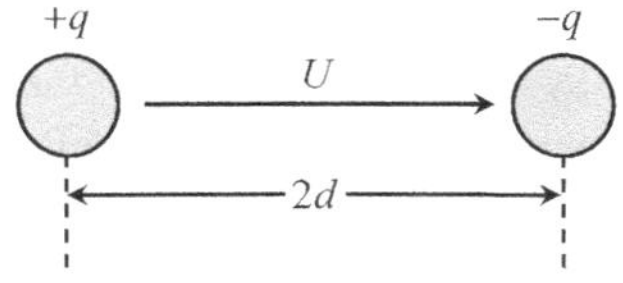

Figure P8.1 Cross-sectional view of a two-wire transmission line

(a) Find the per-unit-length capacitance of the two-conductor line configuration, $C = q/U$. Obtain an expression for the electric field intensity E_S observed at the conductors surface as a function of U.

Electromagnetic Foundations of Electrical Engineering J. A. Brandão Faria

(b) Define the minimum distance between conductors, $2d$, such that E_S does not exceed three-quarters of the dielectric strength of air ($E_d = 20\,\text{kV/cm}$) when $U = U_{3/4} = 10\,\text{kV}$ is applied.

(c) Taking into account the above values for d and U, determine C and q numerically.

2. As shown in Figure P8.2, take into consideration the fact that a ground plane is located a distance h beneath the two-conductor system analyzed in 1.

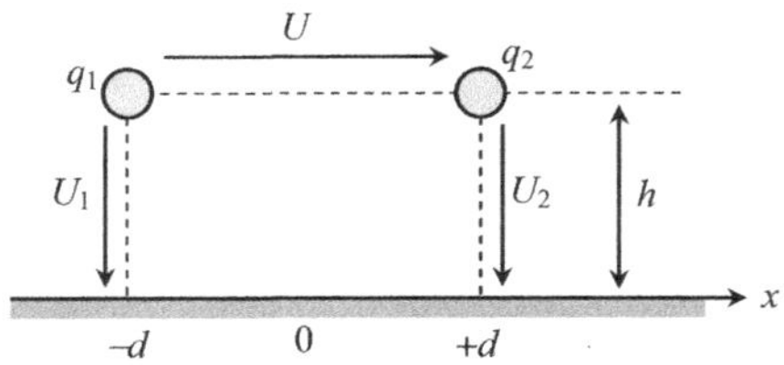

Figure P8.2 Cross-sectional view of a two-wire transmission line above ground

(a) Using the superposition principle and the method of images determine analytically the matrix of potential coefficients per unit length, which describes the capacitive coupling among system conductors

$$\begin{bmatrix} U_1 \\ U_2 \end{bmatrix} = \begin{bmatrix} S_{11} & S_{12} \\ S_{21} & S_{22} \end{bmatrix} \begin{bmatrix} q_1 \\ q_2 \end{bmatrix}$$

(b) Assume that $q_1 = -q_2 = q'$ and $U_1 = -U_2 = U/2$.

(b_1) Obtain the capacitance $C' = q'/U$ and show that it always exceeds the capacitance C of the isolated system in Figure P8.1.

(b_2) The higher the distance h, the closer to an 'isolated system' the configuration in Figure P8.2 will resemble. Find h so that C and C' differ by only 1%.

(b_3) Using the above value for h, compute the capacitance matrix $[C]$ of the multiconductor system.

Part II Uniform Transmission-Line System

Assume that the two-conductor system is sufficiently far away from the ground plane ('isolated system'). Take the geometrical configuration analyzed in Part I, where conductors are assumed lossless.

The two-conductor system defines a uniform transmission line of length $l = 18.75\,\text{m}$ whose aim is to connect a generator (emitter station) to a load (antenna). To simplify things, assume that the input impedance of the antenna is real, $R_0 = 500\ \Omega$. See Figure P8.3.

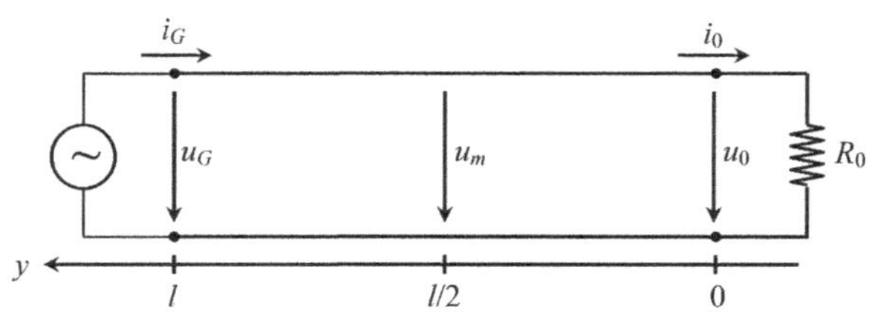

Figure P8.3 Uniform two-wire transmission line of length l, terminated by a resistive load R_0

1. Compute the following line parameters: propagation velocity v, time delay $\tau = l/v$, per-unit-length inductance L and characteristic wave resistance R_w.
2. Time-domain analysis. The generator voltage $u_G(t)$ is a single rectangular pulse, starting at $t = 0$, with duration equal to τ, with 100 V magnitude.
   (a) Line voltages $u_G(t)$, $u_m(t)$ and $u_0(t)$ are monitored with the help of three oscilloscopes positioned at $y = l$, $y = l/2$ and $y = 0$. Plot graphically the monitored voltages.
   (b) Is there any relationship between the voltages analyzed above? Which?
3. Frequency-domain analysis. Consider now that $u_G(t) = U_G \cos(\omega t)$, where $U_G = U_{3/4}$, and $f = 20\,\text{MHz}$ (HF band).
   (a) Find the wavelength, the load reflection coefficient and the standing wave ratio.
   (b) Compute the complex amplitudes of the line voltages and currents at both ends of the line. Obtain the active power delivered to the load.
   (c) Analyze the voltage and current standing wave patterns, identifying their maxima and minima.
   (d) Check that dielectric breakdown phenomena will occur. To avoid them you would need to reduce the generator voltage. How much would you need to reduce U_G?

Part III Exponential Line (Frequency-Domain Analysis)

A possible solution to circumvent the dielectric breakdown problem, without changing the generator voltage, consists of inserting a line-matching section just before the load, as depicted in Figure P8.4.

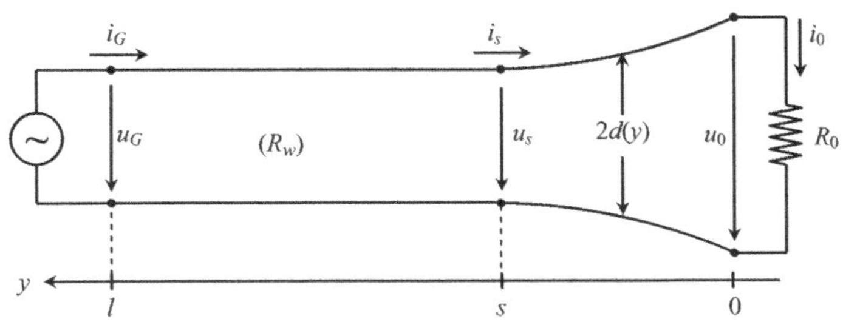

Figure P8.4 Uniform two-wire transmission line, cascaded with an exponential line of length s terminated by a resistive load R_0

The matching section, of length s, is made of the same two conductors that are employed in the previous uniform line (radius r_0), but the lateral distance between conductors varies continuously with y. For the case of the exponential line, the variation $d(y)$ is such that both the per-unit-length inductance and capacitance vary exponentially with y, that is

$$L(y) = L_0 e^{-2\alpha y}, \; C(y) = C_0 e^{+2\alpha y}, \quad \text{for } s \geq y \geq 0$$

where L_0, C_0 and α are parameters to be designed so as to meet the following conditions:

$$\sqrt{L_0/C_0} = R_0, \; L(s) = L, \; C(s) = C$$

where parameters L and C concern the uniform line section already analyzed in Part II.

1. Find the wave solution of the propagation equations for an exponential line. That is, determine $\overline{U}(y)$ and $\overline{I}(y)$, such that the following governing equations are obeyed:

$$\frac{d}{dy}\overline{U}(y) = j\omega L(y)\,\overline{I}(y), \quad \frac{d}{dy}\overline{I}(y) = j\omega C(y)\,\overline{U}(y), \quad \text{for } s \geq y \geq 0$$

   Show that the wave solution includes two exponential complex functions of the type $\exp[(-\alpha \pm j\beta)y]$, where $\beta = \sqrt{\omega^2 LC - \alpha^2}$.
2. Show that the length s can be designed so as to obtain $\overline{U}(s)/\overline{I}(s) = R_w$, thus assuring that the first line section (uniform line) is operating in a matched regime.
3. Compute the smaller value for s that meets the requirement in 2. Compute the α and β parameters. Determine the lateral distance between conductors near to the load terminals.
4. Take into account the design parameters of the exponential line, and consider the generator voltage originally given in II.3.

   (a) Evaluate the complex amplitudes of the line voltages and line currents at $y = l$, $y = s$ and $y = 0$. Find the active power delivered to the load.
   (b) Draw a sketch of the new standing wave patterns.
   (c) Taking into account the variation law for $d(y)$, draw a graphical representation of the conductors' electric field intensity E_S as a function of the y coordinate. Check if the dielectric breakdown problem has been satisfactorily solved.

Project P9

Linear Variable Differential Transformer

In this project you will analyze a linear variable differential transformer (LVDT) from the viewpoint of its application as a position and velocity detector.

The magnetic circuit of the LVDT is shown in Figure P9.1. The magnetic circuit, containing three air gaps, is made of iron, and operates in its linear zone (saturation absent). Neglect the magnetic reluctances of the iron parts except the one concerning the central core. The latter, with a square cross-section S_1, is characterized by its length h_1 and permeability μ_1. Neglect dispersion phenomena.

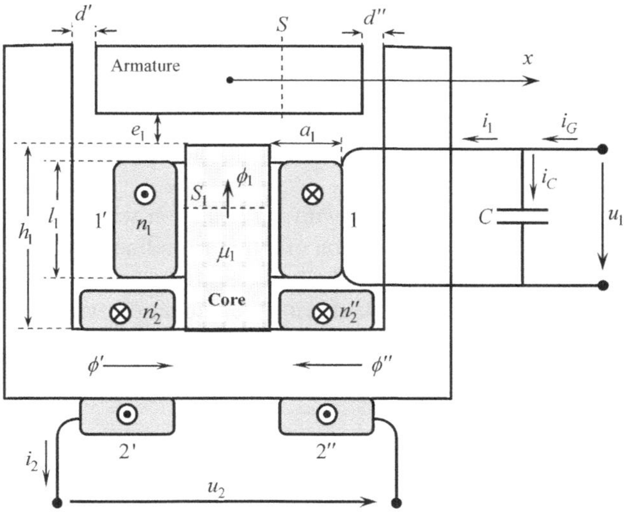

Figure P9.1 Linear variable differential transformer's magnetic circuit. The armature on the top can move along x

The armature part, on the top, is allowed to move along the x-axis.

The transformer includes two windings: the primary or excitation winding with n_1 turns, and the secondary or detection winding with n_2 turns equally split into two symmetrical halves. The detection winding is left open ($i_2 = 0$).

Data: $e_1 = 0.5\,\text{mm}$, $S_1 = 1\,\text{cm}^2$, $S = 0.5\,\text{cm}^2$, $d = 0.5\,\text{mm}$, $h_1 = 3\,\text{cm}$, $l_1 = 1.8\,\text{cm}$, $\mu_1 = 300\mu_0$, $n_2' = n_2'' = n_2/2 = 200$.

Part I Centered Armature at Rest ($d' = d'' = d$)

Assume that the core's magnetic induction field is $B_1 = 100\,\text{mT}$ when the excitation current is set at $i_1 = I_1 = 0.1\,\text{A}$.

1. Find the intensity of the **B** and **H** fields at the air gaps and at the core. Determine the number of turns n_1 of the primary winding.
2. Determine the induction coefficients $(L_{11})_0$, $(L_M)_0$ and $(L_{22})_0$.
3. The excitation winding of length l_1 is made of copper wire ($\sigma_{\text{Cu}} = 5.65 \times 10^7\,\text{S/m}$) of section $S_w = 0.05\,\text{mm}^2$. Assume that the area filling factor is $\eta = 80\%$. Evaluate the winding thickness a_1, as well as the winding internal resistance r_1.

Part II Off-Centered Armature ($d' = d + x$, $d'' = d - x$)

Keep assuming that $i_2 = 0$ and $i_1 = I_1 = 0.1\,\text{A}$. Analyze how the magnetic fluxes ϕ' and ϕ'' depend on x for small shifts around the central position of the armature. Utilize a Taylor series expansion for the results:

$$\phi'(x) = \phi_0' + a'x + b'x^2 + c'x^3 + \cdots$$

$$\phi''(x) = \phi_0'' + a''x + b''x^2 + c''x^3 + \cdots$$

1. Determine the expansion coefficients, up to third order, in terms of the system's given parameters.
2. Obtain an expression for the flux linked with the detection winding, $\psi_2(x)$. Estimate the maximum value of x/d so that the deviation of $\psi_2(x)$ from linearity does not exceed 10%.
3. Determine $L_M(x)$ in the form $L_M(x) = (L_M)_0 + k_M x$.
4. Obtain an expression for the flux linked with the excitation winding, $\psi_1(x)$. Estimate the maximum value of x/d so that the deviation of $\psi_1(x)$ from $\psi_1(0)$ does not exceed 10%. Determine L_{11}.

Part III Position Sensing

The current in the excitation winding is a 50 Hz harmonic current, $i_1(t) = I_1 \cos(\omega t)$, with $I_1 = 0.1\,\text{A}$.

1. For a small shift in the armature position (allowing you to neglect terms higher than first order in ϕ' and ϕ''), find the excitation and detection voltages, $u_1(t)$ and $u_2(x, t)$.
2. Evaluate the amplitudes of u_2 and u_1 for the case $x/d = \pm 0.2$.
3. Find the LVDT sensitivity, $S_x = \partial U_2/\partial x$, expressed in V/mm.

Part IV Velocity Sensing

A direct current, $I_1 = 0.1\,\mathrm{A}$, flows in the excitation winding. The moving armature vibrates along the x-axis according to $x(t) = X_{\max}\cos(\omega t)$, with $\omega = 2\pi f$, $f = 50\,\mathrm{Hz}$ and $X_{\max} = 0.1\,\mathrm{mm}$.

1. Again neglecting terms higher than first order in $\phi'(x)$ and $\phi''(x)$, find the voltage at the detection winding terminals.
2. Determine the LVDT velocity sensitivity $S_v = (U_2)_{\max} / (X_{\max} f)$.

Part V Primary Winding Protection and Transient Analysis

With the goal of protecting the primary winding against switching operations, a capacitor of capacitance C was inserted at the primary winding terminals (see Figure P9.1).

1. Consider the LVDT operating as a position sensor (Part III). Determine C so that the magnitude of the generator's current i_G is minimal. Obtain the complex amplitudes of u_1, i_1, i_C and i_G. Illustrate the results using a phasor diagram.
2. The generator is switched off when $i_1(t) = I_1$. Determine the transient evolution of $i_1(t)$ and $u_1(t)$. Show that a damped periodic regime will occur. Characterize the transient regime through its damping factor and angular frequency.

Project P10

Magnetic Actuator and Sensor Device

In this project you will analyze a magnetic circuit which can be used both as an actuator and as a position sensor.

Consider the magnetic circuit depicted in Figure P10.1, which is made of iron parts of negligible magnetic reluctance. The circuit includes two equal air gaps of small thickness δ. Inductor 1, with internal resistance r_1, is made of two symmetrical halves, each with N_1 turns. Inductor 2, with N_2 turns and negligible resistance, can move up and down in the air gap region, $0 < x < a$; its mean position is given by $x_{av} = \left(x_{N_2} + x_1\right)/2$ and its axial length is $l = x_{N_2} - x_1$, where x_1 and x_{N_2} respectively denote the x coordinates of the first and last winding turns.

Data: $a = 3\,\text{cm}$, $b = 1\,\text{cm}$, $\delta = 2\,\text{mm}$, $l = 1\,\text{cm}$, $r_1 = 20\ \Omega$, $N_2 = 100$, $N_1 = 600$.

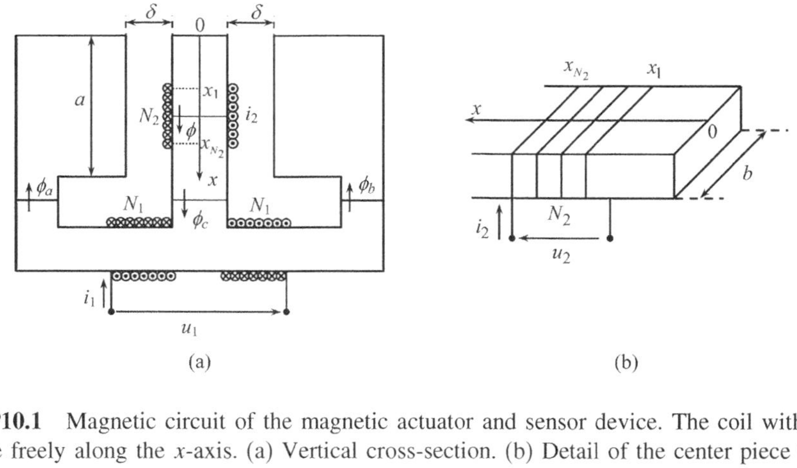

Figure P10.1 Magnetic circuit of the magnetic actuator and sensor device. The coil with N_2 turns can move freely along the x-axis. (a) Vertical cross-section. (b) Detail of the center piece where the coil with N_2 turns is wound around

Part I Magnetic Circuit Analysis

1. Take $i_2 = 0$ and $i_1 = I_1$.
   (a) Determine the magnetic field intensity $H(x)$ in the air gap region, from $x = 0$ to $x = a$.
   (b) For x in the range $[0;\ a]$, find the magnetic flux $\phi(x)$ passing through the iron core where inductor 2 is wound around.
   (c) Obtain ϕ_a, ϕ_b and ϕ_c.
   (d) Find the linkage fluxes ψ_1 and ψ_2 and obtain the analytical expressions for the induction coefficients L_{11} and L_M.
2. Take $i_1 = 0$ and $i_2 = I_2$. Answer the same questions as in 1, obtaining L_{22}.
3. Using the information $N_2 = 100$, $N_1 = 600$:
   (a) Compute L_{11}.
   (b) Compute the extreme values of L_{22} and L_M for the allowed position of inductor 2, whose range is $l/2 < x_{av} < a - l/2$.

Part II Actuator Analysis

Assume now that electric currents i_1 and i_2 are made to flow in both inductors.

Determine, as a function of i_1 and i_2, the global magnetic force exerted on inductor 2 which can drive it into movement along x. Show that the magnetic force is independent of the particular position of the actuated inductor.

Part III Magnetic Induction Phenomena

Assume that inductor 2 is stopped from moving, blocked at the rest position $x_{av} = 1\,\text{cm}$.

1. Determine the numerical values of L_{22} and L_M.
2. Assume that $i_2 = 0$ and that $i_1(t)$ has the ramp shape given in Figure P10.2(a).
   (a) Determine the time evolution of both voltages $u_1(t)$ and $u_2(t)$.
   (b) Determine the driving force exerted on inductor 2.
3. Assume that inductor 2 is short-circuited ($u_2 = 0$), and that $u_1(t)$ has the shape of a positive pulse, as shown in Figure P10.2(b).
   (a) Determine the time evolution of both currents $i_1(t)$ and $i_2(t)$.
   (b) Determine the driving force exerted on inductor 2. Obtain its value for $t = t_0$.

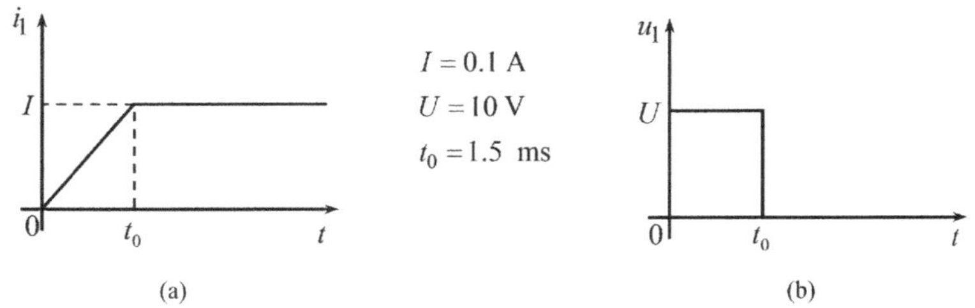

Figure P10.2 (a) Current i_1 against time (ramp shape). (b) Voltage u_1 against time (rectangular pulse)

Part IV AC Analysis

Consider the two inductor system as an ordinary transformer. Assume as above that inductor 2 is short-circuited and blocked at its rest position $x_{av} = 1\,\text{cm}$.

The sinusoidal voltage of the generator that drives inductor 1 (primary winding) is given by

$$u_1(t) = \sqrt{2}\,(U_1)_{rms}\cos(\omega t)$$

where $(U_1)_{rms} = 30\,\text{V},\ \omega = 2\pi f,\ f = 50\,\text{Hz}$.

1. Determine the complex amplitudes of the primary and secondary currents.
2. Find the active and reactive powers at the terminals of inductor 1.
3. In order to minimize the generator's driving current a capacitor of capacitance C is placed in parallel with the primary winding. Determine C and $\overline{I}_G$.
4. Determine the time-averaged value of the magnetic force F_{av} exerted on inductor 2.

Part V Sensor Analysis

The sensing winding (inductor 1) is left open, $i_1 = 0$, while inductor 2 is subjected to a vibration movement described by $x_{av}(t) = x_0 + X_2\cos(\omega t)$. A direct current $i_2 = I$ is made to flow in inductor 2.

1. Show that the voltage in the sensing winding $u_1(t)$ is sinusoidal with the same frequency of vibration movement. Show that its amplitude is proportional to X_2 but it is independent of x_0. Show that $u_1(t)$ and the sinusoidal component of $x_{av}(t)$ are in phase quadrature.
2. Consider that $I = 100\,\text{mA}$, $X_2 = 5\,\text{mm}$ and $\omega = 10\,\text{krad/s}$. Find the amplitude U_1 of the sensing voltage $u_1(t)$, and determine the sensitivity $\partial U_1/\partial X_2$.

Project P11

Overhead-Line Protection by Ground Wires

In this project you will understand the reason why overhead power lines contain not only phase conductors but also one or two additional wires, bonded to ground, for line protection.

Part I Flat Three-Phase Line Analysis

Figure P11.1 shows a cross-sectional view of a flat three-phase line configuration made of three cylindrical conductors of radii $r = 2\,\text{cm}$ at equal height above the soil $h = 15\,\text{m}$; the horizontal distance between conductors is $d = 5\,\text{m}$. The line length is $l = 1\,\text{km}$.

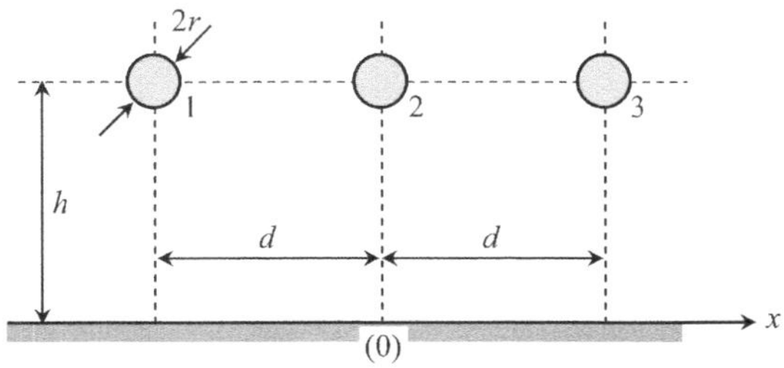

Figure P11.1 Cross-sectional view of a flat three-phase overhead line

Electromagnetic Foundations of Electrical Engineering J. A. Brandão Faria

1. Making use of the method of images, determine the potential coefficients matrix of the line $[S]$, and by inversion find the corresponding capacitance coefficients matrix $[C]$.
2. Consider that in the space occupied by the conductor system a vertical external electric field $\mathbf{E}_{ext}$ exists that is caused by the presence of a strong positively charged cloud positioned above the line conductors. For simplification purposes assume that such a field is approximately uniform in the analysis region; take its intensity as $E_{ext} = 250\,\text{V/cm}$.
   Assume that the line is inactive, with $U_1 = U_2 = U_3 = 0$. Determine the electric charges Q_1, Q_2 and Q_3 on the phase conductors.
   Note that, by using the superposition principle, the voltage U_k can be obtained by summing two contributions: a contribution $(U_k)_{\text{cloud}}$ with zero charge on the conductors, and a contribution $(U_k)_Q$ caused by conductor charges but taking $\mathbf{E}_{ext} = 0$.
3. Estimate the maximum value of the electric field occurring at the surface of the line phase conductors. Comment on possible problems related to air breakdown phenomena.

Part II Ground Wire Analysis

Figure P11.2 is identical to Figure P11.1 except for the inclusion of a fourth conductor (the ground wire) placed above the central phase conductor, at a height $h_4 = 19\,\text{m}$ above the soil. The radius of the cylindrical ground wire is $r_4 = 1\,\text{cm}$. The ground wire is bonded to the soil, that is $U_4 = 0$.

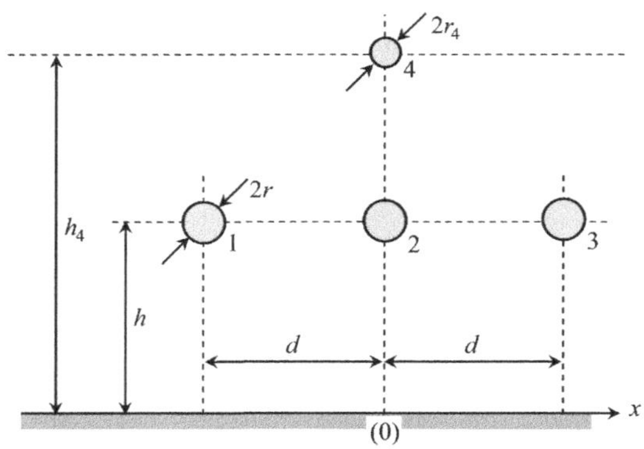

Figure P11.2 Cross-sectional view of a flat three-phase overhead line, including a protection ground wire (conductor 4)

1. For the new configuration determine the new 4×4 potential coefficients matrix of the line $[S]'$, as well as the corresponding capacitance coefficients matrix $[C]'$.
2. Compute the conductor charges Q_1, Q_2, Q_3 and Q_4 considering the same conditions referred to in I.2.
3. Repeat the breakdown phenomena analysis as in I.3.
4. Find the percentage reduction η of the electric field intensity on the phase conductors due to the protecting screening action of the ground wire:

$$\eta_k = \left| \frac{(E_k)_{\text{without ground wire}} - (E_k)_{\text{with ground wire}}}{(E_k)_{\text{without ground wire}}} \right| \times 100\%, \quad \text{for } k = 1, 2, 3.$$

Part III AC Analysis

Assume that the three-phase line plus ground wire configuration (charged cloud absent) is driven by a 50 Hz three-phase balanced generator (neutral grounded) – see Figure P11.3:

$$u_1(t) = \sqrt{2}\, U_{rms} \cos(\omega t + 2\pi/3)$$
$$u_2(t) = \sqrt{2}\, U_{rms} \cos(\omega t)$$
$$u_3(t) = \sqrt{2}\, U_{rms} \cos(\omega t - 2\pi/3)$$

The rms value of the phase-to-phase voltage is $U_\Delta = 120\,\text{kV}$. The line is left open at its far end.

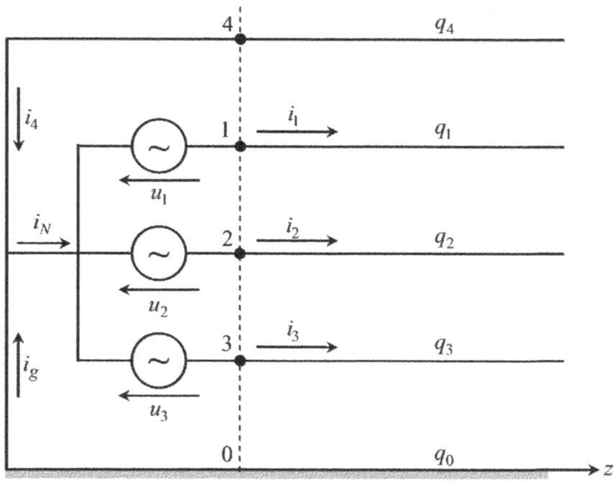

Figure P11.3 Schematic longitudinal view of the overhead line, driven by a three-phase generator, with the neutral grounded

1. Obtain the complex amplitudes of the line voltages u_1, u_2, u_3, line charges q_1, q_2, q_3, q_4, and line currents i_1, i_2, i_3. In addition determine the complex amplitude of the neutral current i_N, breaking it down into its components i_4 and i_g.
2. Find the time-averaged electric energy stored along the line length. Check the result by using the complex Poynting theorem.
3. Find the place where, and the time instant when, the electric field intensity is at its maximum value. Determine $E_{\max}$.

Project P12

Power Line Carrier Communication

In addition to the principal objective of delivering energy, most power line structures are also being used as communication channels, transmitting high-frequency signals for delivering voice and various data services.

In this project we propose that you analyze not only the electromagnetic field of a three-phase line and its 50 Hz steady-state regime, but also the high-frequency regime concerning the issue of signal carrier communications.

As shown in Figure P12.1, the line is made of three parallel cylindrical conductors, all of the same radius $r_0 = 2.5$ mm. The configuration is rotationally symmetric; conductor axes are identified by their (x_k, y_k) coordinates, with $k = 1,\ 2,\ 3$. The distance between conductors is $d = 5$ cm. The surrounding medium is the air ($\varepsilon = \varepsilon_0,\ \mu = \mu_0$).

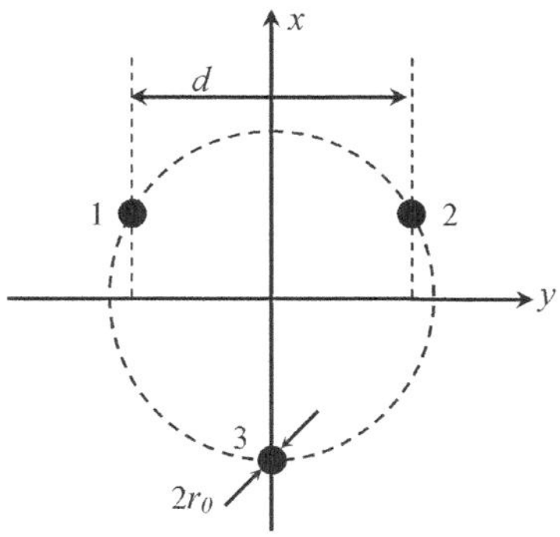

Figure P12.1 Cross-sectional view of a symmetrical three-wire line made of cylindrical conductors

Electromagnetic Foundations of Electrical Engineering J. A. Brandão Faria

Part I Electric Field Analysis

1. Take conductor 3 as the reference conductor, that is $V_3 = V_0 = 0$, $q_3 = q_0 = -(q_1 + q_2)$, where q_1, q_2 and q_3, denote the per-unit-length conductor charges.

   (a) Assume the system charges are $q_1 \neq 0$ and $q_2 = 0$. Determine analytically the expressions for the potential functions V_1 and V_2 on the conductors 1 and 2.
   (b) Repeat the procedure for the case $q_2 \neq 0$ and $q_1 = 0$.
   (c) Find the per-unit-length matrix of potential coefficients $[S]$.

2. Determine numerically the per-unit-length capacitance matrix $[C]$ as well as the corresponding equivalent scheme of partial capacitances.
3. Consider that the line is driven in odd-mode operation, that is $q_1 = -q_2 = q$, $V_1 = -V_2 = U/2$, where U is the voltage applied between conductors 1 and 2.

   (a) Find the odd-mode equivalent capacitance $C = q/U$.
   (b) Find the maximum applied voltage U_{max} in order to prevent air breakdown phenomena from occurring (take $E_d = 25\,\text{kV/cm}$).

Part II Magnetic Field Analysis

Currents i_1, i_2 and i_3 flow across line conductors. Keep considering odd-mode operation, that is $i_1 = -i_2 = I$, $i_3 = 0$ (see Figure P12.2).

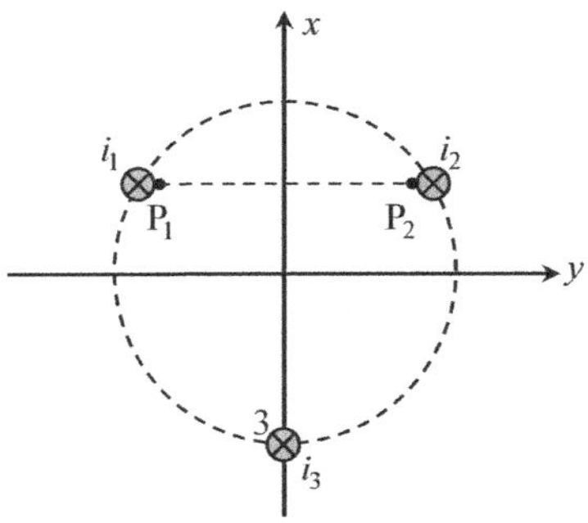

Figure P12.2 Cross-sectional view of the symmetrical three-wire line showing conductor currents. The unit-length rectangle, whose smaller side is the segment P_1P_2, is utilized in the evaluation of the flux linkage

1. Determine analytically the magnetic field $H(y)$, external to the conductors, for points belonging to the plane $x = x_1 = x_2$.
2. Determine the per-unit-length magnetic flux ψ linked with the circuit constituted by conductors 1 and 2, using a rectangular surface whose lateral size corresponds to the segment connecting P_1 to P_2.
3. Using the above result obtain analytically and numerically the self-inductance relative to the conductor pair 1 and 2, $L = \psi/I$.
4. Check the existing relationship between L and C (I.3(a)).

Part III Harmonic Steady-State Analysis

Figure P12.3 shows the connection between a generator and a load, using the three-conductor set previously analyzed in Parts I and II.

The longitudinal length of the distribution line is $l = 175\,\text{m}$.

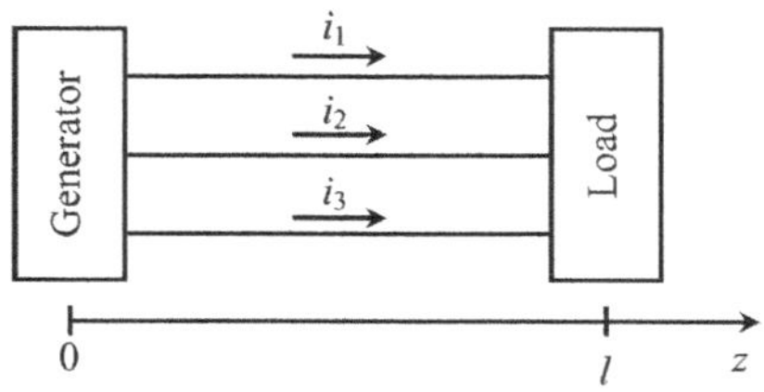

Figure P12.3 Three-wire line, of length l, linking a generator to a load

At the sending end, $z = 0$, the system is driven by a generator set (Figure P12.4(a)) whose time-varying voltages are given by

$$u_a(t) = u'_a(t) + \frac{1}{2}u''(t), \quad u_b(t) = u'_b(t) - \frac{1}{2}u''(t), \quad u_c(t) = u'_c(t)$$

Voltages u' define a balanced three-phase system of frequency $f = 50\,\text{Hz}$ and $U'_{rms} = 230\,\text{V}$:

$$u'_a(t) = \sqrt{2}\,U'_{rms}\cos(\omega t), \quad u'_b(t) = \sqrt{2}\,U'_{rms}\cos(\omega t - 2\pi/3), \quad u'_c(t) = \sqrt{2}\,U'_{rms}\cos(\omega t - 4\pi/3)$$

Voltage u'' between conductors 1 and 2 (odd-mode excitation) is a high-frequency sinusoidal carrier, $f_0 = 1.5\,\text{MHz}$, used for communication purposes:

$$u''(t) = \sqrt{2}\,U''_{rms}\cos(\omega_0 t)$$

where $U''_{rms} = 5\,\text{V}$.

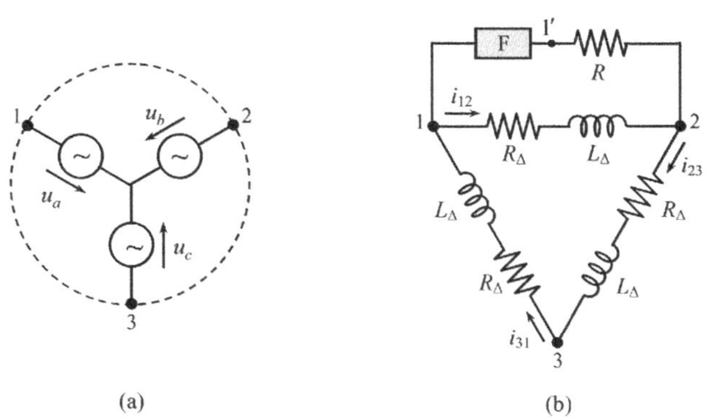

Figure P12.4 (a) Generator voltages. (b) Load arrangement, where F denotes a filtering block

At the receiving end, $z = l$, three delta-connected loads exist (Figure P12.4b), where $R_\Delta = 50\ \Omega$ and $L_\Delta = 91.9\,\text{mH}$.

The block F is a filter device which should be designed so as to behave as an open circuit at 50 Hz and as a short circuit at 1.5 MHz. The resistor R (in Figure P12.4(b)) is to be interpreted as the receiving equipment for the high-frequency signal carrier.

Consider the simplifying approach that the distribution line is lossless.

By invoking the superposition principle, the 50 Hz regime and the 1.5 MHz regime will be analyzed separately.

1. Determine analytically the admittance of the filtering block shown in Figure P12.5.
2. Choosing $C_1 = 10\,\mu\text{F}$, find L_1 and L_2 so as to achieve the project's goals.

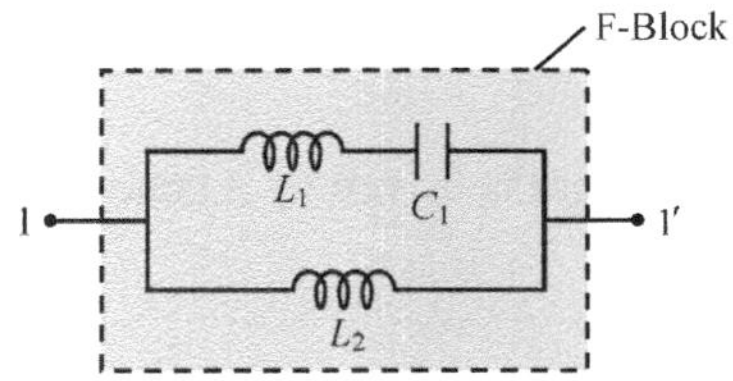

Figure P12.5 Composition of the filtering block

Part III.1 The 1.5 MHz Regime

Take $u'_a = u'_b = u'_c = 0$ and $u'' \neq 0$.

1. Show that the analysis of the problem can be conducted using the transmission-line structure in Figure P12.6. To that end, clarify the meaning of both u_G and $\overline{Z}_L$.

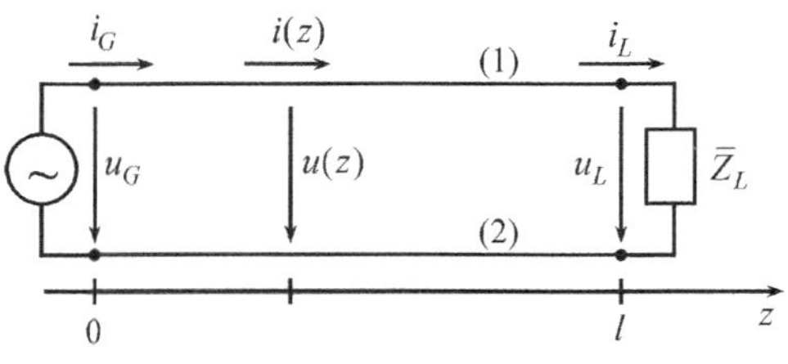

Figure P12.6 Equivalent two-wire line for odd-mode operation

2. Using the per-unit-length odd-mode parameters C and L obtained in Parts I and II, evaluate the characteristic wave resistance of the line R_w, the propagation velocity v, as well as the wavelength λ associated with f_0.
3. Take $R = R_w$ (receiving equipment turned on).

   (a) Determine $\overline{Z}_L$, $\overline{\Gamma}$ and the SWR. Show that, in practical terms, the line is a matched one.

(b) Determine the complex amplitudes of the voltages and currents at the sending and receiving ends of the line.
(c) Evaluate the high-frequency active power guided by the line.

4. Take $R = \infty$ (receiving equipment turned off).

(a) Determine $\overline{Z}_L$, $\overline{\Gamma}$ and the SWR. Show that, in practical terms, the line is left open.
(b) Determine the complex amplitudes of the voltages and currents at the sending and receiving ends of the line.
(c) Analyze the standing wave pattern and identify the extremal points of the rms voltage and current along the line.

Part III.2 The 50 Hz Regime

Take $u' \neq 0$ and $u'' = 0$.

1. Justify why you do not need now to resort to wave theory formalisms.
2. Obtain the voltage and current complex amplitudes of the delta-connected loads (Figure P12.4(b)).
3. Obtain the complex amplitudes of the line currents i_1, i_2 and i_3 (Figure P12.3).
4. Evaluate the instantaneous power delivered to the load.

Project P13

Pseudo-Balanced Three-Phase Lines

In this project you will come to see that an overhead unbalanced three-phase transmission line may behave as a balanced system provided that the geometry of the line is properly designed. In addition you will also examine the main propagation features of the ground mode at high frequencies, where skin-effect phenomena are taken into account.

Part I Capacitance and Inductance Matrices

Consider a three-conductor overhead line above ground characterized by a per-unit-length capacitance matrix

$$[C] = \begin{bmatrix} C_{11} & C_{12} & C_{13} \\ C_{12} & C_{22} & C_{23} \\ C_{13} & C_{23} & C_{33} \end{bmatrix}$$

1. Find the relationship between the capacitance coefficients in such a way that, in spite of the naturally unbalanced character of the structure, the system behaves as a balanced one; that is, the following conditions are simultaneously ensured:

   Condition 1: If $U_1 = U_2 = U_3$ then $Q_1 = Q_2 = Q_3$.
   Condition 2: If $U_1 = -U_3$ and $U_2 = 0$ then $Q_1 = -Q_3$ and $Q_2 = 0$.
   Condition 3: If $U_1 = U_3 = -U_2/2$ then $Q_1 = Q_3 = -Q_2/2$.

2. Show that condition 3 is a direct consequence of conditions 1 and 2.
3. From the preceding results determine the array type composition for the capacitance coefficients matrix $[C]$ and for the potential coefficients matrix $[S]$.
4. Comment on the type of geometry that the line must necessarily obey. Determine the particularities of the partial capacitances among the system conductors.

5. Figure P13.1 shows a cross-sectional view of a three-phase line configuration with $h_1 = 25\,\text{m}$ and $d = 6\,\text{m}$, made of thin cylindrical conductors of radii $r_1 = r_2 = r_3 = 1\,\text{cm}$. Determine the height h_2 of conductor 2 so that the line behaves as a balanced line, according to the conditions defined in 1.
6. Obtain the numerical values for the entries of the per-unit-length potential coefficients matrix $[S]$, capacitance coefficients matrix $[C]$ and external inductance coefficients matrix $[L_e]$. Obtain the partial capacitances of the structure.

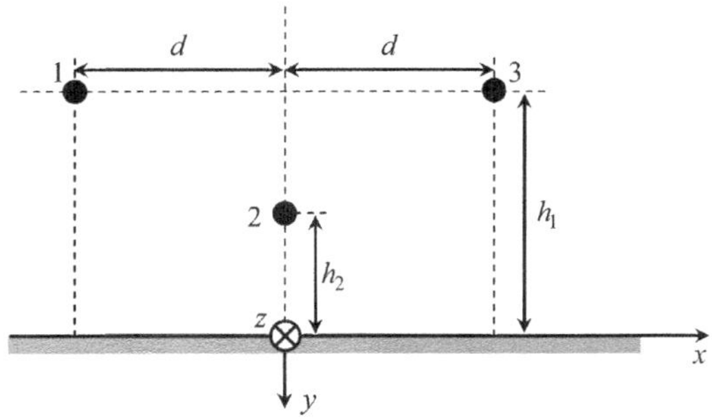

Figure P13.1 Cross-sectional view of a three-conductor overhead line above a ground plane

Part II Lossless Line Analysis

The transmission line previously designed is now employed for communication purposes using a 1 MHz carrier. The line is excited by a ground mode, with $u_1 = u_2 = u_3 = u$, and $i_1 = i_2 = i_3 = i/3$. Consider the line connections shown in Figure P13.2, where $u_G = U_G \cos \omega t$, with $U_G = 10\,\text{V}$ and $R_G = 75\ \Omega$. The line length is $l = 530\,\text{m}$.

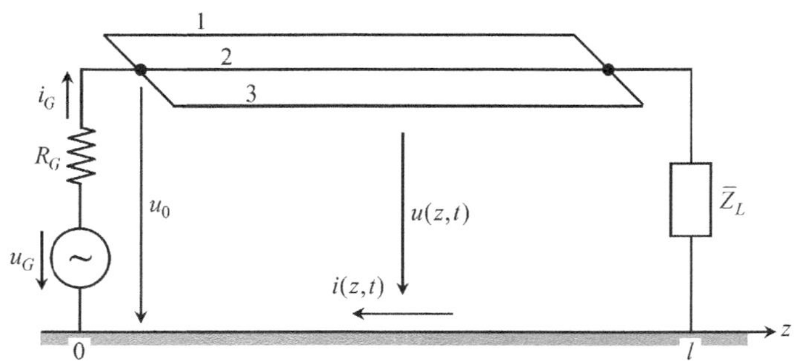

Figure P13.2 Schematic longitudinal view of the pseudo-balanced three-conductor line for ground-mode operation

1. The relationship between the phasors $\overline{U}(z)$ and $\overline{I}(z)$ is described by the usual equations

$$\frac{d\overline{U}}{dz} = -j\omega L\overline{I} \text{ and } \frac{d\overline{I}}{dz} = -j\omega C\overline{U}$$

Making use of the $[L_e]$ and $[C]$ matrices established in I.6, determine the ground-mode equivalent p. u. length parameters L and C. Obtain the mode propagation velocity v as well as its corresponding characteristic wave resistance R_w.
2. Assume that the line is matched, $\overline{Z}_L = R_w$. Obtain the complex amplitudes of the electric and magnetic field vectors $\mathbf{E}(z, x)$ and $\mathbf{H}(z, x)$ at the ground surface $y = 0$.
3. Obtain the voltage and current standing wave patterns for the case of the line being open at its far end, $\overline{Z}_L = \infty$.
4. Find $\overline{Z}_L$ such that $i_G = 0$. For that situation determine the load reflection coefficient $\overline{\Gamma}$ and the standing wave ratio.

Part III Lossy Line Analysis

Keep considering the line configuration in Figure P13.2. Assume that the overhead conductors are characterized by permeability μ_0 and conductivity $\sigma_c = 5.7 \times 10^7$ S/m, and that the soil is characterized by permeability μ_0 and conductivity $\sigma_s = 0.2$ S/m. For the given operating frequency, assume that the skin effect is strong.

1. Find an analytical expression for the correction $\overline{Z}_C$ to the longitudinal impedance of the line associated to the perturbation caused by the skin effect in the overhead conductors.
2. Assume that the distribution of ground currents and the distribution of the magnetic field at the soil surface are very similar to those observed in the lossless line case. However, for an imperfect soil, the ground currents give rise to a longitudinal $\mathbf{E}$ field which is responsible for ground losses. Consider the following simplifying assumptions as far as the fields inside the ground are concerned:

$$\partial/\partial z \ll \partial/\partial y,\ \ \partial/\partial x \ll \partial/\partial y,\ \ \overline{\mathbf{E}} = -\overline{E}\,\vec{e}_z,\ \ \overline{\mathbf{H}} = -\overline{H}\,\vec{e}_x \text{ and } \overline{\mathbf{H}}_{(y=0)} = \overline{\mathbf{H}}^{\text{lossless case}}_{(y=0)}$$

   (a) Using the phasor-domain Maxwell equations establish a pair of differential equations governing the electromagnetic field penetration in the ground. Find the complex amplitude of the longitudinal electric field and particularize it for the ground surface, $y = 0$.
   (b) Determine the complex power $\overline{P}$ corresponding to the flux of the complex Poynting vector across the ground surface, per unit axial length.
   (c) Determine the correction $\overline{Z}_S = \overline{P}/(I)^2_{rms}$ to the longitudinal impedance of the line associated to the perturbation caused by the skin effect in the soil.
   (d) Obtain, numerically, the values for the perturbations $\overline{Z}_C$ and $\overline{Z}_S$. Subsequently, compute the per-unit-length longitudinal impedance of the line, $\overline{Z} = j\omega L_e + \overline{Z}_C + \overline{Z}_S$.
   (e) Obtain the wave parameters that characterize the behavior of the lossy line, that is the propagation constant, the attenuation constant, the phase constant, the phase velocity and the characteristic wave impedance.

3. The line is left open at its far end. Compute the complex amplitudes of the line voltages and line currents at both line ends.

Project P14

Screened High-Voltage Three-Phase Installation

A high-voltage three-phase line feeds a symmetrical load made of three star-connected *RL* circuits, whose assembly is screened from external perturbations by means of a metallic enclosure. As shown in Figure P14.1, the wall of the screen has three circular holes open where the feeding line conductors go through. The set of triangle-connected capacitors C_Δ is used for power factor compensation purposes.

Figure P14.2(a) shows the detailed constitution of the magnetic circuit pertaining to each of the *RL* loads. The magnetic circuit includes an air gap of thickness $\delta = 5\,\text{mm}$ and cross-section $S_0 = 250\,\text{cm}^2$. The shaded iron parts of length h and section S_{Fe} are described by the nonlinear magnetic characteristic outlined in Figure P14.2(b). For simplification

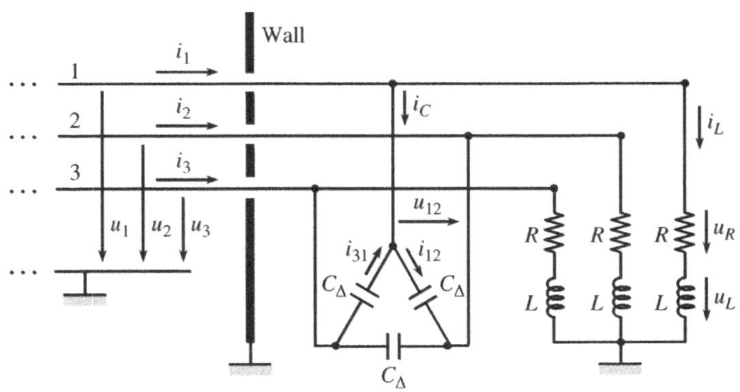

Figure P14.1 A high-voltage three-phase line crosses a screening wall to feed a balanced star-connected *RL* load. The purpose of the capacitors is to ensure power factor compensation

Electromagnetic Foundations of Electrical Engineering J. A. Brandão Faria

purposes the remaining iron parts of the magnetic circuit are assumed to have negligibly small reluctances.

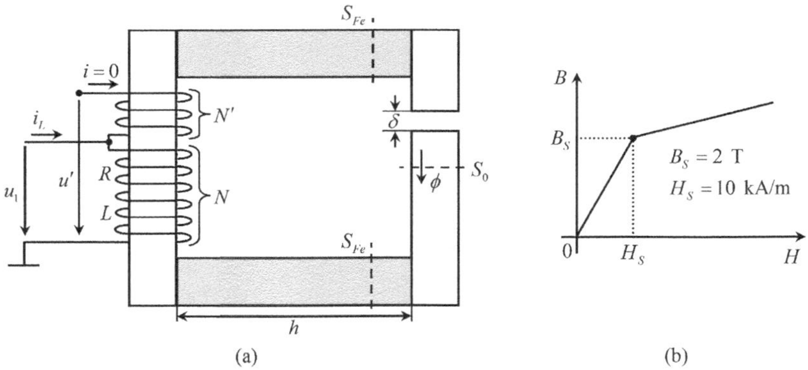

Figure P14.2 (a) Magnetic circuit corresponding to the *RL* load. (b) Nonlinear magnetic characteristic $B(H)$ of the shaded parts of the magnetic circuit

The goals of this project are:

- To properly solve the problem of the screen crossing by the high-voltage feeding conductors.
- To properly design the magnetic circuit in order to avoid saturation effects.
- To properly design the compensation capacitors.

Data: The 50 Hz three-phase voltages are given by $u_k(t) = U \cos[\omega t - 2\pi(k-1)/3]$, for $k = 1,\ 2,\ 3$, and $U = 100\,\text{kV}$.

The insulation material used for wall-crossing purposes is characterized by $\varepsilon = 3\varepsilon_0$ and by a dielectric strength of $E_d = 21\,\text{MV/m}$.

The wall thickness is $l_4 = 20\,\text{cm}$.

The high-voltage feeding conductors radii are $r_0 = 15\,\text{mm}$.

Part I Wall-Crossing Design (Feed-Through Capacitor)

Let r_0 and r_4 be the feeding conductor radius and the hole in the wall radius respectively. The simplest solution to pass the conductor through the hole without causing dielectric breakdown is to use a very large hole. To avoid this brute force solution we can employ the configuration depicted in Figure P14.3, where the insulation medium is subdivided into several regions by employing a set of cylindrical coaxial metallic sheets (three in our case).

The cylindrical metallic sheets of radii $r_1,\ r_2$ and r_3 are equally spaced; their axial lengths (to be designed) are $l_1,\ l_2$ and l_3, respectively.

1. Neglecting fringing field effects, determine an analytical expression for the voltage between consecutive metallic sheets, when $u_k = (u_k)_{\max} = U$.
2. Determine the maximum electric field intensity $E_{\max}$ in each dielectric region. Establish the condition for the invariance of $E_{\max}$.

3. By enforcing $E_{max} = \frac{1}{3} E_d$, determine:
   (a) The radius r_4 of the hole in the wall.
   (b) The radii and lengths of the three cylindrical coaxial metallic sheets.
4. Obtain numerical values for the partial capacitances between the consecutive metallic sheets. Find the overall capacitance C' between each high-voltage conductor and the wall.
5. Obtain the voltages between consecutive metallic sheets.
6. Obtain the electric charge in the external surfaces of the metallic sheets.
7. Plot the radial evolution of the electric field intensity and potential function, $E(r)$ and $V(r)$, for $r_0 < r < r_4$, when $u_k = U$.

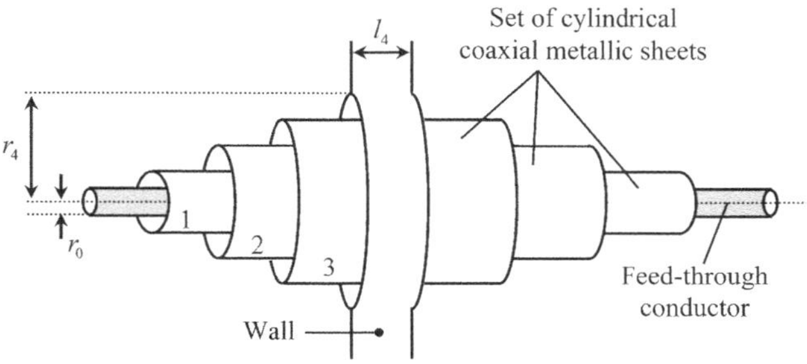

Figure P14.3 The wall crossing by the feeding conductor is accomplished by employing a set of cylindrical coaxial metallic sheets filled with an insulation medium with permittivity $\varepsilon = 3\varepsilon_0$. The wall thickness is $l_4 = 20\,\text{cm}$

Part II Inductor Design

Assume that the amplitude of the inductive voltage u_L in the inductor with N turns is 95 % of the overall applied voltage u_1. Consider that the magnetic induction field in the air gap is to reach a maximum value $B_0 = 1.5\,\text{T}$.

Neglect dispersion phenomena.

1. Obtain the complex amplitudes of u_R, u_L and u_1. Draw the corresponding phasor diagram.
2. Determine ϕ_{max}. Determine the maximum value of the magnetic flux linked with the winding with N turns. Compute N.
3. Knowing that $N' = N/2$, determine the complex amplitude $\overline{U}'$.
4. Assuming that Joule losses in the inductor are $P_J = 60\,\text{kW}$, determine the resistance R and the inductance L of the inductor's winding. Obtain $i_L(t)$.
5. Find the minimal cross-sectional area S_{Fe} of the shaded iron parts so as to prevent saturation from occurring. Determine h as well (see Figure P14.2).

Part III Power Factor Compensation

In order to minimize the magnitude of the feeding currents i_1, i_2 and i_3, a set of triangle-connected capacitors of capacitance C_Δ is placed in parallel with the RL circuits (Figure P14.1).

1. Determine the complex power $\overline{P}$ in play in the RL circuits. Check that

$$\Re\left(\overline{P}\right) = 3 \times P_J = 180\,\text{kW}.$$

   Check that $\Im\left(\overline{P}\right) = 3 \times 2\omega\,(W_m)_{av}$. Evaluate the time-averaged magnetic energy stored in each inductor, $(W_m)_{av}$.
2. Obtain the complex amplitude of the phase-to-phase voltage $u_{12}(t)$. Noting that power factor compensation corresponds to a resonance situation, determine C_Δ based on energy balance considerations.
3. Determine the complex amplitudes of i_{12}, i_{31}, i_C and i_1.
4. Check the importance of the currents (displacement currents) in the concentric capacitor bushings analyzed in I (wall crossing).

Project P15

Shielded Three-Phase Cable Analysis

Figure P15.1 shows a cross-sectional view of a symmetrical three-phase cable, made of three equally distant cylindrical conductors, of radius $r_0 = 2.5\,\text{mm}$, enclosed by a cylindrical conducting shield. The cable insulation medium is characterized by a relative permittivity $\varepsilon_r = 2.25$ and dielectric strength $E_d = 30\,\text{MV/m}$. The cable section is 1 km long, and its partial capacitances are known, $\hat{C}_{10} = 23\,\text{nF}$ and $\hat{C}_{12} = 3\,\text{nF}$.

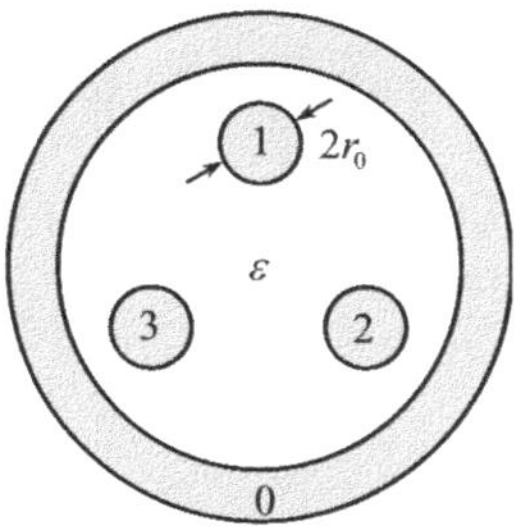

Figure P15.1 Cross-sectional view of a shielded symmetrical three-phase cable

Part I Electrostatic Field Analysis

1. Obtain the capacitance coefficients matrix $[C]$ of the cable.
2. Assume that initially all the conductors are discharged. Afterwards, a constant voltage $U_1 = 10\,\text{kV}$ is applied between conductor 1 and the shield.

   (a) Determine all the voltages and charges in the conductor system.
   (b) Estimate the maximum intensity of the electric field inside the cable.
   (c) What would be the threshold voltage U_1 leading to dielectric breakdown?

3. Next, the voltage generator is removed, and after that the three internal conductors are connected together. Recompute all the system voltages and charges, and estimate the new maximum intensity of the electric field inside the cable.

Part II Experimental Determination of Cable Capacitances

The capacitances of the cable were determined experimentally by resorting to resonance tests using a variable frequency sinusoidal voltage generator. Figure P15.2 shows, as an example, the circuit arrangement for the determination of $\hat{C}_{10}$. The inductor placed in parallel with the generator is characterized by its resistance $r = 300\ \Omega$ and inductance $L = 100\,\text{mH}$.

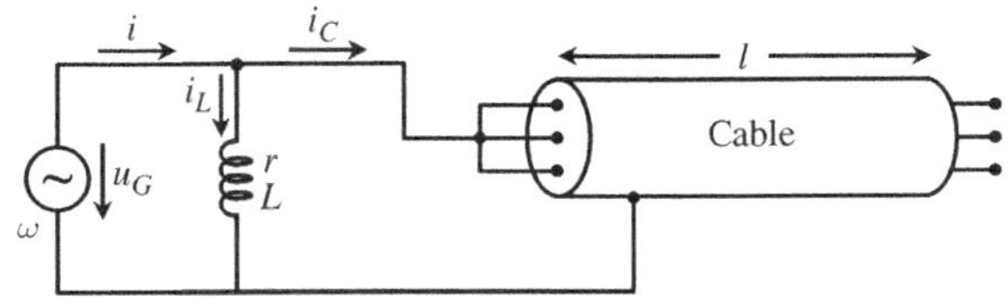

Figure P15.2 Resonant circuit used for the determination of cable capacitances

1. Obtain analytically the input admittance $\overline{Y}$ at the generator terminals, and determine the resonance condition of the circuit.
2. Let the generator frequency be adjusted to resonance $f = f_0$; denote by R_0 the resonance-equivalent resistance at the generator terminals.

   (a) Show that

$$\hat{C}_{10} = \frac{L}{3rR_0} \text{ and } f_0 = \frac{\sqrt{r(R_0 - r)}}{2\pi L}.$$

   (b) Compute R_0 and f_0 for the present case where $\hat{C}_{10} = 23\,\text{nF}$.
   (c) Determine the wavelength λ associated to f_0 and check if cable analysis can be conducted using standard circuit approaches.

3. Consider $u_G(t) = \sqrt{2}\ (U_G)_{rms} \cos(\omega_0 t)$, with $(U_G)_{rms} = 10\,\text{V}$ and $\omega_0 = 2\pi f_0$. Find the complex amplitudes of the circuit currents, i, i_L and i_C.

Part III Transmission-Line Analysis

Consider the situation examined in II (cable left open at its far end, $z = l$). Making use of transmission-line theory, for the limit case of short-length lossless lines, obtain the evolution of conductor voltages (u_1, u_2, u_3) and conductor currents (i_1, i_2, i_3) inside the cable, along its longitudinal coordinate.

Part IV Three-Phase System Steady-State Analysis

Figure P15.3 shows a symmetrical three-phase network where the cable under analysis is connected in parallel to a set of three identical resistors characterized by $R = 50\,\mathrm{k\Omega}$. The cable is left open at is far end. The load neutral conductor is bonded to the cable shield.

The system is driven by a 50 Hz three-phase generator whose voltages are given by:

$$u_k(t) = \sqrt{2}\, U_{rms} \cos[\omega t - 2\pi(k-1)/3], \text{ for } k = 1, 2, 3.$$

The rms value of the phase-to-phase voltage is known: $U_\Delta = 380\,\mathrm{V}$.

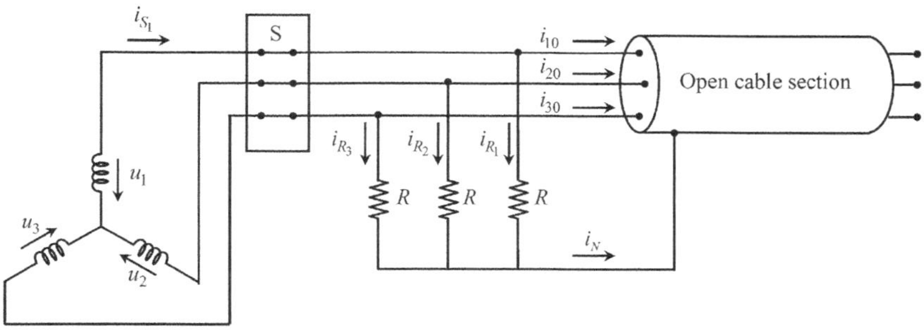

Figure P15.3 Symmetrical three-phase network where the cable under analysis is connected in parallel to a set of three resistors

1. Write the complex amplitudes of the phase-to-neutral voltages $\overline{U}_1, \overline{U}_2, \overline{U}_3$. Making use of the capacitance matrix $[C]$ obtained in I, determine the complex amplitudes of the cable electric charges $\overline{Q}_1, \overline{Q}_2, \overline{Q}_3$ and cable currents $\overline{I}_{10}, \overline{I}_{20}, \overline{I}_{30}$.
2. Determine the neutral current $\overline{I}_N$.
3. Verify that $\overline{Q}_k/\overline{U}_k = C =$ constant; find C. Note that the preceding result allows you to replace the network in Figure P15.3 by the equivalent network in Figure P15.4.

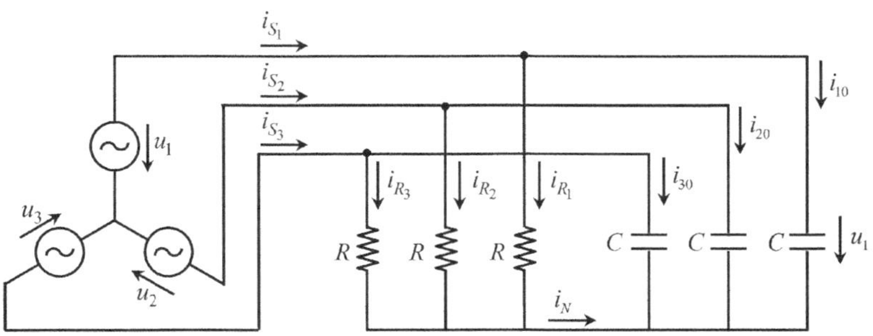

Figure P15.4 Equivalent circuit for the configuration shown in Figure P15.3

4. Determine the complex amplitudes of the resistor currents i_{R_k}, as well as the generator currents i_{S_k}.

5. Determine the active and reactive powers in play at the three-phase generator terminals. Check the results obtained by utilizing the complex Poynting theorem.

Part V Transient Analysis

Assume that the steady-state regime analyzed in Part IV is established for a long time. At $t = t_0$, when u_1 goes through zero, from negative to positive values, the three-phase switch S is opened.

1. Obtain the differential equations governing the cable discharge process.
2. Define the relevant initial conditions of the problem.
3. Determine the time evolution of the cable conductor voltages, for $t > t_0$.
4. Check the consistency of your results by doing an energy balance (compute the initial electric energy stored inside the cable, and compute the total Joule losses in the resistors).
5. Repeat the problem for the case when the neutral conductor is absent.

Project P16

Three-Route Microwave Splitter

Figure P16.1 shows a top view of a microwave three-route splitter which consists of a circular ring metallization of mean radius $r = 10.61\,\text{mm}$, to which four microstrips are connected. The ring and the strips are on a dielectric substrate placed over a conducting ground plane. Figure P16.2 shows one cross-sectional view of the structure.

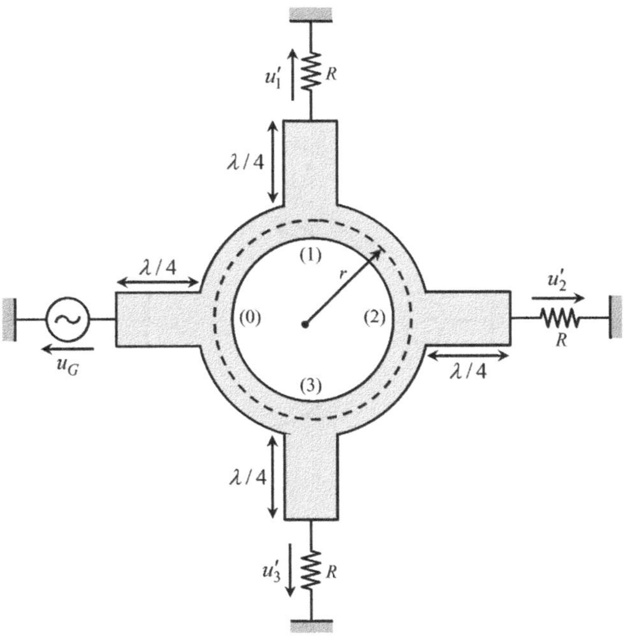

Figure P16.1 Three-route microwave splitter consisting of a circular ring, four microstrip ports, three resistor loads and a voltage generator

Electromagnetic Foundations of Electrical Engineering J. A. Brandão Faria

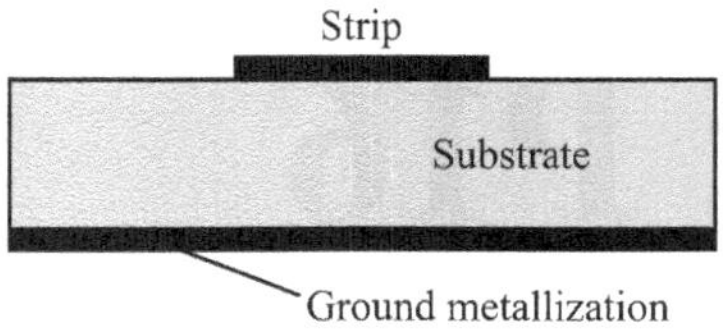

Figure P16.2 Cross-section showing upper strip, substrate and ground plane metallization

In order to neglect bending effects, the ring width is assumed to be much smaller than the ring radius.

The sinusoidal voltage generator connected to the input port 0 provides an active power P_0 which is split by the output ports 1, 2 and 3, which feed three identical loads (symbolized by three identical resistors whose resistance is $R = 75\ \Omega$).

The goal of this project is to analyze the resonating properties of the ring itself, and, in addition, to properly define the necessary properties of the connecting microstrips, so that the output powers P_1, P_2 and P_3 are all equal, for a specified working frequency $f_0 = 3\,\text{GHz}$.

The whole transmission system is assumed to be lossless.

Part I Transmission-Line Ring Analysis

Consider the situation where the ring microstrip connections are still absent (Figure P16.3).

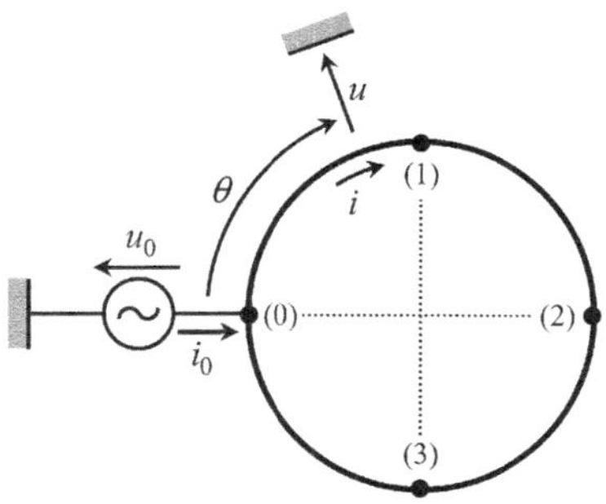

Figure P16.3 Circular ring driven by a voltage generator, with microstrip ports absent

Consider the steady-state harmonic regime corresponding to a sinusoidal voltage signal applied directly to port 0: $u_0(t) = \sqrt{2}\,(U_0)_{rms}\ \cos(\omega t)$.

1. Taking into account the boundary conditions $\overline{U}_{\theta=0} = \overline{U}_{\theta=2\pi} = \overline{U}_0$, determine analytically the complex amplitudes of the line voltage and line current as a function of the peripheral angle θ along the ring.
2. Particularize the preceding results for $\theta = 0$ (port 0) and $\theta = \pi$. Determine the complex amplitude of the current provided by port 0, i_0.
3. Determine the input admittance of the ring at port 0, $\overline{Y}_0 = \overline{I}_0/\overline{U}_0$, as a function of r/λ.

4. In order to determine the wave parameters of the ring line, the operating frequency was swept from 0 to 4 GHz, the input admittance being measured. A first zero was obtained at $f_0 = 3\,\mathrm{GHz}$ (resonance frequency), and at 4 GHz the admittance reading was $\overline{Y}_0 = j34.64\,\mathrm{mS}$. Determine the phase velocity v and the characteristic wave resistance R_w.
5. Consider $f = f_0$. Take $(U_0)_{rms} = 5\,\mathrm{V}$. Evaluate $\overline{U}(\theta)$ and $\overline{I}(\theta)$ for $\theta = \pi/2,\ \pi$ and $3\pi/2$; that is, at the points where ports 1, 2 and 3 will be connected.

Part II Quarter-Wave Transformer Analysis

Quarter-wave transmission lines have the special property of exhibiting real input impedance when the line is terminated by a purely resistive load (the time-averaged electric and magnetic energies inside the quarter-wave line perfectly compensate each other).

1. Find the relationship between the input and output voltages and currents in a quarter-wave line.
2. For a load impedance $\overline{Z}_L = R_L$, determine the line input impedance.

Part III Loaded Ring Line

Assume that three identical resistors R are directly connected at ports 1, 2 and 3 – see Figure P16.4. The excitation voltage at port 0 is $u_0(t) = \sqrt{2}\,(U_0)_{rms}\cos(\omega_0 t)$, with $f_0 = 3\,\mathrm{GHz}$.

Note that the configuration in Figure P16.4 is equivalent to the one shown in Figure P16.5, where the line is segmented into four identical sections, with characteristic wave resistance R_w, of length equal to $\lambda_0/4$.

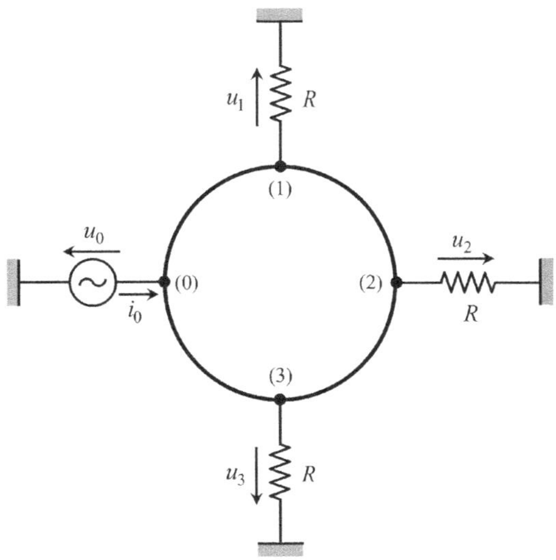

Figure P16.4 Circular ring driven by a voltage generator, with three resistors directly connected to the output ports

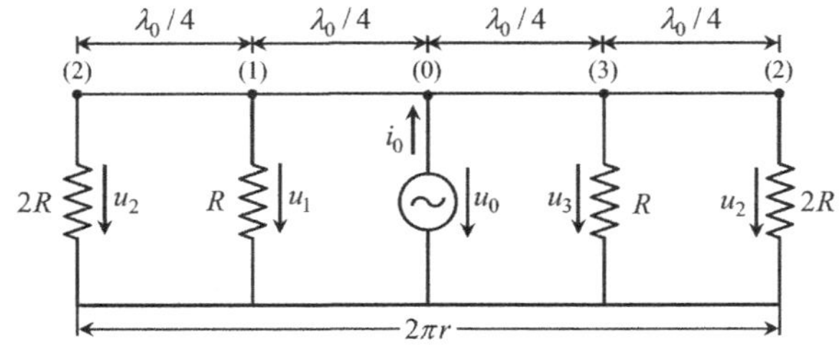

Figure P16.5 Transmission-line structure equivalent to the ring configuration in Figure P16.4

1. Determine analytically the complex amplitudes of the voltages u_1, u_2 and u_3 at the output ports.
2. Determine analytically the active powers delivered to port 1, 2 and 3.
3. Take $(U_0)_{rms} = 5\,\text{V}$, and assume that $R = 75\ \Omega$. Find numerical answers to the questions posed in 1 and 2. Check that the power delivered to the three ports is not equally split among them.
4. Determine the necessary condition to achieve equal power splitting and, for that situation, determine the input impedance at port 0.

Part IV Global Analysis

At last consider the original drawing in Figure P16.1, where the 75 Ω resistor loads, and the generator, are connected to the ring line by means of four microstrips, all of them a quarter-wavelength long. The widths of the microstrips are designed so that their characteristic resistances R_{w_1}, R_{w_2} and R_{w_3} meet the requirement of equal power splitting analyzed in III. The characteristic resistance R_{w_0} of the strip at port 0 is defined so that such a port is perfectly matched.

The generator voltage is given by $u_G(t) = \sqrt{2}\,5\ \cos(\omega_0 t)$ V, with $f_0 = 3\,\text{GHz}$.

1. Determine R_{w_1}, R_{w_2}, R_{w_3} and R_{w_0}. Are the microstrips' metallization widths of the ports smaller or larger than the ring line metallization width?
2. The four circular segments of the ring line are $\lambda_0/4$ long. The four microstrips at the four ports are also a quarter-wavelength long. Does this mean that the physical lengths of the microstrip lines are the same?
3. Compute the complex amplitudes of the ring-to-ground voltages u_0, u_1, u_2, u_3, as well as the load voltages u'_1, u'_2, u'_3.
4. Compute the active power delivered to each load, as well as the active power at the generator terminals.

Project P17

Transmission-Line System with Balun Transformer for Even- to Odd-Mode Conversion

In this project a transmission-line system consisting of two sections of two-conductor lines above ground, linked by a balun transformer, is analyzed – see Figure P17.1. The first line section of length l_E uses even-mode operation whereas the second line section of length l_O uses odd-mode operation. The aerial cylindrical conductors, of radius r_0, are a distance $2d$ apart and are placed at height h above the ground (with $r_0 \ll d, h$).

Data: $r_0 = 5\,\text{mm}$, $d = 40\,\text{cm}$, $h = 4\,\text{m}$.

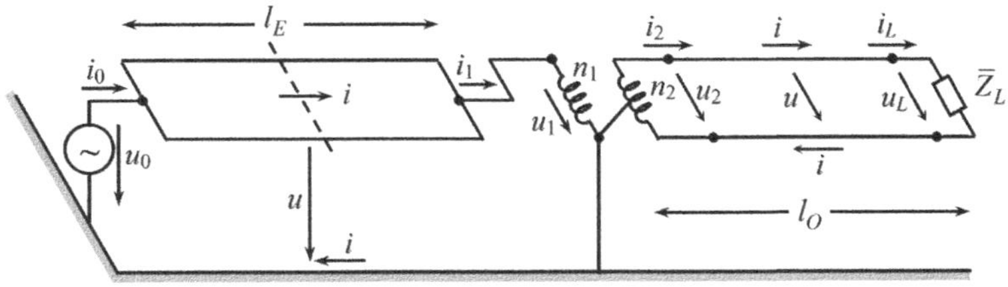

Figure P17.1 Perspective view of the global transmission-line system, with balun transformer, for even- to odd-mode conversion

Part I Electrostatic Field Analysis

1. Even-mode analysis. Consider the cross-sectional view shown in Figure P17.2(a) where q denotes the per-unit-length line charge $q = Q/l$.

   (a) Find the scalar potential function V at a point P(d, y).
   (b) Find the voltage U between the points a and b.

Electromagnetic Foundations of Electrical Engineering J. A. Brandão Faria

(c) Determine the per-unit-length even-mode capacitance of the line, $C_E = q/U$.
(d) Take $U = 1\,\text{kV}$. Evaluate the electric charge q and the electric energy W_e stored per unit length of the line.

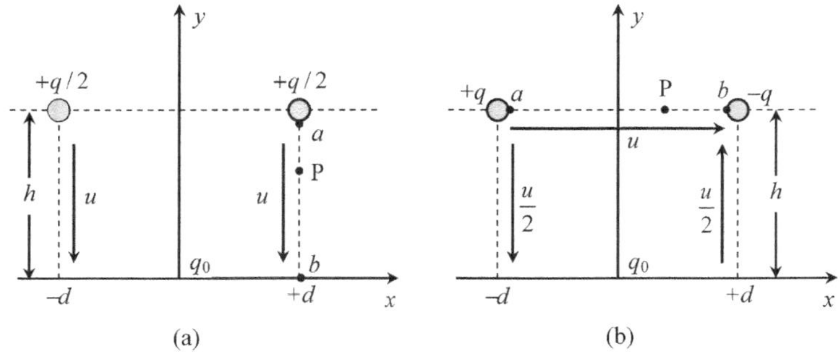

Figure P17.2 Cross-sectional view of the two parts of the transmission-line system. (a) Transmission line of length l_E, supporting even-mode propagation. (b) Transmission line of length l_O, supporting odd-mode propagation

2. Odd-mode analysis. Consider the cross-sectional view shown in Figure P17.2(b) where q denotes the per-unit-length conductor charge $q = Q/l$.

   (a) Find the scalar potential function V at a point $P(x, h)$.
   (b) Find the voltage U between the points a and b.
   (c) Determine the per-unit-length odd-mode capacitance of the line, $C_O = q/U$.
   (d) Take $U = 1\,\text{kV}$. Evaluate the electric charge q and the electric energy W_e stored per unit length of the line.

3. The relationship between line voltages and line charges can be represented by an equivalent scheme of partial capacitances – see Figure P17.3. Using the results in 1 and 2, find the values of C_{10}, C_{20} and C_{12}.

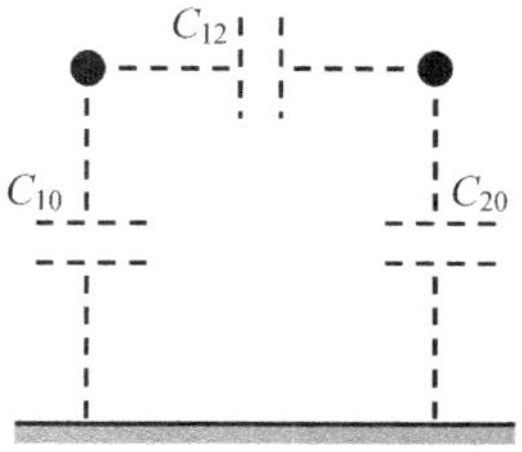

Figure P17.3 Scheme of partial capacitances

Part II Balun Transformer

Figure P17.4 depicts the detailed constitution of the balun transformer's magnetic circuit used to link the two line sections. The two high-permeability ferrite magnetic cores are identical. Neglect dispersion phenomena. Consider a time-harmonic regime characterized by $f = 1\,\text{MHz}$.

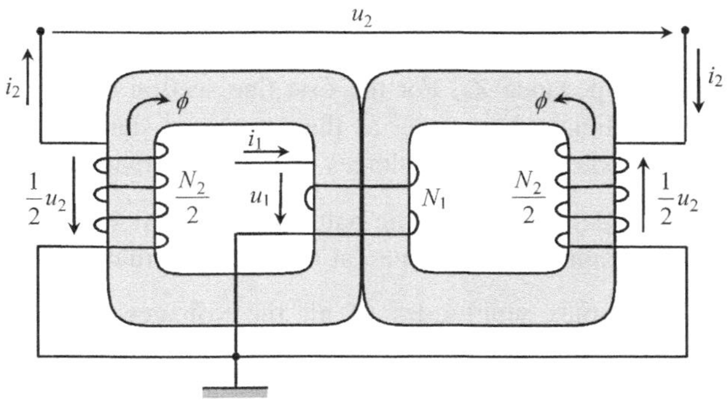

Figure P17.4 Balun transformer

1. To start with, assume that the balun transformer is an ideal transformer.
   (a) By application of Faraday's induction law and Ampère's law, find the relationship between $\overline{U}_1$ and $\overline{U}_2$, as well as between $\overline{I}_1$ and $\overline{I}_2$.
   (b) Determine the turns ratio N_1/N_2 such that $\overline{Z}_1 = \eta\overline{Z}_2$, with $\eta = 0.477\,56$, where $\overline{Z}_1 = \overline{U}_1/\overline{I}_1$ and $\overline{Z}_2 = \overline{U}_2/\overline{I}_2$.
2. Now consider the actual transformer imperfections: $r_1 \neq 0$, $r_2 \neq 0$ and $(R_m)_{\text{core}} \neq 0$. Assume that $r_1/r_2 = N_1/N_2$, with N_1/N_2 defined in 1(b).
   (a) Establish the relationship between the transformer inductances, L_{11}, L_{22} and L_M.
   (b) With the secondary winding open ($i_2 = 0$), at $f = 1\,\text{MHz}$, the transformer input impedance was measured, $\overline{Z}_1 = R_1 + jX_1$, with $R_1 = 10\ \Omega$, $X_1 = 2\,\text{k}\Omega$. Evaluate the transformer characteristic parameters, r_1, r_2, L_{11}, L_{22} and L_M.
   (c) Draw the 'T'-shaped equivalent circuit of the balun transformer.

Part III Transmission-Line System

Take the transmission system shown in Figure P17.1. The working frequency, $f = 1\,\text{MHz}$, is such that $l_O = 2l_E = \lambda/4$ (where λ denotes the wavelength).

Assume that both line sections are lossless.

The generator voltage is given by $u_0(t) = \sqrt{2}\ (U_0)_{rms}\cos(\omega t)$, with $(U_0)_{rms} = 100\,\text{V}$.

1. For the line section that supports even-mode propagation, determine the corresponding phase velocity v_E, the per-unit-length inductance L_E, the characteristic wave resistance R_{w_E}, as well as its physical length l_E.
2. For the line section that supports odd-mode propagation, determine the corresponding phase velocity v_O, the per-unit-length inductance L_O, the characteristic wave resistance R_{w_O}, as well as its physical length l_O.

3. The second line section of length l_O is terminated on a matched load, $\overline{Z}_L = R_{w_O}$.
   (a) Take the ideal model for the balun transformer analyzed in II.1. Show that by using the chosen value for the parameter eta, $\eta = 0.477\,56$, the first line section of length l_E also behaves as a matched line.
   (b) Take the more realistic model for the balun transformer analyzed in II.2. Determine the transformer's impedance $\overline{Z}_1$. For the first line section of length l_E, evaluate the load reflection coefficient $\overline{\Gamma}$, as well as the associated standing wave ratio. Obtain the complex amplitudes of all the voltages and currents marked in Figure P17.1.
4. Make use again of the ideal model for the balun transformer, but now consider that the second line section of length l_O is left open at the load terminals ($i_L = 0$).
   (a) Recompute the complex amplitudes of all the voltages and currents marked in Figure P17.1.
   (b) Draw the standing wave patterns for the rms voltage and current along the transmission system.

Project P18

Transmission-Line System with Transformer-Stage Matching

In this project a transmission-line system which includes a transformer stage for matching purposes is analyzed and designed. Figure P18.1 shows the overall transmission system consisting of a high-frequency generator, an aerial two-conductor symmetrical line, a transformer and a load. The generator is described by an ideal voltage source u_G and by its internal resistance R_G. The two-conductor symmetrical line, of length l, is made of two cylindrical conductors of diameter ϕ, whose axes are separated by d.

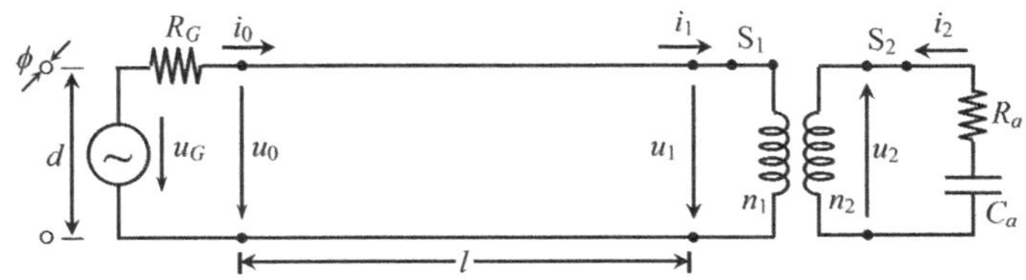

Figure P18.1 Transmission-line system with transformer-stage matching

The transformer stage, with n_1 turns in the primary winding and n_2 turns in the secondary winding, is loaded with a resistor R_a and a capacitor C_a.

Part I Transformer-Stage Analysis

1. In order to determine the characteristic features of the transformer stage, two experiments have been conducted. Firstly, with the secondary winding open, a sinusoidal voltage was applied to the primary winding. Conditions were reversed in the second experiment; with

the primary winding open, a sinusoidal voltage was applied to the secondary winding. Figure P18.2 shows oscilloscope readings concerning one half-period of the monitored voltages and currents.

Write the phasor equations corresponding to both experiments and based on them determine the windings' resistances r_1 and r_2, as well as the self- and mutual inductances L_{11}, L_{22} and L_M. Evaluate the magnetic coupling factor k between transformer windings.

Obtain the voltage $u_1(t)$ corresponding to the second experiment.

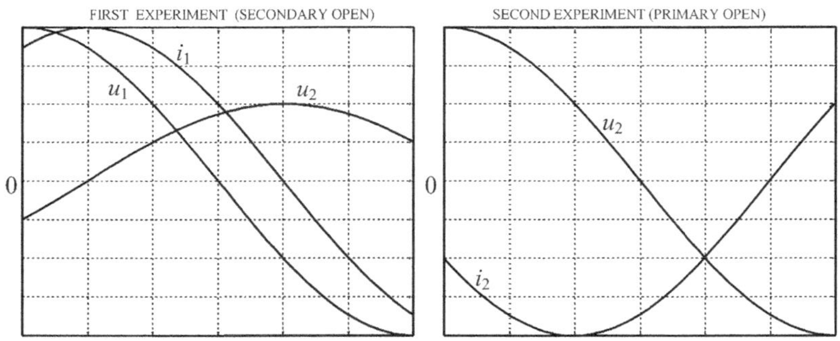

Figure P18.2 Oscilloscope readings concerning transformer measurements. Horizontal time scale: 1 div. = 50 ns. Vertical scale for voltages: 1 div. = 25 V. Vertical scale for currents (first experiment): 1 div. = 216.5 mA. Vertical scale for currents (second experiment): 1 div. = 125.0 mA

2. Draw an equivalent circuit for the actual transformer containing an ideal transformer with a turns ratio of 1:1.
3. Let $\omega = 2\pi f$ be the operating angular frequency. Analyze the input impedance $\overline{Z}_1$ of the loaded transformer,

   (a) Show that $r_1 < \Re\{\overline{Z}_1\} < r_1 + \dfrac{(\omega L_M)^2}{r_2}$

   (b) Show that resonance conditions, $\overline{Z}_1 = \Re\{\overline{Z}_1\}$, can only be met for $f > r_2/(\pi k^2 L_{22})$.
   (c) Particularize the results in (a) and (b) for the case $f = 10\,\text{MHz}$.

4. The design goal is to obtain $\overline{Z}_1 = R_1 = 600\ \Omega$. Determine analytically the appropriate load parameters R_a and C_a, as a function of ω, which allow the specified goal to be achieved.
5. Consider that the transmission system is operating at 10 MHz, with a required voltage in the secondary winding given by $u_2 = \sqrt{2}\,(U_2)_{rms}\cos(\omega t)$, where $(U_2)_{rms} = 1\,\text{kV}$.

   (a) Compute the values of R_a and C_a.
   (b) Making use of the equivalent circuit obtained in 2, evaluate the complex amplitudes of the voltages and currents concerning both transformer windings.
   Verify that $\overline{U}_1/\overline{I}_1 = R_1 = 600\ \Omega$.
   (c) Compute the active and reactive powers at the primary winding terminals. Check the result by using the complex Poynting theorem.

Part II Transmission-Line Analysis

Consider the working conditions established in I.5. The (lossless) transmission line is matched when the switches S_1 and S_2 are closed.

1. Find the phase velocity v and the characteristic wave resistance R_w of the line. In addition, determine the per-unit-length inductance and capacitance of the line. Determine the ratio of the geometrical parameters d and ϕ. Obtain the physical length of the line knowing that it corresponds to one-third of the operating wavelength.
2. Determine the necessary value of the generator's internal resistance R_G such that, in any circumstance, the incident wave voltage at the generator terminals is always half of u_G.
3. Compute the complex amplitude of u_G such that the design goal:
   $u_2 = \sqrt{2}\,(U_2)_{rms}\cos(\omega t),\ \ (U_2)_{rms} = 1\,\text{kV}$, is achieved.
   Obtain the complex amplitudes of $u_0(t)$ and $i_0(t)$.
4. Consider the voltage u_G obtained in 3. Analyze the following two cases:

   case I: switch S_1 open;
   case II: switch S_1 closed, switch S_2 open.

   (a) For both cases find the load reflection coefficient and the standing wave ratio.
   (b) For both cases obtain the rms values of the line voltage and line current at both ends of the line. Determine the maximum and minimum rms values of the line voltage and line current.
   (c) Which case is the worst scenario as far as dielectric breakdown phenomenon is concerned? Assuming that $d = 10\,\text{cm}$, find the maximum intensity of the electric field. At which particular point of the line is the maximum $\mathbf{E}$ field observed?
   (d) For case I, determine the per-unit-length time-averaged force actuating on the line conductors. What would that force be if the line were matched?

Project P19

Two-Way Loudspeaker Analysis

Figure P19.1 represents a simplified version of an electric circuit with discriminating properties as far as audio frequencies are concerned (20 Hz to 20 kHz).

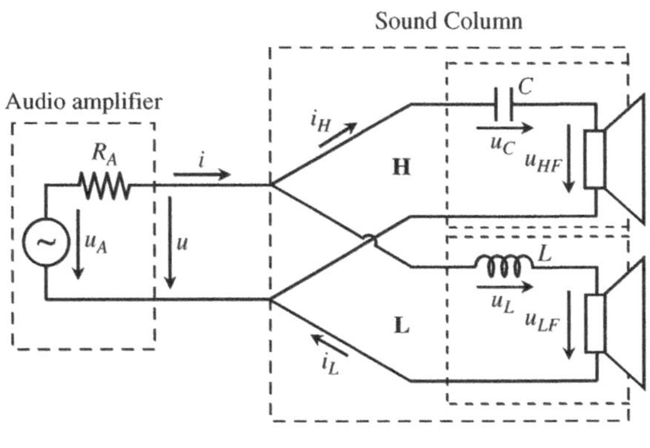

Figure P19.1 A two-way loudspeaker sound column driven by an audio amplifier

The final stage of an audio amplifier is connected to a sound column provided with two loudspeakers, a driver for lows and a tweeter for highs. The output impedance of the amplifier is $R_A = 5\ \Omega$.

The conversion from electric power to acoustic power in both loudspeakers is simulated through an equivalent resistance R (Figure P19.2). Assume that the driver and the tweeter have been properly designed so as to have the same equivalent resistance R, whose value remains approximately constant over the whole audio-frequency range.

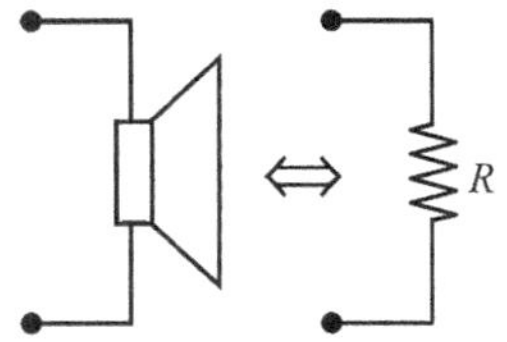

Figure P19.2 A simplified model of the loudspeaker using an equivalent resistance

Part I Steady-State Analysis

1. Write the time-domain and phasor-domain equations governing the circuit.
2. Determine the input impedance $\overline{Z}$ at the amplifier terminals as a function of the design parameters, R, L and C.
3. Determine analytically the appropriate parameters R, L and C, so that the following specifications are fulfilled:
   - equal amounts of power are delivered to both loudspeakers at a given specified frequency f_x (the so-called crossover frequency);
   - the input impedance of the sound column is real, $\overline{Z} = Z$, and independent of frequency;
   - the impedance Z is chosen so as to maximize the amplifier output power.
4. Evaluate numerically R, L, C and Z for the following crossover frequencies: 1 kHz, 2 kHz and 3 kHz.
5. Choose the crossover frequency as $f_x = 1\,\text{kHz}$. Consider that the output voltage of the amplifier is given by $u(t) = \sqrt{2}\ U_{rms} \cos(\omega t)$, with $U_{rms} = 10\,\text{V}$. Analyze the system response for two different audio frequencies $f_L = f_x/3$ (low frequency) and $f_H = 3f_x$ (high frequency).
   (a) Obtain the complex amplitudes of the sinusoidal voltages and currents marked in Figure P19.1.
   (b) Evaluate the active and reactive powers at the amplifier output, and at the input terminals of both the H- and L-blocks. Check the results using the complex Poynting theorem.

Part II Transfer Functions

Consider the design parameters obtained for the case $f_x = 1\,\text{kHz}$.

1. Determine the transfer functions

$$\overline{T}_H(f) = \frac{\overline{U}_{HF}}{\overline{U}}, \quad \overline{T}_L(f) = \frac{\overline{U}_{LF}}{\overline{U}}$$

2. Represent graphically the magnitude characteristics $|\overline{T}_H| = T_H$ and $|\overline{T}_L| = T_L$ against frequency in the range 20 Hz to 20 kHz (use a logarithmic scale on the frequency axis).
3. Elaborate your conclusion concerning the behavior of the H- and L-blocks.

Part III Transient Regime Analysis

Assume that the amplifier is switched on at $t = 0$. The internal voltage of the amplifier is described by

$$\begin{cases} u_A(t) = 0 \text{ for } t < 0 \\ u_A(t) = U_A \cos(\omega t) \text{ for } t > 0 \end{cases}$$

where $U_A = \sqrt{2}\, 20\,\text{V}$, $\omega = 2\pi f$, $f = 3\,\text{kHz}$.

1. Determine the transient response regarding $i(t)$, $u_{HF}(t)$ and $u_{LF}(t)$, for $t > 0$.
2. Discuss the possibility of choosing a particular time instant t_0 such that the steady-state regime of the loudspeaker voltages establishes itself immediately (free regime absent).

Project P20

Variable Reluctance Transformer

In this project we ask you to analyze the functioning of a two-window transformer which includes a variable air gap in the central leg.

Consider the transformer depicted in Figure P20.1, whose primary and secondary windings have the same number of turns, $N_1 = N_2 = N = 500$. The magnetic reluctance of the shaded iron parts is negligibly small. The vertical legs, of length $l = 15\,\text{cm}$, section $S = 5\,\text{cm}^2$, are characterized by a relative permeability $\mu_r = 200$. A small air gap of thickness δ is included in the central leg.

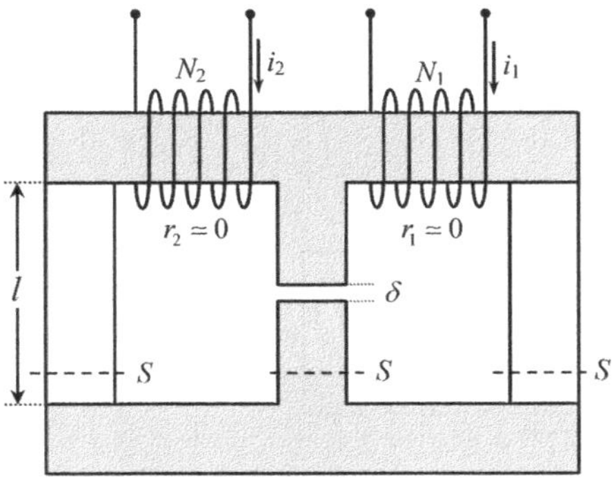

Figure P20.1 Magnetic circuit of the variable reluctance transformer. The magnetic reluctance of the shaded parts is assumed to be negligibly small

Part I Magnetic Circuit Analysis

1. Deduce analytical expressions for the induction coefficients L_{11}, L_{22} and L_M. Compute their values for $\delta = \delta_0 = 2\,\text{mm}$.
2. Determine the magnetic coupling factor k between the inductors, as a function of δ. Write L_{11}, L_{22} and L_M in terms of k.

Electromagnetic Foundations of Electrical Engineering J. A. Brandão Faria

3. Assume that the windings are connected in series, $i_1 = i_2 = I = 1\,\text{A}$. Determine the **H**- and **B**-field intensities in both transformer legs, as well in the air gap. Interpret the results.

Part II Equivalent Circuit

The transformer analysis can be conducted by making use of an equivalent 'T' circuit chain connected to an ideal transformer whose transformation ratio $\upsilon : 1$ is arbitrary.

1. Define the allowed range for the variation of υ.
2. Choosing the upper limit for υ, determine the components of the equivalent circuit shown in Figure P20.2.

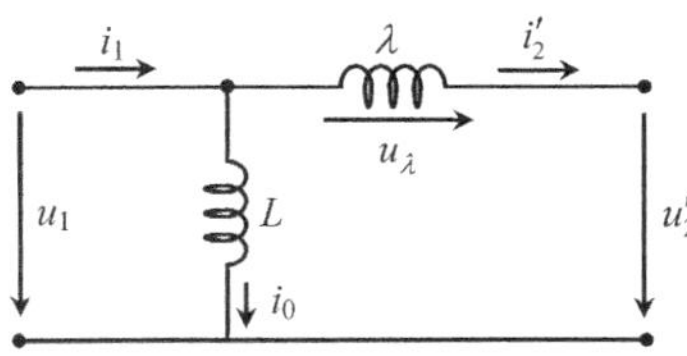

Figure P20.2 The transformer's equivalent circuit

Part III AC Analysis

Consider the circuit connections shown in Figure P20.3. The circuit is driven by a 500 Hz sinusoidal voltage generator whose voltage is given by $u_G(t) = \sqrt{2}\ (U_G)_{rms} \cos(\omega_0 t)$, where $(U_G)_{rms} = 20\,\text{V}$.

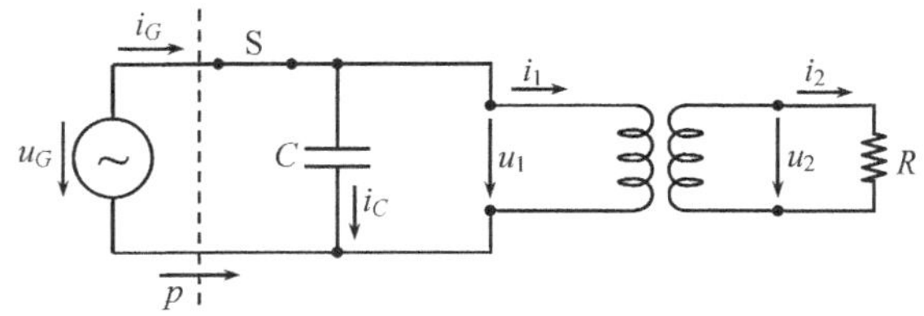

Figure P20.3 The network under analysis includes a generator, a capacitor, the transformer and its resistive load

The design goals for R and C are, on the one hand, to maximize the load active power and, on the other hand, to minimize the generator current amplitude.

1. Redraw Figure P20.3 employing the equivalent circuit analyzed previously.
2. Plot in the complex plane the locus described by the tip of phasor $\overline{I}_1$ when R varies from 0 to ∞.
3. Determine R and C as a function of the magnetic coupling factor between inductors.
4. Determine the input impedance measured at the generator terminals.
5. Determine the complex amplitudes of all the voltages and currents marked in Figure P20.3 and in Figure P20.2.

6. Apply the complex Poynting theorem in order to check the above results.
7. Recompute the design parameters R and C, for the cases $\delta = \delta_0/2 = 1\,\text{mm}$ and $\delta = 2\delta_0 = 4\,\text{mm}$.

Part IV Frequency-Domain Analysis

Take the R and C parameters obtained for the case $\delta = \delta_0$. Let the working frequency vary continuously.

1. Show that the voltage transfer function $T(f) = (U_2)_{rms}/(U_1)_{rms}$ can be written as

$$T(f) = \frac{k}{\sqrt{1 + (f/f_0)^2}}$$

   where k is the magnetic coupling factor between inductors.
2. So far, the inductor's resistances have been neglected. Now take $r_1 = r_2 = 10\ \Omega$. Determine the new transfer function $T(f)$ and plot it against the frequency. Identify graphically the cutoff frequencies for which $T = T_{\max}/\sqrt{2}$.

Part V Transient Regime

The steady-state harmonic regime ($f = f_0$) analyzed in III is the end result of a transient regime started at a given instant. Keep assuming that inductor resistances are approximately zero.

1. Consider that the switch-on instant is $t = 0$. Determine the transient regime for the primary and secondary currents.
2. What would happen if a different switch-on instant were chosen?

Part I
A Brief Overview

Introduction

Ever since Maxwell wrote his *Treatise on Electricity and Magnetism* the subject area of electromagnetism has been evolving at a strong pace. Despite its age, this area of knowledge is both alive and active: not only have its applications exploded in interest, but also new powerful computational methods have been developed for electric and magnetic field evaluation, and, what is more, its theoretical tools have also evolved.

As far as the theoretical aspects are concerned, modern trends point towards the substitution of traditional vector analysis by new, powerful and concise mathematical tools based on geometric algebra formulations[1] (exterior calculus, graded algebra of multivectors, graded algebra of differential forms, etc.). The most salient feature of these new modern tools is that, in addition to an elegant and concise view of electromagnetic phenomena, they provide a perfect match between the laws of electromagnetism and Einstein's theory of relativity.

Geometric algebra formulations of electromagnetism are adequate for advanced courses in the second and third cycle of studies of the Bologna framework. However, for the first cycle of studies, the most familiar vector calculus (Appendix A) formulation is recommended:

$$\text{Maxwell's equations}\begin{cases}\text{curl }\mathbf{E} = -\dfrac{\partial \mathbf{B}}{\partial t}\\ \text{div }\mathbf{B} = 0\\ \text{curl }\mathbf{H} = \mathbf{J} + \dfrac{\partial \mathbf{D}}{\partial t}\\ \text{div }\mathbf{D} = \rho\end{cases} \tag{I.1}$$

The electromagnetic field equations in (I.1) will be utilized throughout this book for the macroscopic description of a variety of electromagnetic phenomena which concern current electrical engineering problems.

The equation set in (I.1) is an axiomatic one; it cannot be mathematically deduced from any source. In fact, Maxwell's equations are a result of experimental research accumulated until the end of the nineteenth century.

[1] In terms of geometric algebra, all the properties of the electromagnetic field can be condensed into a single equation: $\nabla F = J$.

What is truly astonishing is that not a single flaw has been detected in these equations since then. On top of that, Maxwell's equations have passed Einstein's relativity challenge intact and unscathed. Indeed, if you consider two distinct reference frames (x, y, z, t) and (x', y', z', t') where the second is moving with a relative speed $\mathbf{v}$, the new equations for the second referential read

$$\begin{cases} \text{curl}' \ \mathbf{E}' = -\dfrac{\partial \mathbf{B}'}{\partial t'} \\ \text{div}' \ \mathbf{B}' = 0 \\ \text{curl}' \ \mathbf{H}' = \mathbf{J}' + \dfrac{\partial \mathbf{D}'}{\partial t'} \\ \text{div}' \ \mathbf{D}' = \rho' \end{cases} \tag{I.2}$$

showing that the electromagnetic field's governing laws conserved its original shape, despite the change of referential. Note that the primed and unprimed quantities in (I.1) and (I.2) are related through the Lorentz transformation (Appendix B).

Quantum theory, another modern physics development, is mainly devoted to the analysis of microscopic phenomena; nonetheless, Maxwell's equations are not divorced from it. The wave–matter duality principle still allows a link to be established between the two formalisms.

As we will see later (in Part IV), the amount of energy carried by a sinusoidal electromagnetic wave of frequency f, crossing a surface S, during one oscillation period $T = 1/f$, is given by

$$W_T = \int_0^T \left(\int_S (\mathbf{E} \times \mathbf{H}) \cdot \mathbf{n} \, dS \right) dt \tag{I.3}$$

where $\mathbf{n}$ is the unit normal to S directed along the wave propagation direction.

From a quantum theory viewpoint, such energy is evaluated as

$$W_T = N_p h f \tag{I.4}$$

where h is the universal Plank constant, and N_p is the number of photons crossing S during one time period T.

By equating (I.3) and (I.4) you will find the link between the macroscopic field and quantum views:

$$N_p = \frac{1}{hf} \int_0^T \left(\int_S (\mathbf{E} \times \mathbf{H}) \cdot \mathbf{n} \, dS \right) dt$$

These remarks are intended to show you that, almost 150 years from their establishment, Maxwell's equations are still current, and do not conflict with modern physics developments.

1

Basic Field Vectors

1.1 The Electric and Magnetic Field Vectors

A set of four vectors is needed to describe electromagnetic field phenomena. These are:

the electric field vector, **E** (units: V/m, volt per meter)
the magnetic induction vector, **B** (units: T, tesla)
the electric displacement vector, **D** (units: C/m^2, coulomb per square meter)
the magnetic field vector, **H** (units: A/m, ampere per meter)

Among these, the first two have special physical significance, since they can be determined experimentally and measured.

If you place a static charged particle $Q > 0$ in a region where **E** is to be determined, you will see that a force $\mathbf{F}_e$ is exerted on the charge (Figure 1.1(a)) from which **E** is obtained:

$$\mathbf{E} = \frac{1}{Q}\mathbf{F}_e \tag{1.1}$$

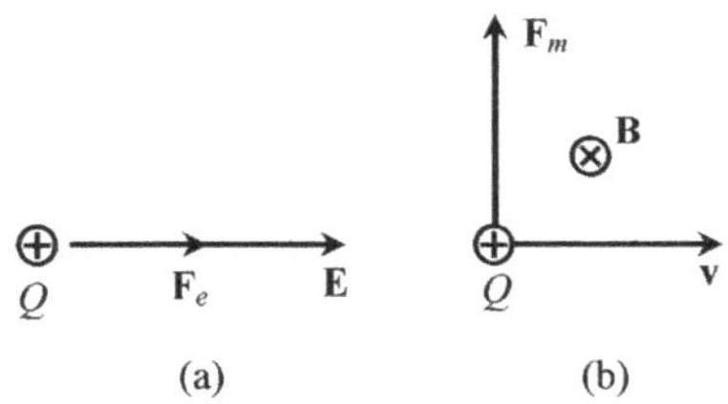

Figure 1.1 Actuating forces on a charged particle. (a) Electric force. (b) Magnetic force

Now consider that the same charged particle is moving with a prescribed velocity **v** in a region where **B** exists. You will notice that the trajectory of the moving particle may start changing due to the presence of a new force, $\mathbf{F}_m = Q(\mathbf{v} \times \mathbf{B})$. If you choose the velocity

Electromagnetic Foundations of Electrical Engineering J. A. Brandão Faria

vector **v** perpendicular to **B** (Figure 1.1(b)) you will obtain the intensity of the magnetic induction vector

$$B = \frac{1}{vQ} F_m \tag{1.2}$$

If a moving charged particle is submitted to an electromagnetic field, the two forces will add and you get the so-called Laplace–Lorentz force

$$\mathbf{F} = Q(\mathbf{E} + \mathbf{v} \times \mathbf{B}) \tag{1.3}$$

The fundamental vector pair (**E, B**) constitutes the electromagnetic field. Its intrinsic properties are defined by the following equations:

$$\text{curl } \mathbf{E} = -\frac{\partial \mathbf{B}}{\partial t} \tag{1.4a}$$

$$\text{div } \mathbf{B} = 0 \tag{1.4b}$$

These equations have two very simple implications: the field lines of vector **B** are closed; and, moreover, if **B** happens to be a time-varying field then it will give rise to the presence of an electric field **E** (induction phenomena).

Electromagnetic fields cannot be created from a void. Their existence requires the presence of sources, namely charges, either at rest or moving in space. Charges at rest are usually characterized by their volume density (a scalar field); when they move, that is when currents are present, they are characterized by a current density (a vector field):

charge density, ρ (units: C/m^3, coulomb per cubic meter)
current density vector, **J** (units: A/m^2, ampere per square meter)

The connection between sources and fields is established using the two remaining Maxwell's equations:

$$\text{curl } \mathbf{H} = \mathbf{J} + \frac{\partial \mathbf{D}}{\partial t} \tag{1.5a}$$

$$\text{div } \mathbf{D} = \rho \tag{1.5b}$$

Note that earlier theories of electromagnetism did not include the term $\partial \mathbf{D}/\partial t$ on the right-hand side of (1.5a); its inclusion due to Maxwell was revealed to be absolutely crucial. This is why the equations of electromagnetism are now termed Maxwell's equations.

1.2 Constitutive Relations

At this point you may feel that something is missing. On the one hand, we have a relationship between field sources (ρ, **J**) and the auxiliary fields (**D, H**) and, on the other hand, we have a relationship that dictates the properties of the electromagnetic field (**E, B**). The missing link is the one that relates the pair (**D, H**) to the pair (**E, B**) – see Figure 1.2.

The connections between (**D, H**) and (**E, B**) do not belong to the set of Maxwell's equations. They have to do with the interaction of the electromagnetic field with the material

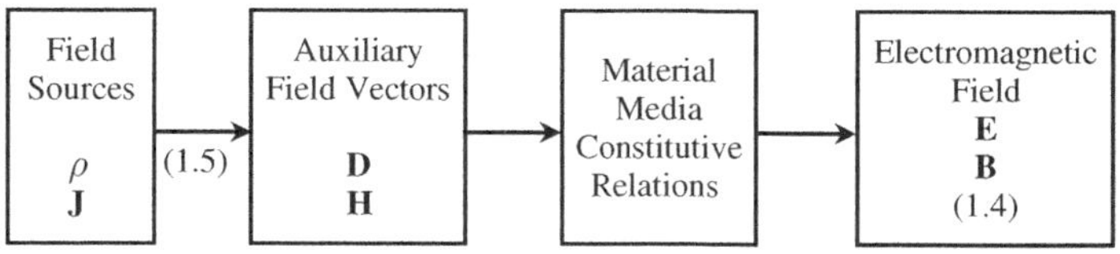

Figure 1.2 Connection among field sources, electromagnetic field vectors and material media properties

media where fields are impressed. Depending on the material medium, analysis of the field–matter interaction can be extremely complicated, requiring, for its understanding, contributions from atomic physics and quantum mechanics. Hopefully, you will learn about this subject in another discipline of your course.

Here, since we are going to deal with rather simple material media with linear isotropic characteristics, a pragmatic heuristic approach of the media macroscopic properties will be adopted. Except for a very few cases, these properties will be described by the following constitutive relations:

$$\mathbf{D} = \varepsilon \mathbf{E}, \quad \mathbf{B} = \mu \mathbf{H} \tag{1.6}$$

where ε and μ denote, respectively, the permittivity and the permeability of the medium. Values for these parameters can be found in tabular form in many books devoted to the study of the electromagnetic properties of materials. In particular, for a vacuum, we have the following fundamental constants:

$$\varepsilon = \varepsilon_0 \approx \frac{1}{36\pi} \times 10^{-9} \text{ F/m (units: farad per meter)}$$

$$\mu = \mu_0 = 4\pi \times 10^{-7} \text{ H/m (units: henry per meter)}$$

from which you can see that

$$\frac{1}{\sqrt{\mu_0 \varepsilon_0}} = c = 3 \times 10^8 \text{ m/s}$$

where c denotes the speed of light in a vacuum.

Note that the actual value for the speed of light is $2.997\,925\ldots \times 10^8$ m/s, thus the correct value for ε_0 is $8.854188\ldots \times 10^{-12}$ F/m (but this is a seldom required fine detail).

1.3 Units and Notation

As you may have noticed in the preceding section, fields and a few other quantities have already been assigned units.

Throughout this book the rationalized International System of Units (SI) is adhered to. Table 1.1 contains a list of the main quantities that appear in this book, as well as their corresponding SI units.

As far as the computation of electromagnetic quantities is concerned, students are reminded that numerical results are incomplete until their units are made explicit.

Table 1.1 Quantities and units

Quantity	Quantity symbol	Unit designation	Unit symbol
Angular frequency	ω	radian per second	rad/s
Capacitance	C	farad	F
Charge density	ρ	coulomb per cubic meter	C/m^3
Conductance	G	siemens	S
Conductivity	σ	siemens per meter	S/m
Current density	$\mathbf{J}$	ampere per square meter	A/m^2
Current intensity	i, I	ampere	A
Electric charge	q, Q	coulomb	C
Electric displacement	$\mathbf{D}$	coulomb per square meter	C/m^2
Electric field	$\mathbf{E}$	volt per meter	V/m
Electric polarization	$\mathbf{P}$	coulomb per square meter	C/m^2
Electric potential	V	volt	V
Energy	W	joule	J
Force	$\mathbf{F}$	newton	N
Frequency	f	hertz	Hz
Inductance	L	henry	H
Length	l	meter	m
Magnetic induction	$\mathbf{B}$	tesla	T
Magnetic field	$\mathbf{H}$	ampere per meter	A/m
Magnetic flux	ϕ, Ψ	weber	Wb
Magnetic voltage	U_m	ampere	A
Magnetization	$\mathbf{M}$	ampere per meter	A/m
Permeability	μ	henry per meter	H/m
Permittivity	ε	farad per meter	F/m
Potential vector	$\mathbf{A}$	weber per meter	Wb/m
Power	p, P	watt	W
Poynting's vector	$\mathbf{S}$	watt per square meter	W/m^2
Reluctance	R_m	henry^{-1}	H^{-1}
Resistance	R	ohm	Ω
Time	t	second	s
Velocity	$\mathbf{v}$	meter per second	m/s
Voltage	u, U	volt	V

Regarding the writing of variables and quantities we use the following notation. Scalar quantities are in italics. Vectors are boldface. Time-invariant quantities are capitalized. Matrix quantities are identified with square brackets. Complex quantities are identified with overbars.

1.4 Fundamental Concepts of Voltage and Current Intensity

In this book the electromagnetic radiation aspects are only touched upon very superficially (Chapter 8); consequently, most of the phenomena we are going to analyze do not need, for their description, more then the key concepts of voltage and current intensity. These two quantities, with which you are certainly familiar, can be measured with the help of voltmeters and ammeters and can be visualized using oscilloscopes.

The concepts of voltage and current intensity, which are based on vector field integration, apply and are valid for any type of regime, either stationary or time varying.

Figure 1.3 Basic definitions of electric voltage (a) and current intensity (b)

As shown in Figure 1.3(a), voltage u between two points a and b is a scalar quantity defined as the line integral of the electric field vector **E** between those points:

$$\text{Voltage} : u = \int_{\overrightarrow{ab}} \mathbf{E} \cdot d\mathbf{s} \tag{1.7}$$

where vector $d\mathbf{s}$ is an infinitesimal element of the path length between a and b.

The reference arrow shown in Figure 1.3(a), usually associated with the definition of u, does not mean that voltage is a vector quantity. The arrow is simply a reminder of the direction of the path length used in the line integral (1.7). Positive voltages signify that the field lines of **E** are predominantly oriented according to the $\overrightarrow{ab}$ path.

A key aspect to be kept in mind is that the evaluation of voltages may depend, or not, on the integration path specification.

As shown in Figure 1.3(b), the current intensity i flowing in a conductor is a scalar quantity defined as a surface integral corresponding to the flux of the current density vector **J** across a conductor section S:

$$\text{Current intensity} : i = \int_{S} \mathbf{J} \cdot \mathbf{n} \; dS \tag{1.8}$$

where dS is an infinitesimal element of area belonging to the section S, and **n** is a unit normal chosen arbitrarily.

The reference arrow shown in Figure 1.3(b), usually associated with the definition of i, does not mean that the current intensity is a vector quantity. The arrow is simply a reminder of the direction of the unit normal **n** used in the surface integral (1.8). Positive current intensities signify that field lines of **J** are predominantly oriented according to **n**.

Again, a key aspect to be kept in mind is that the evaluation of current intensities may depend, or not, on the specification of the conductor section S.

Part II

Stationary Field Phenomena

Introduction

Devoted to stationary fields, the second part of this book covers a wide but specific range of electromagnetic phenomena where all scalar quantities and vector fields are independent of time. Accordingly, from Maxwell's equations in (I.1), by making $\partial/\partial t = 0$, we obtain

$$\begin{cases} \text{curl } \mathbf{E} = 0 \\ \text{div } \mathbf{B} = 0 \\ \text{curl } \mathbf{H} = \mathbf{J} \\ \text{div } \mathbf{D} = \rho \end{cases} \tag{PII.1}$$

This set of equations leads to a particularly simple analysis because electric phenomena and magnetic phenomena are decoupled and therefore can be treated independently.

Part II includes three separate chapters. Chapter 2 is concerned with static electric field phenomena. Currents are absent, $\mathbf{J} = 0$, and electric charges are static in space. Hence, key equations for electrostatics are

$$\begin{cases} \text{curl } \mathbf{E} = 0 \\ \text{div } \mathbf{D} = \rho \end{cases} \tag{PII.2}$$

Chapter 3 is concerned with the intrinsic properties of stationary electric currents. From (PII.1), bearing in mind that div curl $\equiv 0$, we have

$$\begin{cases} \text{curl } \mathbf{E} = 0 \\ \text{div } \mathbf{J} = 0 \end{cases} \tag{PII.3}$$

Chapter 4 is concerned with magnetic fields produced by stationary currents. Therefore, from (PII.1), key equations to be examined are

$$\begin{cases} \text{curl } \mathbf{H} = \mathbf{J} \\ \text{div } \mathbf{B} = 0 \end{cases} \tag{PII.4}$$

2

Electrostatics

2.1 Fundamental Equations

Electrostatic problems are a subclass of stationary field phenomena where electric charges are static in space. This situation ordinarily occurs when conducting bodies in a system are separated by an insulating dielectric medium which prevents the flow of currents.

The fundamental laws governing electrostatic problems are those in (PII.2) together with a constitutive relation concerning the insulating medium behavior. That is,

$$\begin{cases} \operatorname{curl} \mathbf{E} = 0 \\ \operatorname{div} \mathbf{D} = \rho \end{cases} \tag{2.1}$$

and

$$\mathbf{D} = \varepsilon \mathbf{E} \tag{2.2}$$

where ε is the permittivity of the medium (units: F/m, farad per meter).

2.2 Gradient Electric Field, Electric Potential, Voltage, Kirchhoff's Voltage Law

The physical property $\operatorname{curl} \mathbf{E} = 0$ combined with the mathematical identity that the *curl* of the *gradient* of a scalar function is always zero, curl grad $\equiv 0$, leads us to the conclusion that $\mathbf{E}$ is a gradient electric field:

$$\mathbf{E} = \mathbf{E}_g = -\operatorname{grad} V \tag{2.3}$$

where V is the so-called electric potential, scalar potential or, simply, potential.

V is a function of the point in space where its evaluation takes place. A collection of points in space where V assumes a constant value defines an equipotential surface. According to the properties of the *gradient* operator, one immediately concludes, from (2.3), that the field lines of $\mathbf{E}_g$ are always perpendicular to equipotential surfaces.

Electromagnetic Foundations of Electrical Engineering J. A. Brandão Faria

The minus sign in (2.3) is not a mandatory physical requirement of curl $\mathbf{E} = 0$ (one could equally have chosen a plus sign without affecting the coherence of the analysis). The choice of the minus sign is purely conventional and, according to it, the field lines of **E** must point in the direction of decreasing potential (in analogy with the Earth's gravitational field) – see Figure 2.1.

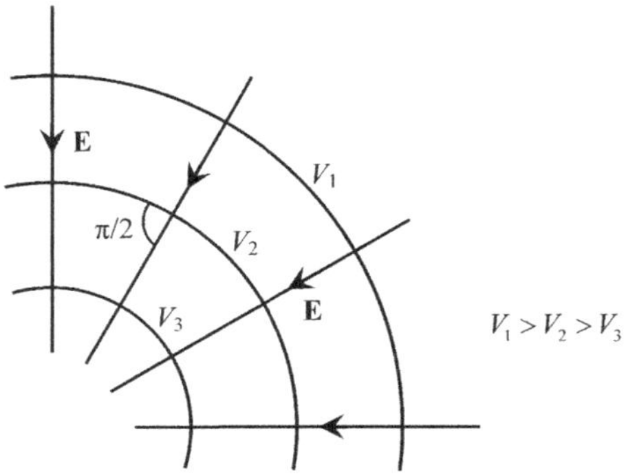

Figure 2.1 Equipotential surfaces and electric field vector

In addition, if in (2.3) we substitute $V' = V + \text{constant}$ for V, we obtain $\mathbf{E} = -\text{grad}\, V'$, therefore showing that the same electric field **E** can be described by different potential functions V and V' differing by an arbitrary constant. The fact that the potential function is not univocally defined leads to the conclusion that such a function is not deeply significant from a physical viewpoint. As a corollary, you can now see that no special meaning should be assigned to the condition $V = 0$.

In order to gain further insight into the properties of the gradient electric field, let us obtain the integral version of curl $\mathbf{E} = 0$. For this purpose we need to utilize Stokes' theorem from vector analysis (a very useful theorem that allows you to transform surface integrals into simpler line integrals):

$$\int_{S_s} \text{curl}\, \mathbf{E} \cdot \mathbf{n}_S \, dS = \oint_{\mathbf{s}} \mathbf{E} \cdot d\mathbf{s} \tag{2.4}$$

where **s** is a simply closed oriented path, vector $d\mathbf{s}$ is an infinitesimal element of the path, S_s is an open surface having the path **s** as its bounding edge, dS is a differential area belonging to S_s, and $\mathbf{n}_S$ is the Stokes unit normal (the direction of the unit normal is related to the path orientation according to the right-hand screw rule) – see Figure 2.2.

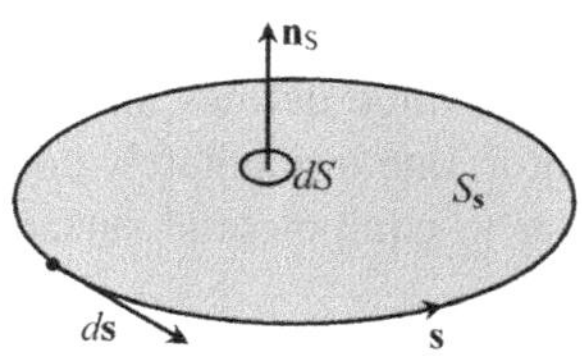

Figure 2.2 Geometrical elements involved in the application of Stokes' theorem

Since curl $\mathbf{E} = 0$, we obtain, from (2.4),

$$\oint_{\mathbf{s}} \mathbf{E} \cdot d\mathbf{s} = 0 \tag{2.5}$$

Therefore, we see that in the framework of stationary regimes, the line integral of the electric field vector along any chosen closed path is always equal to zero or, in other words, such a field is a conservative one. An immediate consequence of (2.5) is that the field lines of **E** are necessarily open curves (they must start and end at different points in space). You can see that this is true by considering the following rationale: if you chose in (2.5) a closed integration path **s** coincident with a hypothetically closed field line of **E**, you would get $(\mathbf{E} \cdot d\mathbf{s}) > 0$ at each and every point of the path, thus implying a positive line integral – an impossible result according to (2.5).

The right-hand side of (2.4) is traditionally known by the name of electromotive force (emf), but that is a misnomer since its units are volt, rather than newton.

At this point let us go back to the definition of voltage between two points a and b, as established in (1.7):

$$U = \int_{\overrightarrow{ab}} \mathbf{E} \cdot d\mathbf{s} \tag{2.6}$$

Does the evaluation of U depend on the specific path going from a to b?

The answer is no, for stationary regimes.

Consider two different paths $\mathbf{s}_1$ and $\mathbf{s}_2$, connecting a to b (see Figure 2.3). Let U_1 and U_2 denote the voltages determined along those two paths. Now, examine the application of (2.5) to a closed path $\widehat{aba}$ containing $\mathbf{s}_1$ and $\mathbf{s}_2$:

$$0 = \oint_{\mathbf{s}} \mathbf{E} \cdot d\mathbf{s} = \int_{\substack{\overrightarrow{ab} \\ (\text{path } s_1)}} \mathbf{E} \cdot d\mathbf{s} + \int_{\substack{\overrightarrow{ba} \\ (\text{path } s_2)}} \mathbf{E} \cdot d\mathbf{s} = \int_{\substack{\overrightarrow{ab} \\ (\text{path } s_1)}} \mathbf{E} \cdot d\mathbf{s} - \int_{\substack{\overrightarrow{ab} \\ (\text{path } s_2)}} \mathbf{E} \cdot d\mathbf{s} = +U_1 - U_2$$

from which you can clearly see that $U_1 = U_2$.

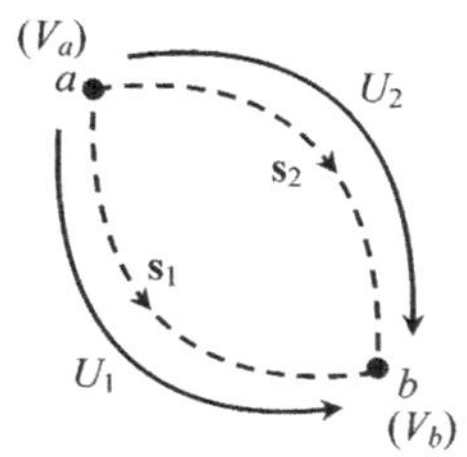

Figure 2.3 For stationary regimes, voltages do not depend on the electric field integration path, $U_1 = U_2 = V_a - V_b$

A different approach to reach this same conclusion consists of substituting –grad V for $\mathbf{E}$ into (2.6). Then we obtain

$$U = \int_{\overrightarrow{ab}} \mathbf{E} \cdot d\mathbf{s} = -\int_{\overrightarrow{ab}} \operatorname{grad} V \cdot d\mathbf{s} = -\int_{\overrightarrow{ab}} \frac{\partial V}{\partial x}dx + \frac{\partial V}{\partial y}dy + \frac{\partial V}{\partial z}dz = -\int_{V_a}^{V_b} dV = V_a - V_b \quad (2.7)$$

showing that, for the evaluation of the voltage U between a and b, you need to know the values of the potential function V only at those two points.

Equations (2.5) and (2.6) lead immediately to another important useful result – Kirchhoff's voltage law (KVL).

By applying (2.5) to the closed path $\widehat{abca}$ depicted in Figure 2.4 we get

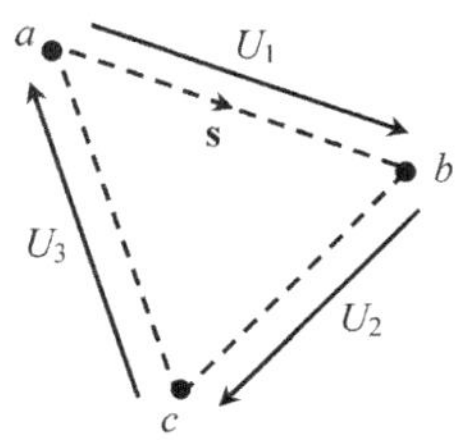

Figure 2.4 Closed circulation path used for the application of KVL

$$0 = \oint_{S} \mathbf{E} \cdot d\mathbf{s} = \int_{\overrightarrow{ab}} \mathbf{E} \cdot d\mathbf{s} + \int_{\overrightarrow{bc}} \mathbf{E} \cdot d\mathbf{s} + \int_{\overrightarrow{ca}} \mathbf{E} \cdot d\mathbf{s} = U_1 + U_2 + U_3$$

or, more generally,

$$\sum_k U_k = 0 \quad (2.8)$$

You should notice, however, that KVL is not, strictly speaking, a general 'law'. Indeed, the result in (2.8) is only valid for stationary phenomena, $\partial/\partial t = 0$. For time-varying regimes (where curl $\mathbf{E} \neq 0$) things are a little more complicated – see Chapter 5.

2.3 Electric Charge, Electric Displacement Vector

We saw earlier that electric field lines cannot be closed; they must start and end at different points in space. Now, we are going to see that field lines start and end at points where positive and negative charges exist, respectively.

Let us find out the integral version corresponding to the local equation div $\mathbf{D} = \rho$.

For this purpose we need to utilize the Gauss theorem from vector analysis (a very useful theorem that allows you to transform volume integrals into simpler surface integrals):

$$\int_V \operatorname{div}\mathbf{D}\,dV = \int_{S_V} \mathbf{D}\cdot\mathbf{n}_o\,dS \tag{2.9}$$

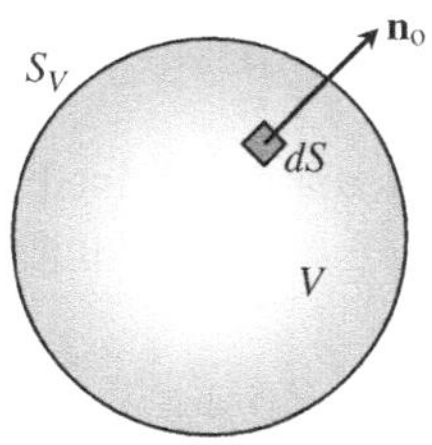

Figure 2.5 Geometrical elements involved in the application of the Gauss theorem

where S_V is a closed surface bounding a region of volume V, dV is a differential volume belonging to V, dS is a differential area belonging to S_V, and $\mathbf{n}_o$ is the outward unit normal – see Figure 2.5. Substituting ρ for div **D** in (2.9) we get

$$\int_{S_V} \mathbf{D}\cdot\mathbf{n}_o\,dS = \int_V \rho\,dV = Q_V \tag{2.10}$$

revealing that the outward flux of the displacement vector across a closed surface is equal to the total electric charge Q_V enclosed by such a surface.

If $Q_V > 0$ then the field lines of **D** must point outward; conversely, if $Q_V < 0$ then the field lines of **D** must point inward.

2.4 Dielectric Media, Permittivity, Polarization, Dielectric Strength

In empty space (vacuum) **D** and **E** are related through

$$\mathbf{D} = \varepsilon_0\,\mathbf{E} \tag{2.11}$$

An insulating material medium (dielectric) is made of electrically neutral atoms immersed in a vacuum. When an **E** field is applied to a dielectric material the cloud of electrons of each atom is displaced from its original position; each atom becomes an elementary electric dipole with its positive and negative centers of charge apart; this phenomenon is called polarization. The stronger the **E**-field intensity the stronger the polarization effect.

The vacuum contribution for **D** is the same as in (2.11); however, an additional contribution **P** must be added due to the polarization effect

$$\mathbf{D} = \varepsilon_0\,\mathbf{E} + \mathbf{P}(\mathbf{E}) \tag{2.12}$$

In (2.12), **P** is the so-called electric polarization vector, a vector that measures the electric dipole moment per unit volume. For isotropic media **P** and **E** are parallel vectors. For linear media **P** and **E** are proportional:

$$\mathbf{P} = \varepsilon_0 \chi_e \mathbf{E} \tag{2.13}$$

where χ_e is a dimensionless positive constant called the electric susceptibility. Substituting (2.13) into (2.12) we obtain

$$\mathbf{D} = \varepsilon_0(1+\chi_e)\mathbf{E} = \varepsilon_0\varepsilon_r\mathbf{E} = \varepsilon\,\mathbf{E} \tag{2.14}$$

The permittivity ε of the insulating media is obtained from ε_0 upon multiplication by the relative permittivity ε_r. Typical values of ε_r concerning a few insulating materials are shown in Table 2.1.

Table 2.1 Permittivity and dielectric strength

Material	ε_r	E_d (kV/cm)
Air	1.0	20–30
Bakelite	4.8	250
Glass	6–10	300
Mica	5.5	2000
Mineral oils	2	150
Paper	3.0	150
Polystyrene	2.6	200
Rubber	2.5–3.0	210

The equation div $\mathbf{D} = \rho$ together with the result in (2.14) can now be used to determine the *divergence* of the electric field vector **E**. For the general case of inhomogeneous dielectrics, where $\varepsilon = \varepsilon(x, y, z)$, we have $\mathrm{div}(\varepsilon\mathbf{E}) = \mathrm{grad}\,\varepsilon \cdot \mathbf{E} + \varepsilon\,\mathrm{div}\,\mathbf{E} = \rho$, from which we find

$$\mathrm{div}\,\mathbf{E} = \frac{\rho - \mathrm{grad}\,\varepsilon \cdot \mathbf{E}}{\varepsilon} \tag{2.15}$$

In the case of homogeneous dielectric media (an assumption we will consider most frequently), the above result simplifies to

$$\mathrm{div}\,\mathbf{E} = \rho/\varepsilon \tag{2.16}$$

The linear behavior of the dielectric medium described by (2.13) and (2.14) does not hold when the medium is subjected to very intense **E** fields; furthermore, if a certain threshold is exceeded, ionization processes will develop and arc discharges (sparking) occur.

The maximum allowed electric field intensity that an insulation material can endure before breakdown takes place is called the dielectric strength, E_d. Typical values of E_d are shown in Table 2.1.

2.5 Conductors in Electrostatic Equilibrium

Contrary to insulating materials, conductors are characterized at the microscopic level by possessing charged particles (electrons) that are free to move under the influence of an electric field. Conductors are said to be in electrostatic equilibrium when all movements of charge have stopped.

In order to ensure that charges do not move, the **E** field inside a conductor must necessarily be zero and, therefore, because $\mathbf{E} = -\mathrm{grad}\, V$, the whole conductor must be an equipotential body and its surface an equipotential surface, $V = \text{constant}$.

Since inside the conductor $\mathbf{E} = 0$, you can see from (2.16) that the volume charge density must also vanish, $\rho = 0$; accordingly, we conclude that in electrostatic regimes, a charged conductor has all its electric charge residing on its boundary surface – see Figure 2.6.

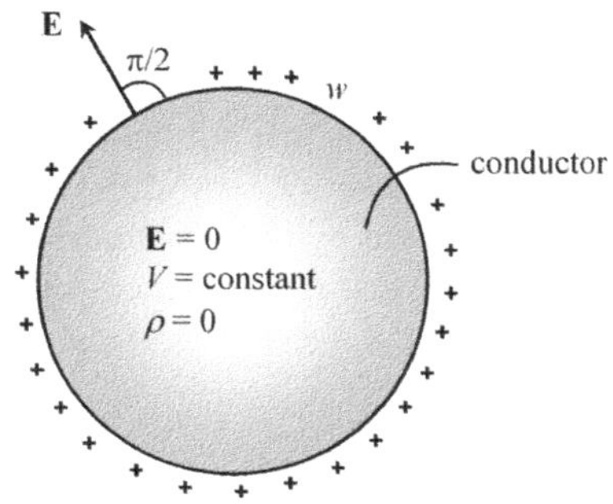

Figure 2.6 In electrostatic equilibrium, charges are distributed on the conductor's surface with a density w

In a system consisting of several conductors immersed in a dielectric medium, the surface charge distribution over a given conductor is such that, together with all the charge distributions of the remaining conductors, it ensures $\mathbf{E} = 0$ inside each and every conductor.

The total charge Q on a conductor is obtained by integrating its surface charge density w (C/m^2) over its boundary surface S_V,

$$Q = \int_{S_V} w\, dS$$

If the surface charge density at a given point on the conductor's surface is known then you can easily evaluate how intense the electric field is in the vicinity of that point (on the dielectric side). This information can be very useful for predicting whether dielectric breakdown is prone to occur.

However, note that, because the conductor's surface is an equipotential, the field lines of **E** are perpendicular to the conductor's surface (the field is entirely specified by its normal component), $\mathbf{E} = E\mathbf{n}$.

In order to evaluate E, consider a small cylinder, perpendicular to the conductor's surface, with the bottom half inside the conductor and the upper half in the dielectric medium. The height of the cylinder is infinitesimal; the area ΔS of the bottom and top ends of the cylinder is also infinitesimal – see Figure 2.7.

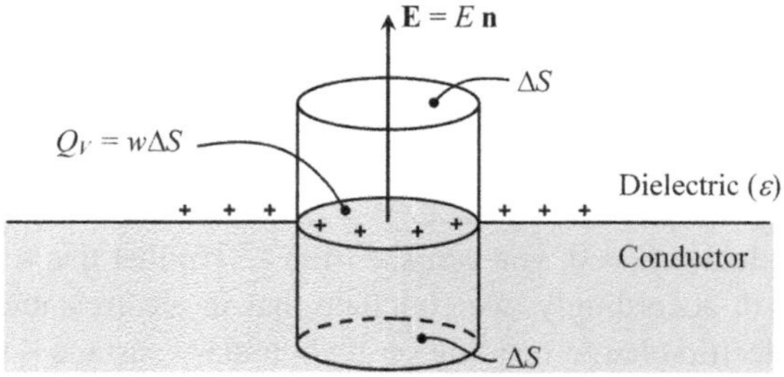

Figure 2.7 Application of the Gauss theorem to the determination of the normal component of the electric field vector in a conductor

The total charge enclosed by the cylinder is just but the charge on the conductor's surface occupying the small circular area ΔS, that is $Q_V = w\,\Delta S$. On the other hand, the outward flux of the displacement vector across the cylinder boundary surface is zero everywhere except at its top end, where it amounts to $\varepsilon E\,\Delta S$. Hence, by making use of the result in (2.10) we find

$$E = \frac{w}{\varepsilon} \tag{2.17}$$

So, if you allow the surface charge density w at a given point of a conductor to exceed the critical limit εE_d (with E_d denoting the dielectric strength) then you must be aware that the dielectric medium runs the risk of breakdown. Ordinarily, this risk is greatly enhanced when needle-shaped conductors are utilized since the charge density tends to increase near the conductors' tips (recall the shape of lightning rods).

At this stage you may be wondering if breakdown phenomena might eventually occur at dielectric zones other than the interfaces with conductors. Well, generally speaking the answer is yes; for instance, remember the case of atmospheric discharges (lightning).

However, under certain circumstances, you may rest assured that the electric field intensity will reach its maximum value nowhere else other than at the surface of a conductor. This is true whenever the dielectric medium is homogeneous and charge free, where, from (2.15), div $\mathbf{E} = 0$.

To prove our point, let us start by evaluating the *Laplacian* of the potential function V:

$$\text{lap } V = \frac{\partial^2 V}{\partial x^2} + \frac{\partial^2 V}{\partial y^2} + \frac{\partial^2 V}{\partial z^2} = \text{div grad } V = -\text{div } \mathbf{E} = 0 \tag{2.18}$$

Next, from vector analysis, we determine the *Laplacian* of the squared field intensity E^2:

$$\text{lap}(E^2) = \text{div grad}(\mathbf{E} \cdot \mathbf{E}) = \text{div grad}(\text{grad } V \cdot \text{grad } V) = 2\,\text{div}(\overbrace{\text{lap } V}^{0}\ \text{grad } V) = 0$$

Thus we see that lap V = lap $E^2 = 0$. The important physical interpretation for this conclusion is that neither V nor E can have maxima or minima inside their space domains – only at their boundaries! (Remember that the necessary condition for an $\mathbb{R}^3$ function to have an extremal point is that its second derivatives with respect to x, y and z must all have the same algebraic sign – which is in clear contradiction with lap = 0.)

2.6 Application Example (Filament of Charge)

The solution of many electrostatic problems where cylindrical conductors are involved is ordinarily based on the analysis of a much simpler problem consisting of a single rectilinear filament of charge of indefinite length. Let us therefore consider a filament of charge, characterized by a per-unit-length charge density q (C/m), immersed in a homogeneous dielectric medium with permittivity ε.

Questions

Q_1 Determine the electric field vector **E** and the potential function V at a generic point P at a distance r from the filament.

Q_2 Find the voltage U_{12} between any two points $P_1(r, \phi_1, z_1)$ and $P_2(r, \phi_2, z_2)$ at a common distance from the charge filament.

Solutions

Q_1 The geometry of the configuration under analysis indicates that a cylindrical reference frame (r, ϕ, z) should be used. The problem geometry together with the need for **E** to be perpendicular to the charge distribution leads us immediately to the conclusion that **E** is a purely radial vector, $\mathbf{E} = E\vec{e}_r$. Moreover, by noting that the problem is insensitive to translation operations along z or rotation operations along ϕ, we can conclude that both **E** and V can only depend on the radial coordinate. In short, $\mathbf{E} = E(r)\vec{e}_r$ and $V = V(r)$.

Now, for the determination of **E** we employ the results in (2.10) and (2.14):

$$\varepsilon \int_{S_V} \mathbf{E} \cdot \mathbf{n}_o \, dS = Q_V$$

At this point you must decide which closed surface S_V should be used to facilitate the surface integration above, where **E** is a radial vector.

As shown in Figure 2.8, the easiest choice consists of a cylindrical surface of radius r and length l coaxial with the charge filament.

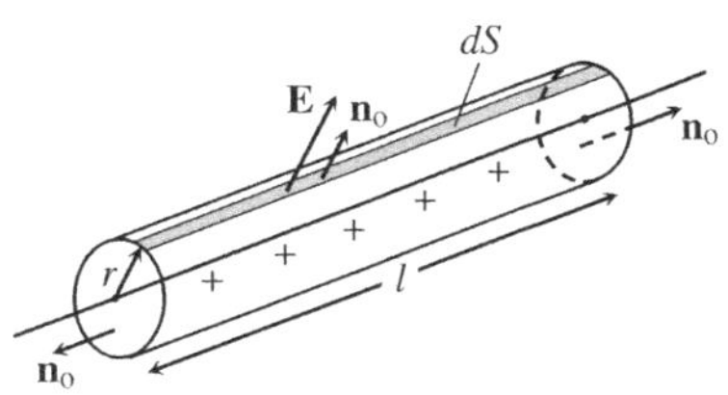

Figure 2.8 Application of the Gauss theorem to the determination of the radial electric field produced by a charge filament

At the cylinder ends the outward normal $\mathbf{n}_o$ is perpendicular to $\mathbf{E}$, and therefore at both ends $\mathbf{E}\cdot\mathbf{n}_o = 0$. On the other hand, along the cylinder's lateral surface $\mathbf{n}_o$ is parallel to the electric field vector, $\mathbf{n}_o = \vec{e}_r$, and consequently $\mathbf{E}\cdot\mathbf{n}_o = E$.

Noting further that $dS = l\,r\,d\phi$ and that the total charge inside the cylinder is $Q_V = ql$, we obtain

$$\varepsilon \int_{S_V} \mathbf{E}\cdot\mathbf{n}_o\, dS = Q_V \rightarrow 0 + 0 + \varepsilon r l \int_0^{2\pi} E\, d\phi = ql$$

Recalling that E does not depend on ϕ, we finally get

$$\mathbf{E} = \underbrace{\left(\frac{q}{2\pi\varepsilon r}\right)}_{E} \vec{e}_r \tag{2.19}$$

As to the evaluation of the potential function V, from (2.3), $\mathbf{E} = -\text{grad}\, V$, we obtain in cylindrical coordinates

$$E\,\vec{e}_r = -\left(\frac{\partial V}{\partial r}\vec{e}_r + \frac{1}{r}\frac{\partial V}{\partial \phi}\vec{e}_\phi + \frac{\partial V}{\partial \phi}\vec{e}_z\right)$$

The above result confirms the conclusion we reached before based on geometry arguments – that is, V depends only on the radial coordinate. Hence,

$$E = -\frac{dV}{dr} \rightarrow \int dV = -\int E dr$$

Substituting $q/(2\pi\varepsilon r)$ for E we find

$$V = \left(\frac{q}{2\pi\varepsilon}\ln\frac{1}{r}\right) + V_0$$

where V_0 is an arbitrary constant. If you decide to set $V = 0$ at an arbitrary point in space $r = r_0$, the above result can be rewritten as

$$V = \frac{q}{2\pi\varepsilon}\ln\frac{r_0}{r} \tag{2.20}$$

Q_2 By noting that points P_1 and P_2 belong to the same equipotential cylindrical surface, we must have $U_{12} = 0$, irrespective of the particular coordinates z_1, z_2, ϕ_1 and ϕ_2.

2.7 Capacitor, Capacitance, Electric Energy

A capacitor is a device consisting of two conductors (called electrodes) separated by an insulating medium. This definition rules out the existence of any other nearby conducting body that may perturb the capacitor's electric field distribution. The analysis of conductor systems involving more than two conductors will be addressed in Section 2.9.

The capacitor electrodes A and B are initially uncharged, but as soon as a voltage $U = V_A - V_B$ is applied between them, using a generator (Figure 2.9), a redistribution of

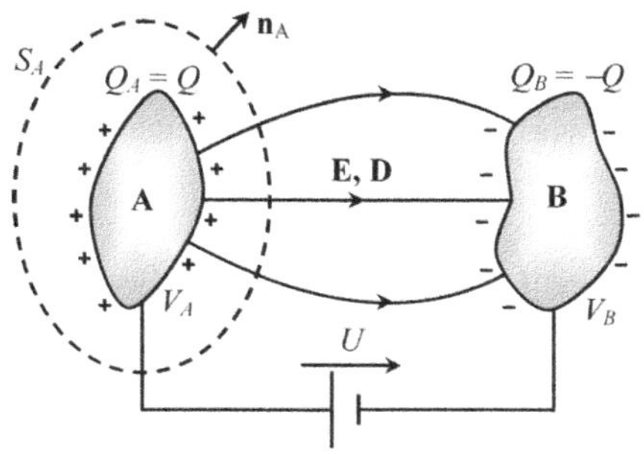

Figure 2.9 Charge and voltage in a capacitor with electrodes A and B

charges takes place, both conductors acquiring symmetrical charges, $Q_A = -Q_B = Q$. Note that the algebraic sign of the capacitor charge Q is the same as the one for the applied voltage U.

Assuming that the capacitor's dielectric material displays a linear behavior, $\mathbf{D} = \varepsilon \mathbf{E}$, then Q will be proportional to U: the higher the applied voltage, the higher the capacitor charge. The proportionality constant relating Q to U, which is positive and only depends on the capacitor's geometry and on the permittivity of the dielectric, is called capacitance C (units: F, farad):

$$C = \frac{Q}{U} = \frac{\int_{S_A} \mathbf{D} \cdot \mathbf{n}_A \, dS}{\int_{\overrightarrow{AB}} \mathbf{E} \cdot d\mathbf{s}} \tag{2.21}$$

Hence you can see that the linear relationship $Q = CU$ between integral quantities is just a consequence of the linear relationship $D = \varepsilon E$ between local fields. You should also realize that if ε is not defined (nonlinear dielectrics) then the concept of capacitance will be inapplicable.

Since the dielectric material is a perfect insulator (zero conductivity, $\sigma = 0$), if you disconnect and remove the voltage generator in Figure 2.9, the electric field distribution in the space surrounding the two electrodes of the capacitor will remain unchanged, the same thing happening to both Q and U. (Note: In reality, because perfect insulators do not exist, Q and U will tend very slowly to zero as time passes; the analysis of this phenomenon will be tackled in Section 6.5.)

After removing the generator, if you decide to connect a light bulb between the capacitor electrodes you will see that the bulb flashes. Where did the released energy come from?

During the charging process of the capacitor, the generator redistributes charges within the system by transporting free electrons from electrode A to electrode B (or, which is formally the same, by moving positive charges from electrode B to electrode A) until electrostatic equilibrium is reached.

The charging process requires the generator to expend a certain amount of energy. Part of it is associated with irreversible dissipation losses (Joule effect) arising from transient currents flowing through the connecting wires. Another part has to do with the fact that the transport of positive elements of charge dq from B to A is opposed by the electric field force

$\mathbf{F}_e = dq\mathbf{E}$ directed from electrode A to electrode B. In order to counteract this opposing force, the generator has to produce additional work.

The differential amount of work produced by the generator to drive dq from B to A is given by

$$dW_G = \int_{\overrightarrow{\mathrm{BA}}} -\mathbf{F}_e \cdot d\mathbf{s} = dq \int_{\overrightarrow{\mathrm{AB}}} \mathbf{E} \cdot d\mathbf{s} = u\,dq \tag{2.22}$$

Summing all dW_G contributions starting from the initial uncharged state ($q = 0$, $u = 0$) up to the final state of equilibrium ($q = Q$, $u = U$) gives the electric energy stored in the capacitor:

$$W_e = \int_{\substack{\text{initial}\\ \text{state}}}^{\substack{\text{final}\\ \text{state}}} u\,dq \tag{2.23}$$

This result for W_e is graphically interpreted in Figure 2.10, for linear and nonlinear dielectric media.

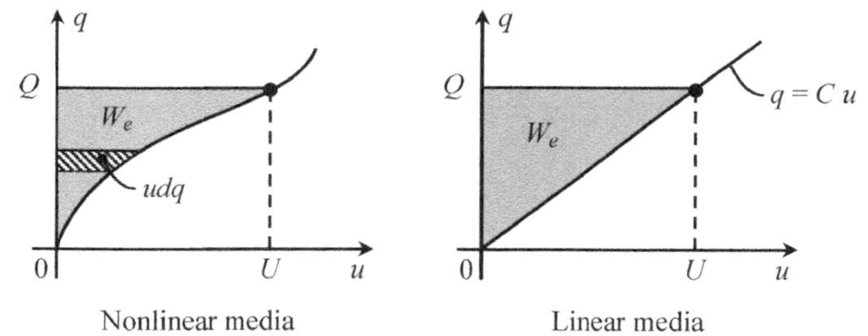

Figure 2.10 Graphical interpretation of the electric energy stored in a two-conductor system, for linear and nonlinear dielectric media

For the ordinary case of linear dielectric media, by using $q = Cu$ in (2.23), we find

$$W_e = C\int_0^U u\,du = \frac{1}{2}CU^2 = \frac{1}{2}\frac{Q^2}{C} = \frac{1}{2}QU \tag{2.24}$$

The work of the generator evaluated according to (2.23) or (2.24) remains stored in the capacitor in the form of electric energy, which remains available for subsequent use, even after the generator is disconnected.

When you plug in an external dissipative load (a flashbulb, for example) to the capacitor terminals, the capacitor discharges. Electrons move from electrode B to electrode A, and, as the energy is transferred to the external load, the energy stored in the capacitor decreases to zero.

But where was the electric energy of the capacitor stored?

In order to answer this question let us consider the simple example of a parallel-plate capacitor (Figure 2.11), a device made of two metallic plates of area S separated by an insulator of very small thickness δ (so that fringing field effects at the corners can be ignored).

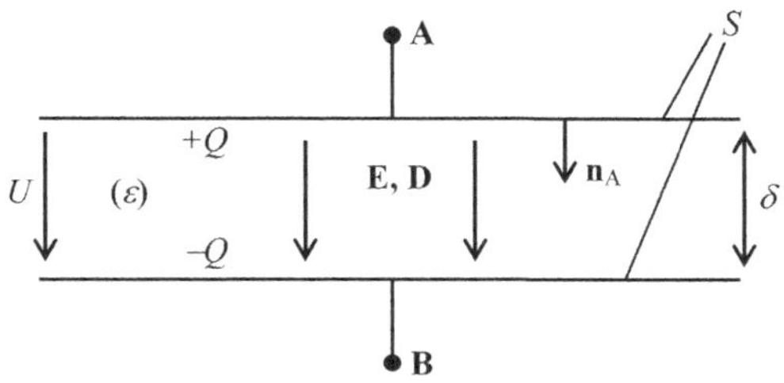

Figure 2.11 Parallel-plate capacitor of small thickness δ (uniform fields)

Inside the device, the electric field lines are straight lines perpendicular to the metallic plates and the field intensity is uniform throughout the dielectric region. Therefore the capacitor voltage and charge can be easily correlated with **E** and **D**:

$$U = \int_{\overrightarrow{AB}} \mathbf{E} \cdot d\mathbf{s} = \int_{\overrightarrow{AB}} E\,ds = E \int_{\overrightarrow{AB}} ds = E\delta \tag{2.25a}$$

$$Q = \int_{S_V} \mathbf{D} \cdot \mathbf{n}_A\,dS = \int_{S} D\,dS = D \int_{S} dS = DS \tag{2.25b}$$

From (2.24) and (2.25) we obtain the electric energy stored:

$$W_e = \frac{1}{2} QU = \left(\frac{ED}{2} \right) S\delta$$

We note now that $S\delta$ represents the volume V_d of the dielectric medium sandwiched between the plates, and consequently the quantity in parentheses must represent a volume energy density (units: $\mathrm{J/m^3}$),

$$\hat{w}_e = \frac{1}{2} ED = \frac{1}{2} \varepsilon E^2 = \frac{1}{2} \mathbf{E} \cdot \mathbf{D} \tag{2.26}$$

So we see that for the parallel-plate capacitor case we can write $W_e = \hat{w}_e\,V_d$. Generalization of the preceding result for other capacitor structures with non-uniform field distributions is simply

$$W_e = \int_{V_d} \hat{w}_e\,dV \tag{2.27}$$

The result in (2.27) clearly suggests that the electric energy is stored in the electric field itself, at each and every point of the dielectric region, with a density given by (2.26). Analogous to Figure 2.10, you will find in Figure 2.12 a graphical interpretation for the computation of $\hat{w}_e$ for linear and nonlinear media.

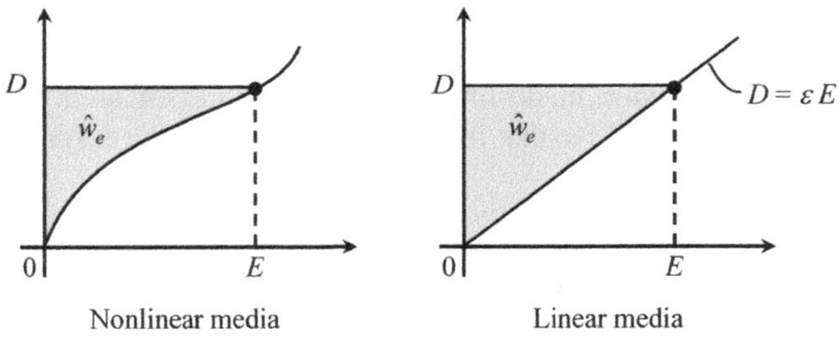

Figure 2.12 Graphical interpretation of the electric energy density stored in the electric field, for linear and nonlinear dielectric media

2.8 Application Example (Two-Wire Transmission Line)

The two-wire line is a ubiquitous conductor arrangement that you may find in most transmission systems, for power delivery, wave guiding and signal communication.

Figure 2.13 represents a cross-sectional view of a two-wire transmission line of length l, made of two parallel identical cylindrical conductors of radius r, whose axes are a distance $2d$ apart. The wires are immersed in air. When a voltage U is applied between the line conductors, these acquire symmetrical electric charges $\pm Q$. Choose $V = 0$ at $x = y = 0$.

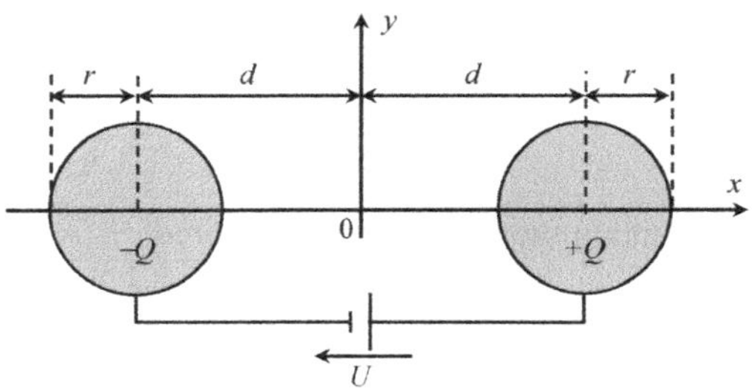

Figure 2.13 Cross-sectional view of a symmetrical two-wire transmission line

Questions

Q_1 Due to conductor proximity effects, the charges are not uniformly distributed over the conductors' surfaces and, for that reason, the computation of the electric field and potential in the space surrounding the transmission line cannot be done by replacing the actual charge distributions with fictitious filaments of charge located at the conductor axes. Depending on how close the conductors are, you will need to locate the fictitious filaments of charge somewhere nearer the periphery of the conductors.

Taking into account that field lines of **E** are circumferential arcs starting and ending at the fictitious filaments of charge, and recalling that field lines are perpendicular to the surfaces of conductors, find with the help of Figure 2.14(a) the correct position for the location of the fictitious filaments of charge.

Q_2 Using the superposition principle (that is, adding the contributions of the positive and negative charge filaments), determine the electric field vector **E** and the potential function V at the generic point P outside the conductors, as shown in Figure 2.14(b). Evaluate the absolute value of the electric field intensity at P.

Q_3 The cross-section of the cylindrical equipotential surface containing a generic point P is defined by a circumference of radius ρ with center at $y = 0$, $x = x_C$. Find ρ and x_C as a function of V.

Q_4 Using the preceding results, determine the two-wire line capacitance $C = Q/U$.

Q_5 With the help of Figure 2.14(c), determine the surface charge density w over the positively charged conductor as a function of its peripheral angle θ.

Q_6 Obtain an expression for the electric field maximum intensity E_{max}. Where does it occur?

Q_7 Consider the frequently used thin-wire approximation, $r/d \ll 1$. Find approximate results for the line capacitance C_{approx} and for the maximum electric field intensity $(E_{max})_{approx}$. Discuss the errors incurred by using such an approximation.

Q_8 Numerical example. Take $d = 2r = 2\,\text{mm}$, $l = 50\,\text{m}$. Assuming that the two-wire line is allowed to operate at 20 % below dielectric breakdown, find numerical values for the line capacitance, allowed line charge and voltage, and electric energy stored. Find the ratio η of the charge density values evaluated at two opposite points, $\theta = 0$ and $\theta = \pi$, on the conductors' surface.

Solutions

Q_1 By analyzing the right-angled triangle in Figure 2.14(a) with sides a, d and r, you can easily find the radius a of the circumferential arc centered at the origin, intersecting the conductors at right angles, and crossing the x axis at the points where the fictitious charge filaments are to be positioned, $x = \pm a$, with

$$a = (d^2 - r^2)^{1/2} \tag{2.28}$$

Q_2 Making use of the results (2.19) and (2.20) previously obtained in Application Example 2.6, by adding the contributions from the positively and negatively charged filaments with charge density $q = Q/l$ you can readily find

$$\mathbf{E} = \mathbf{E}^{(+)} + \mathbf{E}^{(-)} = \frac{Q}{2\pi\varepsilon_0 l}\left(\frac{\vec{e}_1}{r_1} - \frac{\vec{e}_2}{r_2}\right) \tag{2.29}$$

$$V = V^{(+)} + V^{(-)} = \frac{Q}{2\pi\varepsilon_0 l}\ln\frac{r_2}{r_1} \tag{2.30}$$

where $\vec{e}_1$ and $\vec{e}_2$ are unit vectors along the radial directions r_1 and r_2, with origin, respectively, at the positive and negative filaments of charge – see Figure 2.14(b).

The absolute value of $\mathbf{E}$ at P can be evaluated from

$$E = \sqrt{\mathbf{E}\cdot\mathbf{E}} = \frac{Q}{2\pi\varepsilon_0 l r_1 r_2}\sqrt{(r_2\vec{e}_1 - r_1\vec{e}_2)\cdot(r_2\vec{e}_1 - r_1\vec{e}_2)}$$

where $(r_2\vec{e}_1 - r_1\vec{e}_2)\cdot(r_2\vec{e}_1 - r_1\vec{e}_2) = r_1^2 + r_2^2 - 2r_1r_2\,\vec{e}_1\cdot\vec{e}_2$.

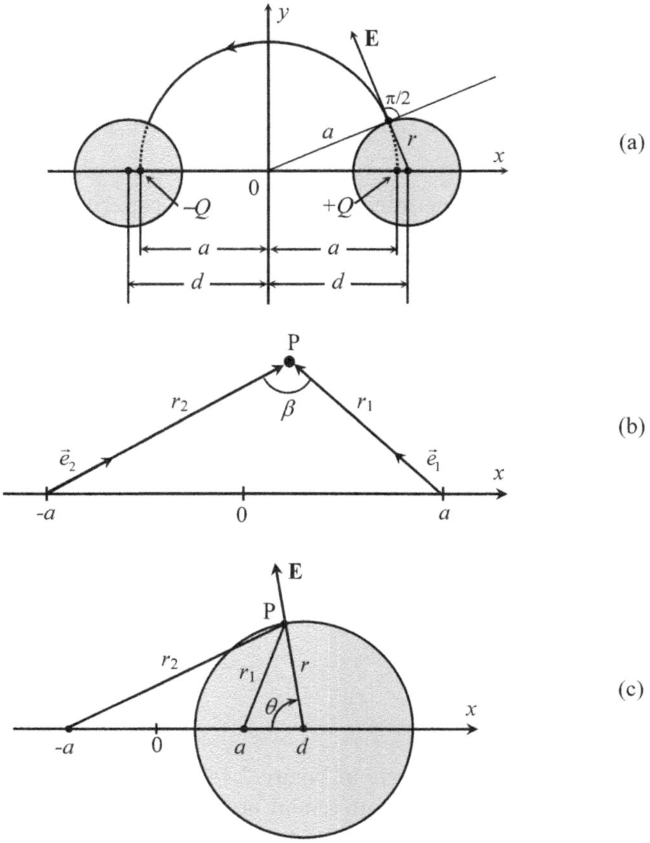

Figure 2.14 (a) Geometrical construction for finding the location of the fictitious charge filaments that replace the actual conductor charges. (b) The electric field at P is the vector sum of two radial electric fields originating from two charged filaments positioned at $x = \pm a$. (c) Geometrical construction for the determination of the surface charge density at P

Noting further, from Figure 2.14(b), that $\vec{e}_1 \cdot \vec{e}_2 = \cos\beta$, and using the law of cosines from trigonometry, you can get

$$(r_2\vec{e}_1 - r_1\vec{e}_2) \cdot (r_2\vec{e}_1 - r_1\vec{e}_2) = r_1^2 + r_2^2 - 2r_1r_2\cos\beta = (2a)^2$$

which yields the useful result

$$E = \frac{Q}{\pi\varepsilon_0 l} \times \frac{a}{r_1 r_2} \tag{2.31}$$

Q_3 Let V be the potential of the equipotential surface containing the point P(x, y). According to (2.30), all the points belonging to that same surface are such that the ratio r_2/r_1 remains constant:

$$\frac{r_2}{r_1} = k_V = \exp\left(\frac{2\pi\varepsilon_0 l}{Q}V\right) \tag{2.32}$$

Next, from Figure 2.14(b), r_2 and r_1 are expressed in terms of the x and y coordinates of P:

$$r_1^2 = (a-x)^2 + y^2; \quad r_2^2 = (a+x)^2 + y^2$$

Hence, the condition for equipotentiality takes the form

$$(a+x)^2 + y^2 = k_V^2(a-x)^2 + k_V^2 y^2.$$

Rearranging terms you find

$$\left(x - a\frac{k_V^2+1}{k_V^2-1}\right)^2 + (y-0)^2 = \left(a\frac{2k_V}{k_V^2-1}\right)^2$$

This is the well-known equation of a circumference in the xy transverse plane, with center at $x = x_C$ and $y = 0$, and radius ρ, with

$$x_C = a\frac{k_V^2+1}{k_V^2-1}; \quad \rho = a\frac{2k_V}{|k_V^2-1|} \tag{2.33}$$

where, as seen before, in (2.32), k_V depends on V, through

$$k_V = \exp\left(\frac{2\pi\varepsilon_0 l}{Q}V\right)$$

The expressions obtained for x_C and ρ are not independent; you can easily check that

$$a = (x_C^2 - \rho^2)^{1/2}$$

and

$$\rho = \begin{cases} k_V(x_C - a) \text{ for } k_V > 1, x_C > 0 \\ k_V(a - x_C) \text{ for } k_V < 1, x_C < 0 \end{cases}$$

The first relationship above being particularized to the conductors' surfaces (equipotential surfaces) gives $a = (d^2 - r^2)^{1/2}$, which confirms the result obtained in (2.28).

Q_4 From (2.32) you can see that the vertical plane of symmetry $x = 0$, where $r_2 = r_1$, is characterized by $V = 0$. The half-space $x > 0$ is filled with positive equipotentials ($k_V > 1$), whereas the half-space $x < 0$ is filled with symmetrical (negative) equipotentials ($k_V < 1$).

Consequently, if U is the voltage between the conductors then the positively charged conductor, of radius r with center at $x = d$, will be characterized by a potential $V = U/2$.

Then you have

$$k_{V=U/2} = k = \exp\left(\frac{\pi\varepsilon_0 l}{Q}U\right) \tag{2.34}$$

On the other hand, by making $x_C = d$ and $\rho = r$ in (2.33) you find $d/r = (k^2+1)/(2k)$ which allows you to determine k:

$$k = \frac{d}{r} + \sqrt{\left(\frac{d}{r}\right)^2 - 1} \tag{2.35}$$

Finally, from (2.34) and (2.35), you can evaluate the capacitance of the two-wire line configuration

$$C = \frac{Q}{U} = \frac{\pi\varepsilon_0 l}{\ln\left(\frac{d}{r} + \sqrt{\left(\frac{d}{r}\right)^2 - 1}\right)} \tag{2.36}$$

Q_5 According to (2.17), the surface charge density over a conductor immersed in a medium of permittivity ε_0 is given by $w = \varepsilon_0 E$. On the other hand, we saw in (2.31) that the electric field intensity at a generic point P in space is given by

$$E = \frac{Q}{\pi\varepsilon_0 l} \times \frac{a}{r_1 r_2}$$

Examining the representation in Figure 2.14(c), and using the trigonometric law of cosines to express r_1 and r_2 as functions of θ, you find

$$r_1^2 = r^2 + (d-a)^2 - 2r(d-a)\cos\theta; \quad r_2^2 = r^2 + (d+a)^2 - 2r(d+a)\cos\theta$$

Taking into account that $a^2 = d^2 - r^2$, evaluation of the $r_1^2 r_2^2$ product yields

$$(r_1 r_2)^2 = 4r^2d^2 + 4r^4\cos^2\theta - 8r^3 d\cos\theta = [2r(d - r\cos\theta)]^2$$

from which we obtain the electric field intensity normal to the conductor surface

$$E = \frac{Q}{2\pi r\varepsilon_0 l} \times \frac{a}{d - r\cos\theta} \tag{2.37}$$

which allows us to determine the surface charge distribution as a function of θ

$$w(\theta) = \frac{Q}{2\pi r l} \times \frac{a}{d - r\cos\theta} \tag{2.38}$$

Note: You can verify that this result is correct by checking that integration of $w(\theta)$ over the whole conductor surface S_C gives the total charge Q:

$$\int_{S_C} w(\theta)\, dS = Q$$

where $dS = lr\, d\theta$, with θ running from 0 to 2π.

Q_6 Taking into account the result in (2.37), you can obtain, for $\theta = 0$,

$$E_{\max} = \frac{Q}{2\pi\varepsilon_0 lr}\frac{a}{d-r} = \frac{Q}{2\pi\varepsilon_0 lr}\sqrt{\frac{d+r}{d-r}} \tag{2.39}$$

Q_7 Considering $d \gg r$, the following approximations apply:

$$C_{\text{approx}} = \frac{\pi\varepsilon_0 l}{\ln(2d/r)} \tag{2.40}$$

$$(E_{\max})_{\text{approx}} = \frac{Q}{2\pi\varepsilon_0 lr}. \tag{2.41}$$

The errors incurred by using the above results are analyzed in Table 2.2.

Table 2.2 Errors as a function of r/d

$\frac{r}{d}$	$\frac{C_{\text{approx}} - C}{C}\%$	$\frac{(E_{\max})_{\text{approx}} - E_{\max}}{E_{\max}}\%$
0.10	−0.08	−09.55
0.20	−0.44	−18.35
0.30	−1.22	−26.62
0.40	−2.65	−34.53
0.50	−5.00	−42.26

As far as the computation of the two-wire line capacitance is concerned, the approximation works very well even when the wires are not very thin (the error is about 5 % for $d/r = 2$). However, the same is not true for the computation of $E_{\max}$ (the error already exceeds 40 % for $d/r = 2$).

Q_8 From (2.36), you obtain $C = 1.055\,\text{nF}$.
By making $E_{\max} = 0.8E_d = 2.4\,\text{MV/m}$, you can obtain from (2.39):

$$Q = 2\pi\varepsilon_0 lrE_{\max}\sqrt{\frac{d-r}{d+r}} = 3.85\,\mu\text{C}$$

From $U = Q/C$, you obtain $U = 3.65\,\text{kV}$.
From $W_e = QU/2$, you obtain $W_e = 7.02\,\text{mJ}$.
From (2.38), you can obtain

$$\frac{w(0)}{w(\pi)} = \eta = \frac{d+r}{d-r} = 3$$

2.9 Multiple Conductor Systems

The analysis of electric coupling phenomena among conductors is of utmost importance in a variety of electrical engineering problems where multiple conductor systems are in play, such as overhead power lines, communication lines and cables, printed circuit boards and integrated circuits for electronic applications, and so on.

The electrostatic analysis of multiple conductor systems cannot be carried out using the trivial concept of capacitance introduced in Section 2.7 (a concept which is only defined for two-conductor linear systems); what is more, if you let yourself be guided by the familiar capacitor equation $Q = CU$, you will certainly make mistakes.

To give you an example of what we are talking about, consider the four-conductor system depicted in Figure 2.15.

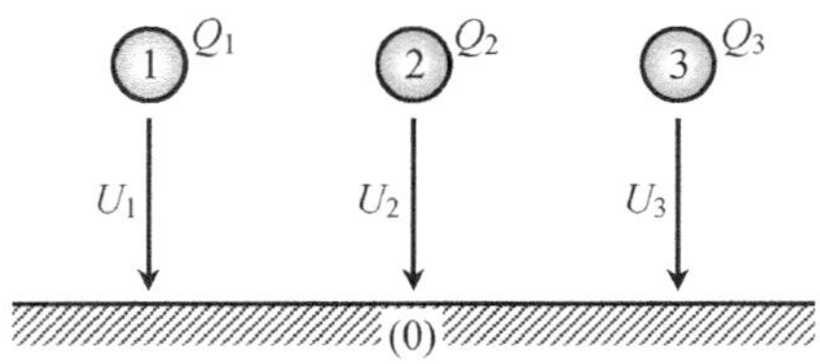

Figure 2.15 Example of a four-conductor system

All the conductors are initially uncharged. Then, with the help of a generator, a positive voltage U_1 is applied between conductors 1 and 0 (ground); further, let conductor 3 be grounded ($U_3 = 0$).

For this situation, because Q_2 remains zero, you would probably guess that $U_2 = 0$; likewise, because U_3 is zero, you would probably guess that $Q_3 = 0$.

Well, both guesses are wrong!

As you will see in the upcoming analysis, voltages and charges are going to show up where you would probably least expect them.

In fact, field lines of **E** are present between conductors 1 and 0, between 1 and 2, between 1 and 3, between 2 and 0, and also between 2 and 3. The existing electric coupling among conductors makes $U_2 > 0$ and $Q_3 < 0$.

Let us next approach this subject in a systematic way.

2.9.1 Capacitance Matrix

Consider a set of several uncharged conductors immersed in an insulating dielectric medium. Let the total number of conductors be $n+1$, and let them be numbered from 0 to n. Assume that the dielectric material exhibits a linear behavior, $\mathbf{D} = \varepsilon\mathbf{E}$.

Next, assume that a set of voltage generators is connected among the conductors giving rise to a redistribution of charges within the system. In electrostatic equilibrium each conductor becomes characterized by a certain charge Q_k and by a certain potential V_k, with k running from 0 to n. Since the total amount of charge of the system must remain invariant (null), we can always write

$$Q_0 = -\sum_{k=1}^{n} Q_k \tag{2.42}$$

On the other hand, because the potential function is not uniquely defined, we may arbitrarily set $V_0 = 0$; in other words, we are selecting conductor 0 as the reference conductor of the system. (Note: You can arbitrarily choose the zeroth conductor.) Although the number of conductors is $n+1$, we can see that the problem we are dealing with is of order n.

The kth conductor voltage is defined as the voltage between conductor k and the reference conductor (0), that is $U_k = V_k - V_0 = V_k$.

Due to the existing electric coupling among conductors, the kth conductor charge Q_k depends not only on U_k but also on the remaining voltages. Since a linear behavior has been assumed for the system, any conductor charge can be written as a linear combination of all conductor voltages

$$Q_k = C_{k1}U_1 + \cdots + C_{kk}U_k + \cdots + C_{kn}U_n \tag{2.43}$$

for $k = 1$ to n. The coefficients C_{kj} are called the capacitance coefficients (units: F). The above result can be put in compact matrix form as

$$\underbrace{\begin{bmatrix} Q_1 \\ \vdots \\ Q_k \\ \vdots \\ Q_n \end{bmatrix}}_{[Q]} = \underbrace{\begin{bmatrix} C_{11} & \cdots & C_{1k} & \cdots & C_{1n} \\ \vdots & & \vdots & & \vdots \\ C_{k1} & \cdots & C_{kk} & \cdots & C_{kn} \\ \vdots & & \vdots & & \vdots \\ C_{n1} & \cdots & C_{nk} & \cdots & C_{nn} \end{bmatrix}}_{[C]} \underbrace{\begin{bmatrix} U_1 \\ \vdots \\ U_k \\ \vdots \\ U_n \end{bmatrix}}_{[U]} \tag{2.44}$$

The square $n \times n$ real matrix $[C]$ is called the capacitance matrix.

The entries of $[C]$ can be determined using various methods. They can be found experimentally by measuring system charges and voltages, they can be found numerically using dedicated computer programs, and in some cases, when very simple geometries are considered, they can also be determined analytically.

Whatever the method, the results you will obtain should be checked against a few general properties that $[C]$ must necessarily obey.

In order to illustrate these general properties of $[C]$, let us consider a system of four conductors (that is, $n = 3$) where one of them completely screens another – see Figure 2.16.

Let $U_1 > 0$ and $U_2 = U_3 = 0$ (Figure 2.16(a)). For this situation, the electric field coupling among conductors gives rise to a set of conductor charges such that $Q_1 > 0$, $Q_2 < 0$, $Q_3 = 0$, $Q_0 < 0$. Then, from

$$Q_1 = C_{11}U_1;\ Q_2 = C_{21}U_1;\ Q_3 = C_{31}U_1;\ Q_0 = -(Q_1 + Q_2 + Q_3) = -(C_{11} + C_{21} + C_{31})U_1$$

we see that

$$C_{11} > 0;\ C_{21} < 0;\ C_{31} = 0;\ \sum_{k=1}^{n} C_{k1} > 0 \tag{2.45a}$$

Next, let $U_2 > 0$ and $U_1 = U_3 = 0$ (Figure 2.16(b)). The conductor charges are such that $Q_1 < 0$, $Q_2 > 0$, $Q_3 < 0$, $Q_0 < 0$. Then, from

$$Q_1 = C_{12}U_2;\ Q_2 = C_{22}U_2;\ Q_3 = C_{32}U_2;\ Q_0 = -(Q_1 + Q_2 + Q_3) = -(C_{12} + C_{22} + C_{32})U_2$$

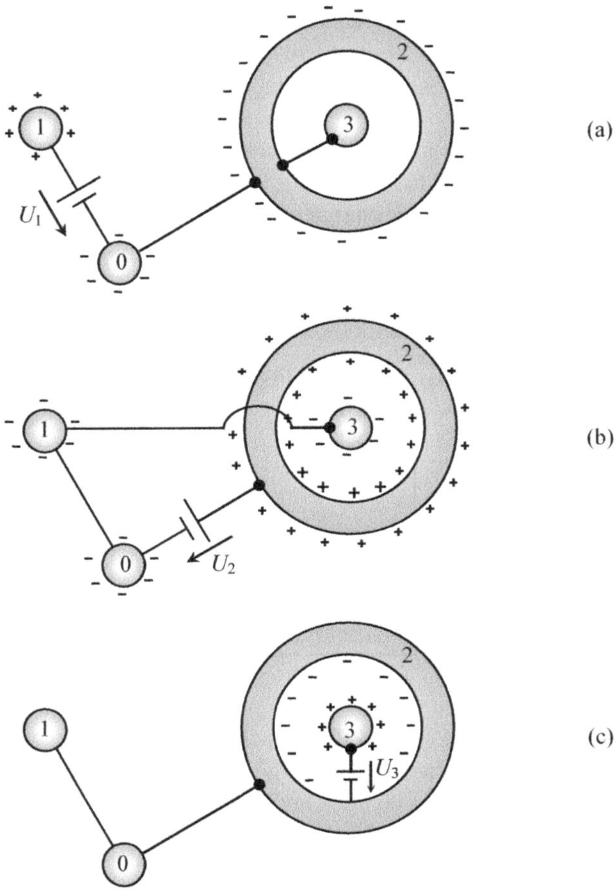

Figure 2.16 Charge distributions in a four-conductor system where conductor 3 is enclosed by conductor 2. (a) $U_1 > 0$, $U_2 = U_3 = 0$. (b) $U_2 > 0$, $U_1 = U_3 = 0$. (c) $U_3 > 0$, $U_1 = U_2 = 0$

we see that

$$C_{22} > 0;\ \ C_{12} < 0;\ C_{32} < 0;\ \ \sum_{k=1}^{n} C_{k2} > 0 \tag{2.45b}$$

Finally, let $U_3 > 0$ and $U_1 = U_2 = 0$ (Figure 2.16(c)). The conductor charges are such that $Q_1 = 0$, $Q_2 < 0$, $Q_3 > 0$, $Q_0 = 0$. Then, from

$$Q_1 = C_{13}U_3;\ \ Q_2 = C_{23}U_3;\ \ Q_3 = C_{33}U_3;\ \ Q_0 = -(Q_1 + Q_2 + Q_3) = -(C_{13} + C_{23} + C_{33})U_3$$

we see that

$$C_{33} > 0;\quad C_{13} = 0;\quad C_{23} < 0;\quad \sum_{k=1}^{n} C_{k3} = 0 \tag{2.45c}$$

From (2.45) the following general properties result:

$$C_{kk} > 0; \quad C_{kj} \leq 0; \quad \sum_{k=1}^{n} C_{kj} \geq 0 \tag{2.46}$$

You can see from (2.46) that only the diagonal entries of matrix $[C]$ are positive, all the remaining entries being zero or negative. Nonetheless, when you sum up all the entries of any chosen column of $[C]$ you cannot obtain a negative number; in other words, matrix $[C]$ is a diagonally dominant matrix.

A few more properties of $[C]$ can be derived based on energy considerations.

As in Section 2.7, the electric energy stored in the dielectric media surrounding the system conductors can be evaluated taking into account the amount of work produced by the voltage generators in order to have the conductors of the system charged to their final state of equilibrium

$$W_e = \frac{Q_1 U_1}{2} + \cdots + \frac{Q_k U_k}{2} + \cdots + \frac{Q_n U_n}{2} = \sum_{k=1}^{n} \frac{Q_k U_k}{2} \tag{2.47}$$

This result can be put in the matrix form

$$2W_e = \begin{bmatrix} Q_1 \cdots Q_k \cdots Q_n \end{bmatrix} \begin{bmatrix} U_1 \\ \vdots \\ U_k \\ \vdots \\ U_n \end{bmatrix} = [Q]^T [U]$$

where superscript T stands for transposition.

Using $[Q] = [C][U]$ from (2.44), we find

$$W_e = \frac{1}{2} [U]^T [C]^T [U] \tag{2.48}$$

Noting further that W_e is an ordinary scalar (1×1 matrix), $W_e = W_e^T$, we may write

$$W_e = \frac{1}{2} [U]^T [C] [U] \tag{2.49}$$

showing that, except for a factor 1/2, the stored electric energy is expressed by a quadratic form of matrix $[C]$.

Regardless of the polarity of the generators (that is, whether the voltages are positive or negative), the process of charging the system necessarily requires that the global work produced by the generators is positive, $W_e > 0$. Then, from (2.49), we can conclude that $[C]$ is a positive definite matrix and, consequently, its principal minors must all be positive.

For example, for $n = 3$,

$$C_{11} > 0; \quad C_{11}C_{22} - C_{12}C_{21} > 0; \quad \det[C] > 0$$

Moreover, since $\det[C] \neq 0$, one can always ensure that the inverse of $[C]$ exists; hence

$$[Q] = [C][U] \rightarrow [U] = [S][Q], \text{ with } [S] = [C]^{-1} \tag{2.50}$$

where $[S]$ is the so-called matrix of potential coefficients.

The quadratic form of (2.49) allows you to determine W_e as a function of conductor voltages; alternatively, if you prefer, you can also determine W_e via a quadratic form involving conductor charges, in which case the matrix of potential coefficients $[S]$ should be used:

$$W_e = \frac{1}{2}[Q]^T [S][Q]$$

At last, we prove the important property that $[C]$ is a symmetric matrix.

The physical fact that the work produced by the generators, to bring the system from its initial uncharged state to its final charged state, does not depend on the order in which the generators are turned on is equivalent to saying that W_e is a state function (that is, a function whose value is entirely specified by the system voltages at its final state of equilibrium).

The generalization of (2.22) for a multiple conductor system is

$$dW_e = \sum_{k=1}^{n} u_k \, dq_k = \sum_{k=1}^{n} A_k \, du_k$$

where

$$A_k = u_1 C_{1k} + \cdots + u_i C_{ik} + \cdots + u_n C_{nk}$$

Since the electric energy W_e is a state function, dW_e must be an exact differential. From a mathematical point of view this requires the following condition to be fulfilled:

$$\frac{\partial A_k}{\partial u_i} = \frac{\partial A_i}{\partial u_k}$$

from which we get

$$C_{ik} = C_{ki} \Leftrightarrow [C] = [C]^T \tag{2.51}$$

showing that the capacitance matrix is symmetric.

At this point you may be tempted to think that the preceding conclusion could have been reached before by simply comparing (2.48) and (2.49), which yields

$$[U]^T \underbrace{\left([C]-[C]^T\right)}_{[\Delta]} [U] = 0$$

but you would be wrong. Indeed, for the above quadratic form to be zero you do not really need to have $[\Delta] = 0$; it suffices that $[\Delta]$ is a skew-symmetric matrix, that is $\Delta_{ik} = -\Delta_{ki}$.

2.9.2 Partial Capacitances Scheme

The utilization of the $[C]$ matrix to describe the electric coupling phenomena among $n+1$ conductors is not the only approach available.

An alternative method – the so-called partial capacitance scheme – consists of modeling the electric interactions among conductors through a network of $n(n+1)/2$ fictitious capacitors.

Physically speaking, each capacitor connecting any two conductors is intended somehow to represent the electric field lines existing between those conductors. The electric energy stored in the multiple conductor system is the sum of the electric energies pertaining to the capacitors of the scheme.

To provide you with an example we can immediately sketch (see Figure 2.17) the partial capacitance scheme corresponding to the conductor configuration previously shown in Figure 2.15.

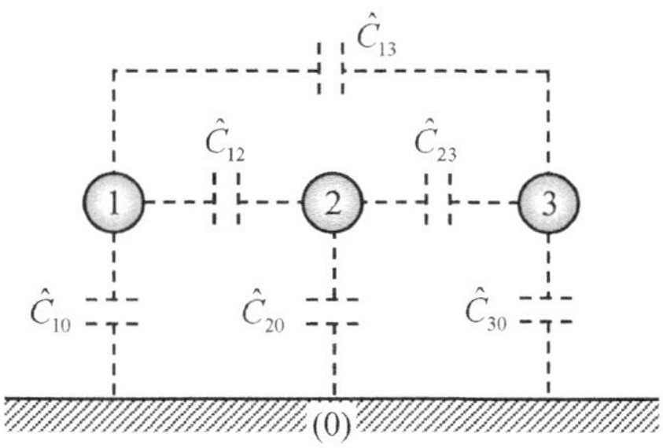

Figure 2.17 Scheme of partial capacitances corresponding to the configuration in Figure 2.15

Note that, to distinguish the capacitance matrix coefficients from the partial capacitances, a 'hat' sign (circumflex) is used in the latter.

In order to establish a relationship between the capacitance matrix formalism in Section 2.9.1 and the proposed partial capacitance scheme, let us obtain expressions for the conductor charges $Q_k (k = 1, 2, 3)$. By analyzing the scheme in Figure 2.17 we find

$$\begin{aligned} Q_1 &= \hat{C}_{10}U_1 + \hat{C}_{12}(U_1 - U_2) + \hat{C}_{13}(U_1 - U_3) \\ Q_2 &= \hat{C}_{20}U_2 + \hat{C}_{12}(U_2 - U_1) + \hat{C}_{23}(U_2 - U_3) \\ Q_3 &= \hat{C}_{30}U_3 + \hat{C}_{13}(U_3 - U_1) + \hat{C}_{23}(U_3 - U_2) \end{aligned}$$

Rearranging terms we get

$$\begin{aligned} Q_1 &= (\hat{C}_{10} + \hat{C}_{12} + \hat{C}_{13})U_1 - \hat{C}_{12}U_2 - \hat{C}_{13}U_3 \\ Q_2 &= -\hat{C}_{12}U_1 + (\hat{C}_{20} + \hat{C}_{12} + \hat{C}_{23})U_2 - \hat{C}_{23}U_3 \\ Q_3 &= -\hat{C}_{13}U_1 - \hat{C}_{23}U_2 + (\hat{C}_{30} + \hat{C}_{13} + \hat{C}_{23})U_3 \end{aligned} \tag{2.52}$$

Comparing the preceding results to the ones obtained from the $[C]$ matrix formalism,

$$\begin{aligned} Q_1 &= C_{11}U_1 + C_{12}U_2 + C_{13}U_3 \\ Q_2 &= C_{21}U_1 + C_{22}U_2 + C_{23}U_3 \\ Q_3 &= C_{31}U_1 + C_{32}U_2 + C_{33}U_3 \end{aligned}$$

and bearing in mind that $[C]$ is a symmetrical matrix, we finally reach the general conclusion that

$$\begin{cases} \hat{C}_{k0} = \sum\limits_{j=1}^{n} C_{kj} \\ \hat{C}_{kj} = -C_{kj} \ (j \neq 0) \end{cases} \tag{2.53}$$

In short, the partial capacitance between conductor k and the reference conductor 0 is obtained by adding the entries of the kth line in matrix $[C]$; the partial capacitance between conductors k and j is symmetrical to the corresponding entry in matrix $[C]$.

We would like to make a recommendation. Although the partial capacitance scheme may seem more appealing from a physical point of view, the fact is that, for multiple conductor systems with a large number of conductors, the $[C]$ matrix formalism is simpler to use and much more effective as far as calculations are concerned.

2.10 Application Example (Electric Coupling in Printed Circuit Boards)

A cross-sectional view of a printed circuit board (PCB) consisting of two conducting lands on the surface of a dielectric board above a reference ground plane is shown in Figure 2.18. The three-conductor system is characterized by a partial capacitance scheme whose capacitances have been experimentally determined.

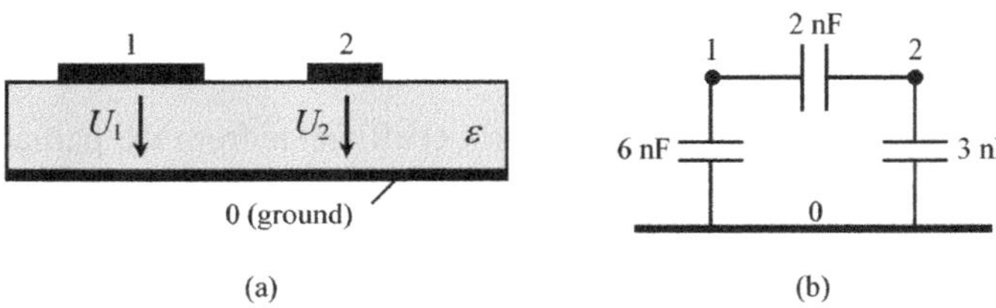

Figure 2.18 Printed circuit board with two lands over a substrate on a ground plane. (a) Cross-sectional view. (b) Partial capacitance scheme

Questions

Q_1 Determine the capacitance matrix $[C]$ of the multiple conductor system.

Q_2 Consider that land 1 is active, with $U_1 = 10\,\text{V}$, and that land 2 is idle ($Q_2 = 0$). Find the resulting values for charge Q_1 and voltage U_2.

Q_3 Next, consider that the voltage generator enforcing U_1 has been disconnected and, afterwards, land 2 has been short-circuited to ground. Determine the resulting charges Q_1 and Q_2 and voltages U_1 and U_2.

Q_4 Compute the electric energy stored for the situations addressed in questions Q_2 and Q_3. How do you justify the fact that $(W_e)_{(3)} < (W_e)_{(2)}$?

Solutions

Q_1 From (2.53) you have

$$C_{12} = C_{21} = -\hat{C}_{12} = -2\,\text{nF};\ C_{11} = \hat{C}_{10} - C_{12} = 8\,\text{nF};\ C_{22} = \hat{C}_{20} - C_{21} = 5\,\text{nF}$$

Hence

$$[C] = \begin{bmatrix} 8 & -2 \\ -2 & 5 \end{bmatrix} \text{nF}$$

Q_2 By making $U_1 = 10\,\text{V}$ in

$$\begin{cases} Q_1 = C_{11}U_1 + C_{12}U_2 \\ 0 = C_{21}U_1 + C_{22}U_2 \end{cases}$$

you find $Q_1 = 72\,\text{nC}$ and $U_2 = 4\,\text{V}$.

Q_3 Charge Q_1 remains trapped and unchanged, $Q_1 = 72\,\text{nC}$.
Land 2 is short-circuited, $U_2 = 0\,\text{V}$.
By making $Q_1 = 72\,\text{nC}$ in

$$\begin{cases} Q_1 = C_{11}U_1 \\ Q_2 = C_{21}U_1 \end{cases}$$

you find $Q_2 = -18\,\text{nC}$ and $U_1 = 9\,\text{V}$.

Q_4 Here

$$(W_e)_{(2)} = (Q_1U_1 + Q_2U_2)/2 = 360\,\text{nJ};\ (W_e)_{(3)} = (Q_1U_1 + Q_2U_2)/2 = 324\,\text{nJ}$$

Land 2, which was initially discharged, acquired a negative charge of -18 nC after its connection to the ground. During the transient regime, free electrons have flowed from ground to land 2 through the connecting conductor; the corresponding transient electric current gave rise to Joule losses totaling 36 nJ.

2.11 Electric Forces and Torques

With the recent development of microelectromechanical systems (MEMS) technology, such as micromachined capacitive actuators, increased attention is being currently paid to the evaluation of electric forces and electric torques exerted on charged conductor devices.

Electric forces $\mathbf{F}_e$ can be evaluated by resorting to two different approaches.

The first is based on the well-known basic result for the local force

$$\mathbf{F}_e = q\mathbf{E} \tag{2.54}$$

which was referred to in Chapter 1. The problem with this method is that, for its application, it requires a detailed knowledge of the distribution of charges on the conductors – a requisite that, except for very simple problems, can hardly be met.

In the second approach, the resultant force exerted on a given conductor is evaluated globally, based on electric energy considerations.

To exemplify the latter very powerful approach let us consider the two-conductor system (capacitor) shown in Figure 2.19 where electrode (0) is fixed and electrode (1) undergoes a

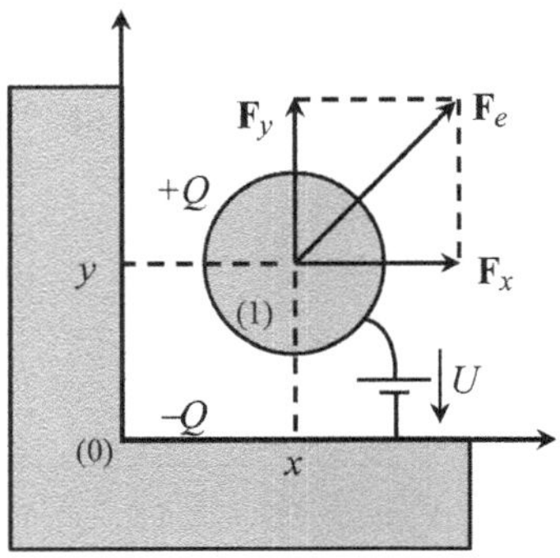

Figure 2.19 Electric forces on a charged conductor part of a two-conductor system (capacitor)

virtual displacement $d\mathbf{s} = dx\,\vec{e}_x + dy\,\vec{e}_y$ associated with the global force $\mathbf{F}_e = f_x\,\vec{e}_x + f_y\,\vec{e}_y$ exerted on it. To put the problem in a more general perspective we will allow the insulating dielectric medium to be nonlinear.

The generator connected to the capacitor electrodes is responsible not only for the electric energy stored in the capacitor, but also for the mechanical work associated with the actuating force. In differential terms the system energy balance is thus described by

$$dW_G = dW_e + dW_{\text{mec}} \tag{2.55}$$

On the one hand, according to (2.22), the generator's work is given by $dW_G = udq$. On the other hand, the mechanical energy is given by $dW_{\text{mec}} = \mathbf{F}_e \cdot d\mathbf{s} = f_x dx + f_y dy$.

Therefore you can see that

$$dW_e = udq - f_x dx - f_y dy \tag{2.56}$$

As illustrated in Figure 2.20, the electric energy of the capacitor depends on the respective charge q and also on the configuration variables x and y

$$W_e = W_e(q, x, y) \tag{2.57}$$

By differentiating (2.57) we find

$$dW_e = \left(\frac{\partial W_e}{\partial q}\right)_{x,y} dq + \left(\frac{\partial W_e}{\partial x}\right)_{q,y} dx + \left(\frac{\partial W_e}{\partial y}\right)_{q,x} dy \tag{2.58}$$

Comparing (2.56) and (2.58) yields

$$f_x = -\left(\frac{\partial W_e}{\partial x}\right)_{q,y} \; ; \quad f_y = -\left(\frac{\partial W_e}{\partial y}\right)_{q,x} \tag{2.59}$$

or, more generally,

$$\mathbf{F}_e = -\operatorname{grad} W_e|_{q\,\text{invariant}}$$

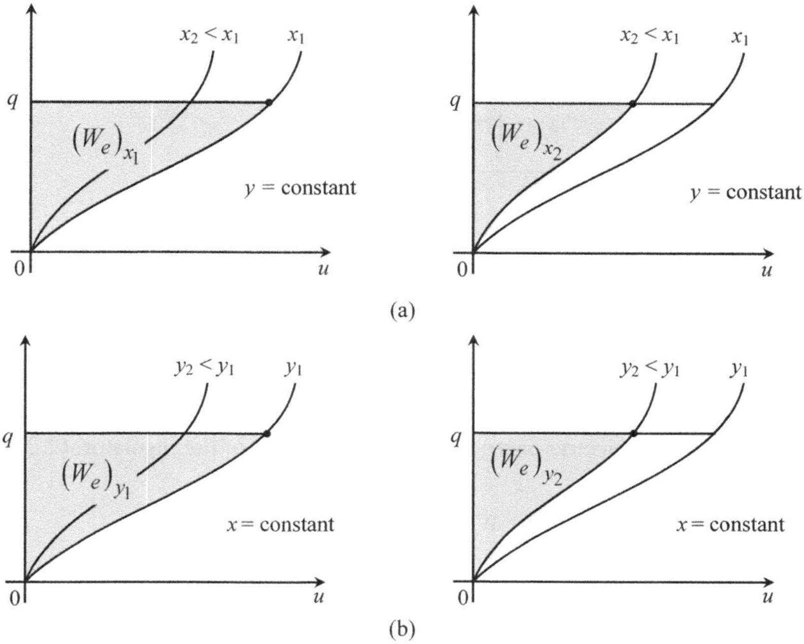

Figure 2.20 Electric energy stored in a capacitor, under constant charge conditions, upon a virtual displacement of one of the electrodes. (a) Displacement along x, keeping y constant. (b) Displacement along y, keeping x constant

Note that the evaluation of the partial derivatives in (2.59) is executed assuming that q is held fixed despite the movement. With constant charge, $dq = 0$, the amount of energy delivered by the generator $dW_G = udq$ is zero, meaning that mechanical work results from a decrease of the electric energy stored.

Hence, you can see from Figure 2.20(a) and Figure 2.20(b) that, if the electric energy stored is decremented ($\Delta W_e < 0$), either x or y is decremented ($\Delta x < 0$ or $\Delta y < 0$). Consequently, from (2.59) you will obtain $f_x < 0$ and $f_y < 0$, which shows, unsurprisingly, that $\mathbf{F}_e$ is an attractive force – the capacitor electrodes are attracted to each other.

To clarify things a little, we present an application example.

Let us turn our attention to the very familiar case of the parallel-plate capacitor (Figure 2.11) we analyzed earlier.

The insulating material is air, which is a linear medium ($D = \varepsilon_0 E$, $Q = CU$). Assume that the separation x between plates is small (by neglecting fringing field effects, both **D** and **E** can be considered uniform fields). For this problem, we have $U = Ex$ and $Q = DS$, where S is the area of each plate where forces are exerted (Figure 2.21).

The attractive force, perpendicular to the plates, is determined from

$$f_x = -\left(\frac{\partial W_e}{\partial x}\right)_Q \tag{2.60}$$

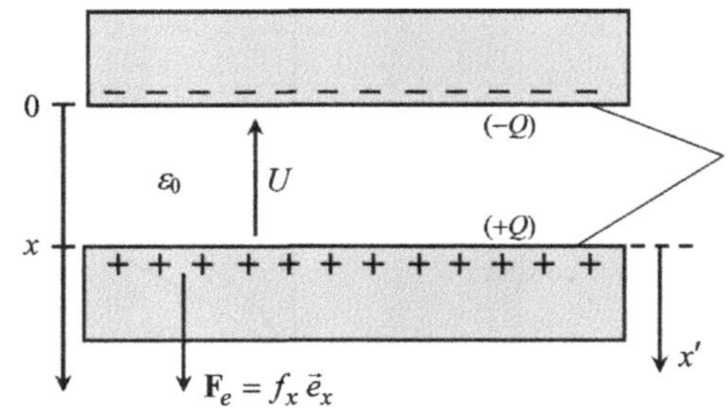

Figure 2.21 Attractive electric force in a parallel plate capacitor ($f_x < 0$)

The electric energy W_e can be expressed either in terms of the capacitor capacitance C, (2.24), or in terms of the electric energy density $\hat{w}_e$ residing in the insulator, (2.27):

$$W_e = \frac{Q^2}{2C} = \hat{w}_e V$$

where V is the volume of the dielectric material ($V = Sx$) sandwiched between the plates.

Using the capacitance formulation we find

$$f_x = -\frac{\partial}{\partial x}\left(\frac{1}{C(x)}\frac{Q^2}{2}\right)_Q = -\frac{Q^2}{2}\frac{d}{dx}\left(\frac{1}{C(x)}\right)$$

but

$$\frac{1}{C(x)} = \frac{U}{Q} = \frac{Ex}{\varepsilon_0 ES} = \frac{x}{\varepsilon_0 S}$$

hence, we have the result

$$f_x = -\frac{Q^2}{2\varepsilon_0 S} \tag{2.61a}$$

Further, noting that $Q = CU$, we have

$$f_x = -\frac{\varepsilon_0 S}{2x^2}U^2 \tag{2.61b}$$

The force is attractive (minus sign). It does not depend on the polarity of the generator (if you reverse the generator terminals the force does not become repulsive!): the smaller the separation between the plates, the stronger the attractive force between them. Another conclusion you should note is that the force acts so as to diminish the plates' separation; put another way, the force acts so as to increase the capacitor capacitance.

Finally, let us explore the formulation based on the electric energy density:

$$f_x = -\frac{\partial}{\partial x}(\hat{w}_e \overbrace{Sx}^{V})_Q = -\hat{w}_e S$$

This result immediately shows that the pressure P (units: N/m^2) exerted on the plates' surface by the attractive force is numerically equal to the electric energy density stored near the plates

$$P = \frac{|f_x|}{S} = \hat{w}_e \tag{2.62}$$

From an engineering point of view, it is very important to acquire an estimate of the maximum pressures achievable by electric forces. Assuming the dielectric medium is air, $\varepsilon = \varepsilon_0$, and taking into account that breakdown electric fields are around 30 kV/cm, we find from (2.62)

$$P = \hat{w}_e = \varepsilon_0 E^2/2 \approx 40\,\text{N/m}^2 \tag{2.63}$$

In absolute terms, this is not very much ($\approx 0.4\,\text{gf/cm}^2$) indeed, but for MEMS applications it is more than sufficient.

So far we have paid attention to linear displacements originating from electric forces. However, problems exist when charged conductors are subjected to rotational movements originating from torques of electrical origin.

The type of formulation for these new problems follows the same rationale we used before. The only other thing required is a change of notation, justified by the fact that for rotational movements the differential mechanical work is expressed as $dW_{\text{mec}} = T_\alpha\, d\alpha$, where α is the angle of rotation and T_α the associated torque (units: Nm).

Therefore, generalizing the results in (2.59),

$$T_\alpha = -\left(\frac{\partial W_e}{\partial \alpha}\right)_q \tag{2.64}$$

2.12 Proposed Homework Problems

Problem 2.12.1

The parallel-plate capacitor in Figure 2.22 includes two dielectric layers characterized by $\varepsilon_1 = 4\varepsilon_0$ and $\varepsilon_2 = 6\varepsilon_0$. Capacitor size is defined by $a = 1\,\text{cm}$ and $\delta = 0.5\,\text{mm}$. A positive voltage U is applied between the plates.

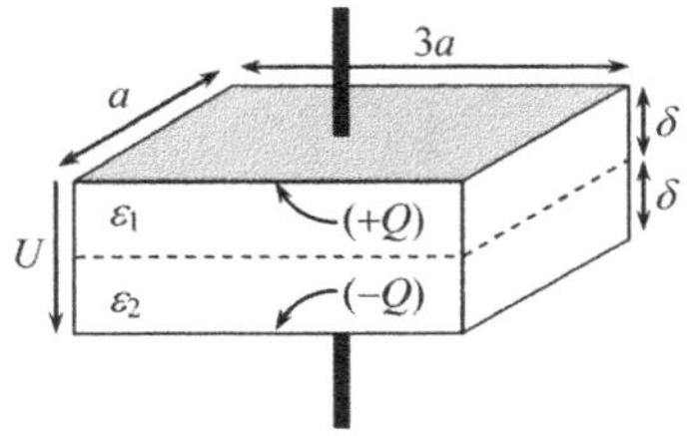

Figure 2.22 Parallel-plate capacitor filled with an inhomogeneous dielectric medium

Q_1 Assuming that the resulting capacitor charge is $Q = 127.3\,\text{nC}$, determine the field intensities D_1, E_1, D_2 and E_2 inside each dielectric layer. Next, find the value of the voltage U between the plates, and determine the capacitance C of the capacitor. Evaluate the electric energy stored in the capacitor W_e and specify its distribution among the dielectric layers.

Q_2 Assume that the insulating layers are characterized by the following values of dielectric strength: $E_{d1} = 200\,\text{kV/cm}$ and $E_{d2} = 100\,\text{kV/cm}$. Determine the maximum voltage $U_{\max}$ that the capacitor can withstand before dielectric breakdown occurs.

Answers

Q_1 $D_1 = D_2 = 424.33\,\mu\text{C/m}^2$; $E_1 = 120\,\text{kV/cm}$; $E_2 = 80\,\text{kV/cm}$; $U = 10\,\text{kV}$; $C = 12.73\,\text{pF}$; $W_e = 636.5\,\mu\text{J}$; $W_{e1} = 381.9\,\mu\text{J}$; $W_{e2} = 254.6\,\mu\text{J}$.

Q_2 $U_{\max} = 12.5\,\text{kV}$.

Problem 2.12.2

Consider an infinitesimal point charge dQ immersed in a homogeneous dielectric medium of permittivity ε.

Q_1 Determine the infinitesimal electric field vector $d\mathbf{E}$ at a distance r' from the charge.

Q_2 Assume that a rectilinear filament of charge oriented along z is made up from an infinite collection of infinitesimal point charges $dQ = qdz$, extending from $z = -\infty$ to $z = +\infty$. Consider a point P at a distance r from the filament of charge – see Figure 2.23. Determine the infinitesimal contribution $d\mathbf{E}_\text{P}$ to the electric field at P originating from two symmetrically positioned point charges.

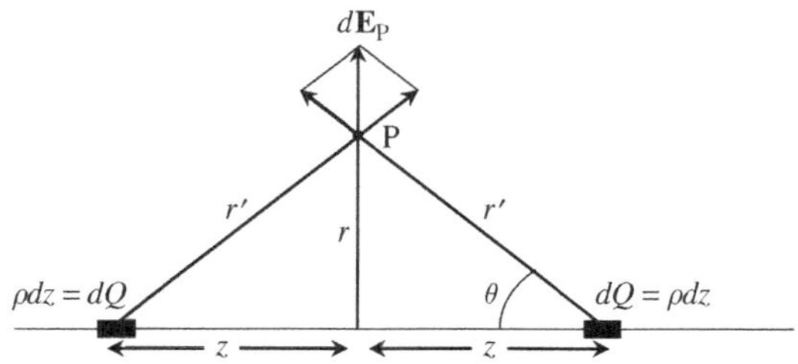

Figure 2.23 Electric field produced by two symmetrically positioned point charges dQ

Q_3 Finally, by summing the contributions from all the charge pairs, show that the total electric field vector at a generic point P a distance r from the charged filament is given by the result obtained in (2.19).

Answers

Q_1 $d\mathbf{E} = dE\vec{e}_{r'}$, where

$$dE = \frac{dQ}{4\pi\varepsilon\,(r')^2}$$

Q_2

$$d\mathbf{E}_P = dE\frac{2r}{\sqrt{r^2+z^2}}\vec{e}_r$$

Q_3

$$\mathbf{E} = \int_{z=0}^{z=\infty} d\mathbf{E}_P = \left(\frac{q}{2\pi\varepsilon r}\right)\vec{e}_r$$

Problem 2.12.3

The structure in Figure 2.24 represents a coaxial cable of length $l = 900\,\text{m}$. The cable consists of two conductors: an external cylindrical shield of inner radius $r_2 = 13.59\,\text{mm}$, and an internal cylindrical conductor whose radius r_1 is unknown. The characteristic parameters of the dielectric medium are $\varepsilon = 2\varepsilon_0$ and $E_d = 200\,\text{kV/cm}$. The cable has a charge Q as a result of the application of a voltage U between cable conductors. Assume that the cable shield is at zero potential.

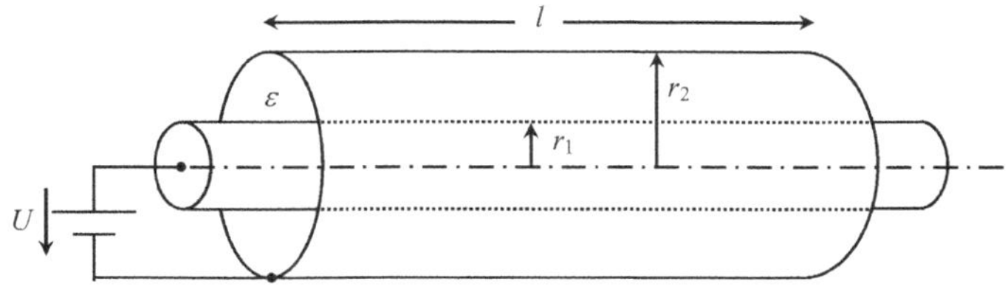

Figure 2.24 Coaxial cable

Q_1 By using the results in Application Example 2.6, determine $\mathbf{E}(r)$ and $V(r)$.

Q_2 Determine the cable capacitance C.

Q_3 Obtain an expression for E_{max} as a function of both U and r_2/r_1. Determine the optimal value for r_1 such that the function $E_{max}(r_2/r_1)$ goes through its minimum.

Q_4 Assuming that r_1 has been optimized, and assuming that the cable is allowed to operate at 20 % below dielectric breakdown, find numerical values for the cable capacitance, cable allowed voltage, cable charge, as well as for the electric energy stored along the whole cable.

Q_5 Consider the cable voltage determined above. Show that if the cable had not been properly designed, choosing for example $r_1 = 2\,\text{mm}$ or $r_1 = 9\,\text{mm}$, then dielectric breakdown would have certainly taken place.

Answers

Q_1

$$\mathbf{E} = \begin{cases} 0, \text{ for } r < r_1 \\ \dfrac{Q}{2\pi\varepsilon l r}\vec{e}_r, \text{ for } r_1 < r < r_2 \\ 0, \text{ for } r > r_2 \end{cases} \qquad V = \begin{cases} U, \text{ for } r < r_1 \\ \dfrac{Q}{2\pi\varepsilon l}\ln\dfrac{r_2}{r}, \text{ for } r_1 < r < r_2 \\ 0, \text{ for } r > r_2 \end{cases}$$

Q_2

$$C = \frac{2\pi\varepsilon l}{\ln(r_2/r_1)}$$

Q_3

$$E_{max} = E_{(r=r_1)} = \frac{U}{r_2}\frac{k}{\ln k}, \text{ with } k = \frac{r_2}{r_1}$$

$$(r_1)_{optimal} = \frac{r_2}{e} = 5\,\text{mm}$$

Q_4 $C = 2\pi\varepsilon l = 100\,\text{nF}$; $U = r_1 \times 0.8 E_d = 80\,\text{kV}$; $Q = 8\,\text{mC}$; $W_e = 320\,\text{J}$.

Q_5 $r_1 = 2\,\text{mm} \rightarrow E_{max} = 208.8\,\text{kV/m} > E_d$; $r_1 = 9\,\text{mm} \rightarrow E_{max} = 215.7\,\text{kV/m} > E_d$.

Problem 2.12.4

Figure 2.25 depicts a single-phase overhead line of length l, where a cylindrical conductor of radius r runs parallel to the ground conductor at height d. Consider the typical situation $r \ll d$.

A voltage U is applied between the conductors giving rise to an electric field distribution in the air around the conductors. Electric fields from transmission lines may affect the

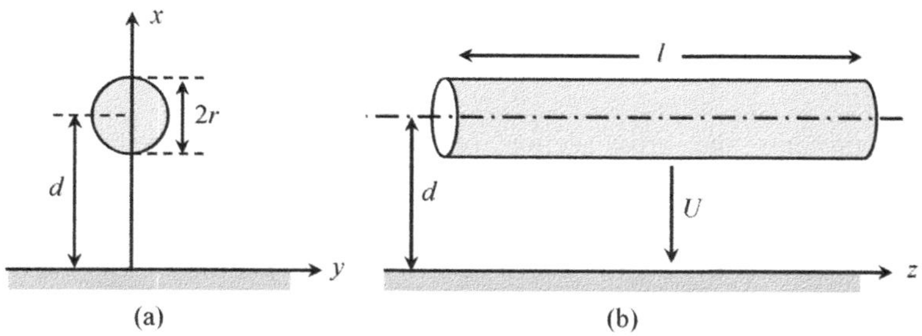

Figure 2.25 Cylindrical conductor above ground. (a) Cross-sectional view. (b) Longitudinal view

operation of electronic appliances and, furthermore, may affect human health. For this reason, limitations have been internationally adopted concerning the exposure to **E** fields ($E_{max} = 5\,\text{kV/m}$).

Q_1 Find the line capacitance C.

Q_2 Determine the charge density distribution on the ground surface, $w(y)$.

Q_3 Determine the electric field along the x axis.

Q_4 Take $r = 1\,\text{cm}$, $d = 10\,\text{m}$, $l = 1\,\text{km}$. Find the maximum voltage U at which the line can be operated such that $E = \frac{1}{4}E_{max}$ at a point 1.8 m above ground (head height).

(Note: The analysis of this problem can be conducted by employing the results already derived in Application Example 2.6. Explain why.)

Answers

Q_1

$$C = 2 \times \frac{\pi\varepsilon_0 l}{\ln\left(\frac{d}{r} + \sqrt{\left(\frac{d}{r}\right)^2 - 1}\right)} \approx \frac{2\pi\varepsilon_0 l}{\ln\left(\frac{2d}{r}\right)}$$

Q_2

$$w(y) = -\frac{CU}{\pi l} \times \frac{d}{y^2 + d^2}$$

Q_3

$$\mathbf{E} = E(x)\vec{e}_x, \; E(x) = \frac{CU}{\pi\varepsilon_0 l} \times \frac{d}{x^2 - d^2}, \; x \in [0, d - r] \cup [d + r, \infty[$$

Q_4

$$C = 8.04\,\text{nF}; \; U = 41.8\,\text{kV}$$

Problem 2.12.5

Figure 2.26 represents the cross-section of a three-conductor system comprising a communications coaxial cable of length l running parallel to a conducting wall (reference conductor). Consider the following data: $l = 10\,\text{m}$, $d = 1\,\text{cm}$, $r_2 = 5\,\text{mm}$, $r_1 = 1\,\text{mm}$, $\varepsilon = 2\varepsilon_0$.

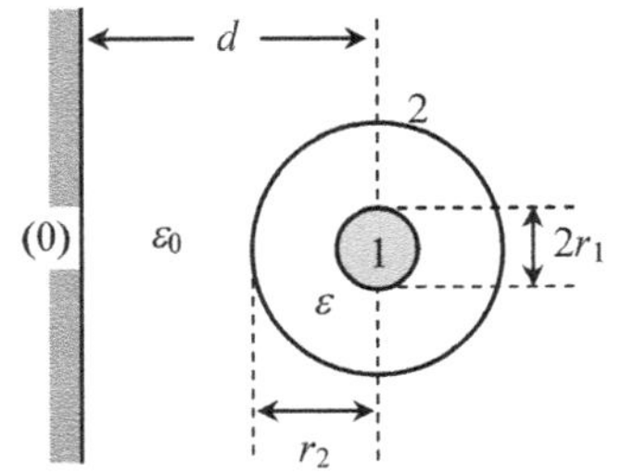

Figure 2.26 Coaxial cable parallel to a conducting ground plane

Q_1 Determine the partial capacitance scheme of the conductor system.

Q_2 Find the corresponding capacitance matrix $[C]$.

Answers

Q_1

$$\hat{C}_{10} = 0;\ \hat{C}_{20} = \frac{2\pi\varepsilon_0 l}{\ln\left(\dfrac{d}{r_2} + \sqrt{\left(\dfrac{d}{r_2}\right)^2 - 1}\right)} = 0.422\,\text{nF};\ \hat{C}_{12} = \frac{2\pi\varepsilon l}{\ln\left(\dfrac{r_2}{r_1}\right)} = 0.690\,\text{nF}$$

Q_2

$$[C] = \begin{bmatrix} 0.690 & -0.690 \\ -0.690 & 1.112 \end{bmatrix}\ \text{nF}$$

Problem 2.12.6

Figure 2.27 shows a cross-sectional view of a typical low-voltage single-phase cable configuration consisting of three identical conductors: phase conductor (1), neutral conductor (2) and ground conductor (0).

The cable configuration exhibits three-fold symmetry. The equivalent phase–neutral capacitance has been measured, $C_{eq} = 150\,\text{nF}$.

Q_1 Determine the partial capacitance scheme of the cable as well as the corresponding capacitance matrix $[C]$.

Q_2 Keeping the ground conductor in its idle state ($Q_0 = 0$), a voltage $U = 200\,\text{V}$ has been applied between the phase and neutral conductors. Evaluate cable charges Q_1 and Q_2, and cable voltages U_1 and U_2.

Q_3 Repeat the preceding calculations assuming that the neutral conductor is grounded ($U_2 = 0$).

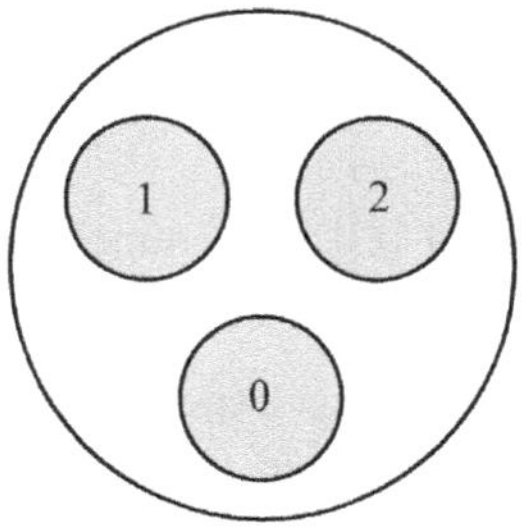

Figure 2.27 Cross-sectional view of an indoor low-voltage single-phase cable, including phase, neutral and ground conductors

Answers

Q_1

$$\hat{C}_{10} = \hat{C}_{20} = \hat{C}_{12} = \frac{2}{3}C_{eq} = 100\,\text{nF};\; [C] = \begin{bmatrix} 200 & -100 \\ -100 & 200 \end{bmatrix}\text{nF}$$

Q_2

$$Q_1 = -Q_2 = 30\,\mu\text{C};\; U_1 = -U_2 = 100\,\text{V}.$$

Q_3

$$U_1 = 200\,\text{V};\; U_2 = 0;\; Q_1 = 40\,\mu\text{C};\; Q_2 = -20\,\mu\text{C};\; Q_0 = -20\,\mu\text{C}.$$

Problem 2.12.7

Figure 2.28(a) shows a cross-sectional view of a three-phase overhead line with three identical conductors, of radius $r = 2\,\text{cm}$, parallel to the ground (reference conductor). The line, of length $l = 10\,\text{km}$, exhibits a vertical plane of symmetry.

The capacitance matrix describing the structure is

$$[C] = \begin{bmatrix} 100 & -x & -5 \\ -x & 80 & -x \\ -5 & -x & 100 \end{bmatrix}\text{nF}$$

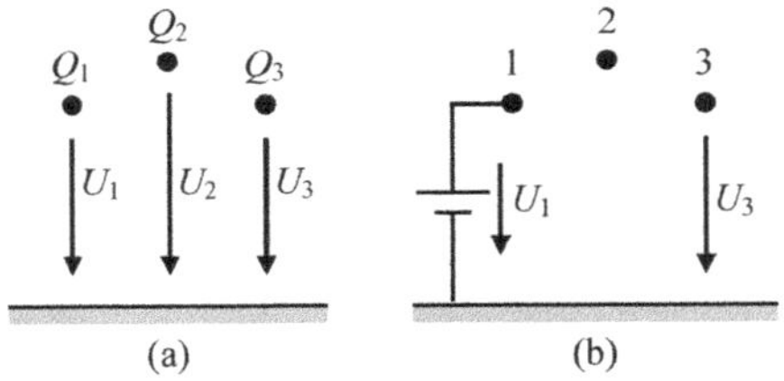

Figure 2.28 Three-phase overhead line configuration. (a) Cross-section. (b) Excitation status

Q_1 Based on the general properties of $[C]$, determine the maximum allowed value for the unknown coefficient x.

Q_2 In order to determine x the following experiment was conducted (Figure 2.28(b)): keeping both conductors 2 and 3 isolated ($Q_2 = Q_3 = 0$), a voltage $U_1 = 3.8\,\text{kV}$ was applied between conductor 1 and the ground. The resulting voltage U_3 was measured, $U_3 = 400\,\text{V}$.

Find x. Determine U_2 and Q_1. Estimate the electric field intensity occurring at the surface of the energized conductor, E_1.

Answers

Q_1 $x < 40$.

Q_2 $x = 20$; $U_2 = 1050\,\text{V}$; $Q_1 = 357\,\mu\text{C}$; $E_1 \approx 32\,\text{kV/m}$.

Problem 2.12.8

In Section 2.11 we analyzed the attractive electric force actuating on the plates of a parallel-plate capacitor (see Figure 2.21). Resorting to energy balance considerations, we reached the conclusion that the intensity of the attractive force was given by $f = Q^2/(2\varepsilon_0 S)$. However, if you try to evaluate f by using the basic local force definition $f = QE$ you will be either disappointed or puzzled.

In fact, from $E = D/\varepsilon_0 = Q/(\varepsilon_0 S)$ you will get $f = Q^2/(\varepsilon_0 S)$, which differs from the correct result by a factor of 2.

The reason for this contradiction lies in the fact that, from a microscopic point of view, the conductor/dielectric interface is not actually an abrupt perfect plane. The so-called surface charge does not really reside at the conductor's surface; it is a volumetric charge ρ extending a little into the interior of the conductor (a few atomic layers) but vanishing very rapidly. A similar thing happens with the electric field. At the conductor surface you have $E_0 = Q/(\varepsilon_0 S)$, yet as you progress to the interior of the conductor, the field intensity decays continuously; it does not drop to zero abruptly.

Let x' be the direction perpendicular to the positively charged capacitor plate oriented towards the inside of the conductor (Figure 2.21). Assume a typical decay law for $\rho(x')$ given by

$$\rho(x') = \rho_0 \, e^{-kx'}$$

where k denotes the rate of decay – an arbitrary parameter that you can choose as large as you wish.

Q_1 Obtain ρ_0.
(Hint: $\int_{V_{\text{plate}}} \rho(x') dV = Q; \quad dV = S dx'$.)

Q_2 By using the equation $\text{div}\, \mathbf{E} = \rho(x')/\varepsilon_0$, where $\mathbf{E} = -E\vec{e}_{x'}$, determine $E(x')$.

Q_3 Evaluate the global electric force f exerted on the charged plate.
(Hint: $f = \int_{V_{\text{plate}}} df_e; \quad df_e = dqE; \quad dq = \rho dV$.)

Q_4 Find the plate attractive pressure P for a capacitor 0.5 mm thick submitted to a 10 V voltage.

Answers

Q_1 $\rho_0 = kQ/S$.

Q_2 $E(x') = E_0 \, e^{-kx'}$.

Q_3 $f = Q^2/(2\varepsilon_0 S)$.

Q_4 $P = 1.77\,\text{N/cm}^2$.

Problem 2.12.9

Consider an aerial two-wire transmission line of indefinite length, whose cylindrical conductors are separated by a distance $x = 1\,\text{cm}$. The radius of both conductors (thin conductors) is $r = 1\,\text{mm}$. The applied voltage between the wires is $U = 200\,\text{V}$.

Q_1 Write the equation for the per-unit-length line capacitance C as a function of x. Compute its value according to the problem data.

Q_2 Determine the intensity of the per-unit-length attractive force f exerted on both line conductors.

Answers

Q_1

$$C(x) = \frac{\pi\varepsilon_0}{\ln(x/r)} = 12\,\text{pF/m}$$

Q_2

$$f = \frac{(CU)^2}{2} \times \frac{d}{dx}\left(\frac{1}{C(x)}\right) = \frac{(CU)^2}{2\pi\varepsilon_0 x} = 10.5\,\mu\text{N/m}$$

3

Stationary Currents

3.1 Fundamental Equations

The topic of stationary currents (also called direct currents) belongs within the subclass of stationary field phenomena. The properties of time-invariant electric currents, associated with free charges moving along closed conductor circuits, are analyzed in this chapter.

The fundamental laws governing stationary current problems are those in (PII.3) together with a constitutive relation concerning conductor media behavior. That is,

$$\begin{cases} \text{curl } \mathbf{E} = 0 \\ \text{div } \mathbf{J} = 0 \end{cases} \tag{3.1}$$

and

$$\mathbf{J} = \sigma \mathbf{E} \tag{3.2}$$

where σ denotes conductor conductivity (units: S/m, siemens per meter).

The equation curl $\mathbf{E} = 0$ has been fully examined in Chapter 2. The properties of the electric field vector $\mathbf{E}$ we studied in electrostatics are exactly the same that you need to keep in mind throughout this new chapter.

However, here – contrary to electrostatics – because currents are allowed to exist ($\mathbf{J} \neq 0$), the electric field vector inside conductors is not zero, $\mathbf{E} = \mathbf{J}/\sigma$, and, consequently, conductors can no longer be considered equipotential bodies. Only in the limit case of perfect conductors ($\sigma \to \infty$) can you use the approximation $\mathbf{E} = 0$ and $V =$ constant.

3.2 Conductivity, Current Density, Electric Circuits

As far as conduction properties are concerned, material media can be coarsely split into two major categories, insulators and conductors. While insulators, like glass, mica, rubber, etc., are characterized by extremely low conductivity values in the range 10^{-8} to 10^{-17} S/m, conductors, like silver, copper, aluminum, etc., have extremely high conductivity values in the range 10^6 to 10^7 S/m. Typical conductivities (at 20 °C) for some common conductors

Table 3.1 Conductor conductivities

Conductor	Conductivity (S/m) at 20 °C
Aluminum	3.5×10^7
Constantan	2×10^6
Copper	5.6×10^7
Gold	4.1×10^7
Graphite	10^5
Iron	10^6
Manganin	2.3×10^6
Silver	6.1×10^7
Tin	9×10^6
Water (sea water)	5.6
Water (tap water)	0.01–0.1

are listed in Table 3.1. Note that the conductivity is a temperature-dependent parameter; conductivities decrease with increasing temperature in the case of metallic conductors.

Free charges inside a conducting medium can move under the influence of impressed electric fields. In the case of good conductors (metals), free charges are electrons and their movement occurs in the direction opposite to **E**; however, from a formal point of view, you can imagine that an equivalent flow of positive charges occurs parallel to **E** (Figure 3.1).

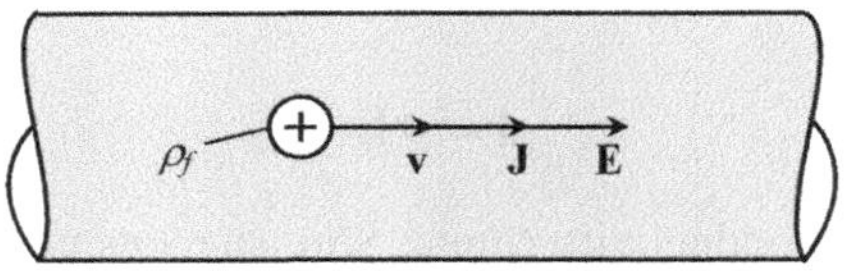

Figure 3.1 Current flow inside a conductor driven by an electric field

Electric currents are free charges in movement. Hence, in order to provide a physical interpretation for the current density vector **J**, we can write

$$\mathbf{J} = \rho_f \mathbf{v} \tag{3.3}$$

where ρ_f (C/m^3) represents the positive free charge per unit volume and **v** denotes the average value of the charge velocity parallel to the impressed **E** field.

For not very intense fields (linear media) the velocity **v**, resulting from random collision processes inside the medium atomic lattice, is proportional to **E**

$$\mathbf{v} = m\mathbf{E} \tag{3.4}$$

where m is the so-called charge mobility. (Note: The electron mobility, for good conductors, can typically be found in the range 10^{-2} to $10^{-3}\,\mathrm{m^2\,V^{-1}\,s^{-1}}$.)

Finally, from (3.2)–(3.4), we obtain $\mathbf{J} = \sigma\mathbf{E} = \rho_f m\mathbf{E}$, from which you can see that $\sigma = \rho_f m$.

At this stage it is worth making an important remark. You have certainly heard before that electrical signals propagate at a velocity close to the speed of light (3×10^8 m/s). Well, this is true. But do not confuse matters: such a velocity has absolutely nothing to do with the velocity that electrons move along conductors!

Considering **E**-field values that may typically occur inside good conductors ($\approx$100 V/m), you can check from (3.4) that the velocity of the electrons is merely about 1 m/s.

Let us now return to (3.1) and focus our attention on the equation div $\mathbf{J} = 0$.

Again using the Gauss theorem – see (2.9) from Chapter 2 – we obtain

$$\int_{S_V} \mathbf{J} \cdot \mathbf{n}_o \, dS = 0 \tag{3.5}$$

This means that, in the framework of stationary regimes ($\partial/\partial t = 0$), the flux of **J** across a closed surface bounding a given volume is always zero or, in other words, the number of **J**-field lines entering a given volume is equal to those leaving it (Figure 3.2).

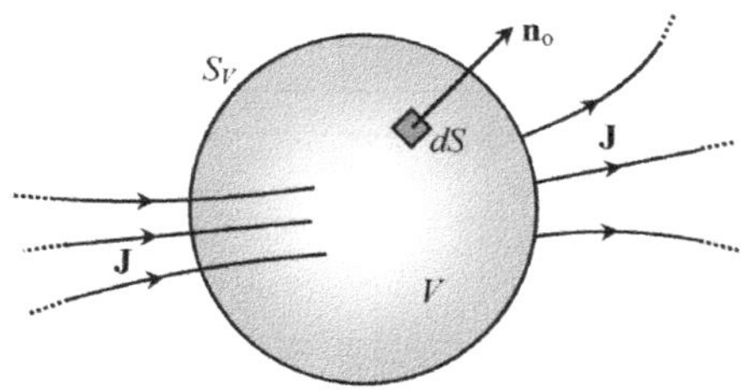

Figure 3.2 The flux of **J** across a closed surface is zero for time-invariant regimes

An obvious consequence of this result is that the field lines of **J** cannot end or start anywhere. In general, any field vector whose *divergence* is zero must have its field lines closed.

An electric circuit is made of simple or multiple conductor connections forming closed loops so as to ensure that **J**-field lines are closed, otherwise one would end up with $\mathbf{J} = 0$ as in electrostatics.

The simplest electric circuit that can be imagined consists of a conductor loop immersed in a dielectric insulating medium. The conductor loop forms a closed tube where free charges can circulate. In most applications, given the huge discrepancy between conductor and insulator conductivities, leakage currents escaping the conductor loop (the circuit) can be considered absolutely negligible.

Although electric circuits are practically perfect tubes for **J**-field lines, you must be aware that the same is not true for the electric field **E**. In fact, **E** exists not only inside the circuit conductors but outside of them as well.

At the conductor/insulator interface (conductor side) you have a purely tangential component for the electric field, $\mathbf{E}_{cond} = \mathbf{E}_t = \mathbf{J}/\sigma$.

At the conductor/insulator interface (insulator side), although $\mathbf{J}=0$, you will not get $\mathbf{E}=0$. The electric field vector on the insulator side can be obtained by adding two orthogonal components

$$\mathbf{E}_{\text{insul}} = \mathbf{E}_{\text{n}} + \mathbf{E}_{\text{t}}$$

According to (2.17), the magnitude of the normal component, which is usually the dominant term, depends on the local surface charge density, $E_{\text{n}} = w/\varepsilon$.

As for the tangential component, it is a simple matter to show (using curl $\mathbf{E}=0$) that it coincides with the one observed inside the conductor.

To prove the continuity of the tangential component of the electric field vector across the conductor/insulator interface, consider the illustration in Figure 3.3 where a closed rectangular infinitesimal path **s** is depicted.

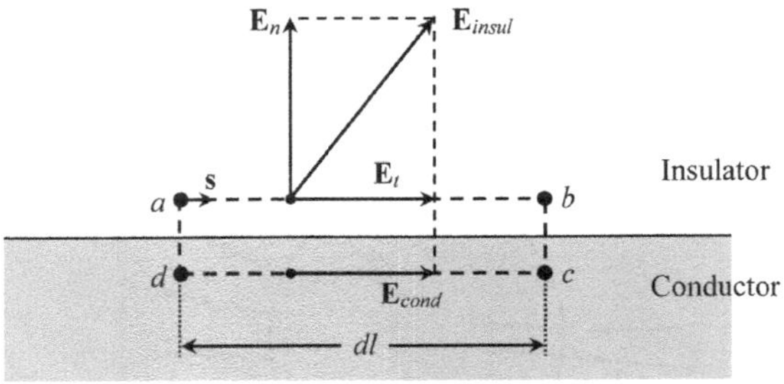

Figure 3.3 The conductor's imperfection ($\sigma \neq \infty$) gives rise to a tangential component of the electric field vector which is continuous at the conductor/insulation interface

Then from

$$\oint_{\mathbf{s}} \mathbf{E} \cdot d\mathbf{s} = 0 = \int_{\overrightarrow{ab}} \mathbf{E}_{\text{insul}} \cdot d\mathbf{s} + \int_{\overrightarrow{cd}} \mathbf{E}_{\text{cond}} \cdot d\mathbf{s} = (E_{\text{t}} - E_{\text{cond}})\, dl$$

we conclude that $E_{\text{t}} = E_{\text{cond}} = J/\sigma$.

3.3 Current Intensity, Kirchhoff's Current Law

The notion of current intensity has already been introduced in Chapter 1.

Current intensity in a conductor, I, is just a simple measure of the flux of **J**-field lines through a conductor cross-section S in a prespecified reference direction **n** (recall Figure 1.3(b))

$$I = \int_{S} \mathbf{J} \cdot \mathbf{n}\, dS \tag{3.6}$$

Not surprisingly, for stationary regimes ($\partial/\partial t = 0$), the current intensity through a conductor immersed in an insulating medium does not depend on the particular cross-section being

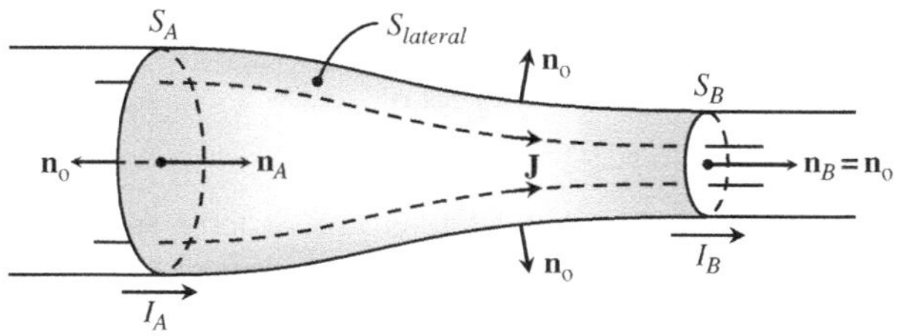

Figure 3.4 Application of the Gauss theorem to show that the current intensity along a conductor remains unchanged, $I_A = I_B$, for time-invariant regimes

considered. Take the conductor volume bounded by S_A, S_B and S_{lateral} shown in Figure 3.4, and make use of the result in (3.5).

Then, taking into account that the external medium is an insulator ($\sigma \approx 0$), you get

$$0 = \int_{S_V} \mathbf{J} \cdot \mathbf{n}_o \, dS = \int_{S_A} \mathbf{J} \cdot \mathbf{n}_o \, dS + \int_{S_B} \mathbf{J} \cdot \mathbf{n}_o \, dS + \overbrace{\int_{S_{\text{lateral}}} \mathbf{J} \cdot \mathbf{n}_o \, dS}^{0}$$

$$= -\int_{S_A} \mathbf{J} \cdot \mathbf{n}_A \, dS + \int_{S_B} \mathbf{J} \cdot \mathbf{n}_B \, dS = -I_A + I_B = 0$$

Hence, you can see that I_A across S_A and I_B across S_B are identical, and for that reason you can drop the unnecessary subscript labels A and B,

$$I_A = I_B = I$$

By the same token, Kirchhoff's current law (KCL) can be obtained. Consider a closed surface S_V which is intersected by several current-carrying conductors – see Figure 3.5.

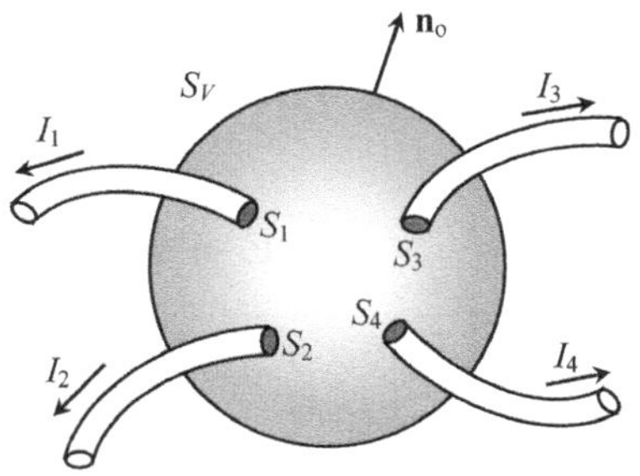

Figure 3.5 Kirchhoff's current law, $\sum I_k = 0$

Then, from (3.5), you get

$$0 = \int_{S_V} \mathbf{J} \cdot \mathbf{n}_o \, dS = \int_{S_1} \mathbf{J} \cdot \mathbf{n}_o \, dS + \int_{S_2} \mathbf{J} \cdot \mathbf{n}_o \, dS + \int_{S_3} \mathbf{J} \cdot \mathbf{n}_o \, dS + \int_{S_4} \mathbf{J} \cdot \mathbf{n}_o \, dS = I_1 + I_2 + I_3 + I_4$$

or, more generally,

$$\sum_k I_k = 0 \tag{3.7}$$

As in Chapter 1, you should notice that KCL is not, strictly speaking, a general 'law'. Indeed, the result in (3.7) is only valid for stationary phenomena, $\partial/\partial t = 0$. For time-varying regimes (where div $\mathbf{J} \neq 0$) things are a little more complicated – see Chapter 6.

3.4 Resistor, Conductance, Resistance, Ohm's Law

To put it simply, a resistor is nothing but a piece of conducting material with two accessible terminals. When a voltage U is applied between the resistor terminals a current of intensity I will flow along the device – see Figure 3.6.

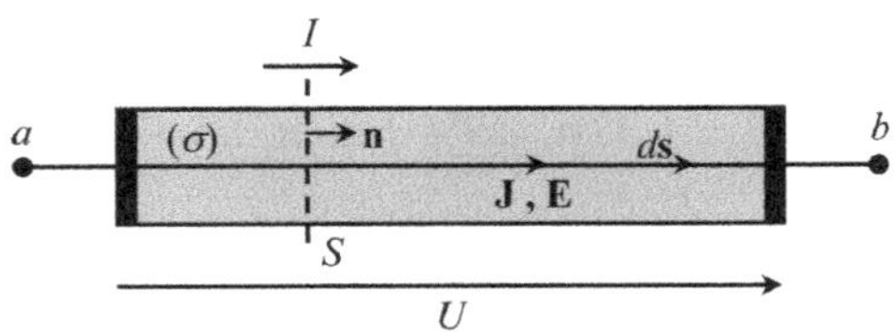

Figure 3.6 Voltage and current in a resistor

If the conducting material behaves as a linear medium, or, put another way, if $\mathbf{J} = \sigma\mathbf{E}$, then I and U will be proportional:

$$I = GU \tag{3.8}$$

The proportionality constant G is called conductance (units: S, siemens). This parameter depends not only on the geometrical configuration of the resistor, but also on the conductivity of the material of which the resistor is made. However, you can see from

$$G = \frac{I}{U} = \frac{\int_S \overbrace{\sigma\mathbf{E}}^{\mathbf{J}} \cdot \mathbf{n}\, dS}{\int_{\overrightarrow{ab}} \mathbf{E} \cdot d\mathbf{s}} \tag{3.9}$$

that G does not depend on the intensity of the electric field in the device.

In many instances, it is often preferred to utilize the inverse of G to describe the resistor's characteristics. The inverse of G is called resistance, $R = 1/G$ (units: Ω, ohm). By employing R, (3.8) translates into

$$U = RI \tag{3.10}$$

which you will certainly recognize as a statement of the familiar Ohm's law.

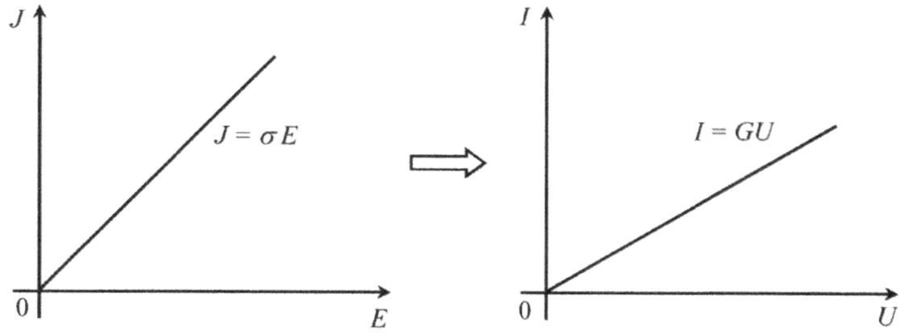

Figure 3.7 The linear relationship $I(U)$ is just a consequence of the linear relationship $J(E)$

At this point two interesting remarks are in order.

The linear relationship between I and U, (3.8), is just a consequence of the assumed linear relationship between **J** and **E**, (3.2) – see Figure 3.7.

Also, if you compare the definition of conductance in (3.9) to the one for the capacitance given in (2.21) you will see the striking analogy between them. This analogy provides you with a very effective means to easily compute C from G, or G from C, whenever you have a capacitor and a resistor with the same geometrical features.

As a simple example, consider a parallel-plate geometrical configuration (Figure 3.8), where in one case the medium sandwiched between the plates is an insulator with permittivity ε (capacitor) and in the other case the medium is a conductor with conductivity σ (resistor).

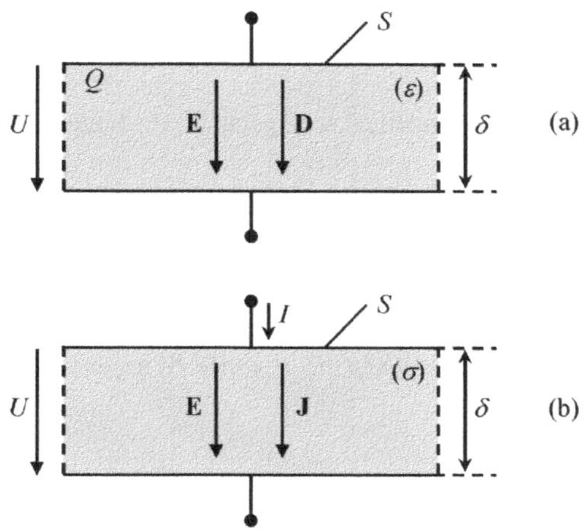

Figure 3.8 Utilization of a parallel-plate structure for showing the analogy between electrostatics and stationary current problems. (a) Capacitor. (b) Resistor

As for the capacitance of the capacitor (see Chapter 2), you get

$$C = \frac{Q}{U} = \frac{DS}{E\delta} = \frac{\varepsilon S}{\delta}$$

Likewise, for the conductance of the resistor, you get

$$G = \frac{I}{U} = \frac{JS}{E\delta} = \frac{\sigma S}{\delta}$$

It then becomes obvious that

$$G = \frac{\sigma}{\varepsilon} C \tag{3.11}$$

3.5 Application Example (The Potentiometer)

Potentiometers are variable resistors that you may find in a large variety of electrical and electronic appliances. These devices are used principally as gain or volume controls, voltage dividers and current controls. Figure 3.9(a) shows the geometrical configuration of a common type of potentiometer and Figure 3.9(b) shows its corresponding equivalent electric circuit.

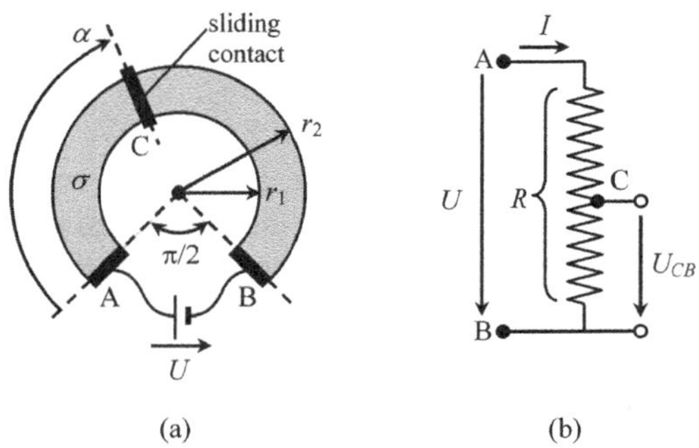

Figure 3.9 Potentiometer. (a) Geometrical configuration. (b) Equivalent circuit representation

The resistor, which includes a sliding contact (terminal C), is made of a thin layer of conducting material of conductivity σ with the shape of a circular crown of width $w = r_2 - r_1$ and thickness t (perpendicular to the drawing plane).

A voltage U is applied between the metallic terminals A and B.

Data: $r_2 = 9\,\text{mm}, \; r_1 = 6\,\text{mm}, \; t = 100\,\mu\text{m}, \; \sigma = 10^3\,\text{S/m}$.

Questions

Q_1 Obtain the equation for the radial dependence of the current density **J**. Using the latter find an expression for the current intensity I.

Q_2 Deduce an expression for the resistor resistance R and compute its numerical value.

Q_3 Evaluate the voltage U_{CB} as a function of both α and U.

Q_4 Show that if the width of the circular crown is small, then the device resistance can be approximately evaluated through $R = l/(\sigma S)$, where l is the average length of the resistor and S denotes its cross sectional area. Taking into account the problem data, estimate the relative error incurred by using such an approximation.

Solutions

Q_1 The electric field lines inside the resistor are circumferential arcs parallel to the circular crown walls. For geometrical reasons the intensity of **E** remains constant along each field line; however, when you jump from field line to field line (that is, when you change r) the field intensity must vary; it is weaker at the outer wall ($r = r_2$) but stronger at the inner wall ($r = r_1$). Hence you can write

$$\text{For } r_1 \leq r \leq r_2 : \mathbf{E} = E(r)\vec{e}_\phi$$

Integration of **E** along a circumferential arc starting at A and ending at B (infinitesimal path length $d\mathbf{s} = r\ d\phi\ \vec{e}_\phi$) yields the applied voltage U between the potentiometer terminals

$$U = \int_{\overrightarrow{AB}} \mathbf{E} \cdot d\mathbf{s} = \int_{\phi=0}^{\phi=\frac{3}{2}\pi} E(r) r\ d\phi = \frac{3\pi r}{2} E(r) \rightarrow E(r) = \frac{2U}{3\pi r} \tag{3.12}$$

From (3.12) you can find the current density field,

$$\mathbf{J} = \sigma \mathbf{E} = \mathbf{J}(r) = \frac{2\sigma U}{3\pi r} \vec{e}_\phi$$

The current intensity I is obtained by evaluating the flux of **J** through the rectangular cross-section S of the resistor

$$I = \int_S \mathbf{J} \cdot \mathbf{n}\ dS, \text{ where } \mathbf{n} = \vec{e}_\phi \text{ and } dS = t\ dr$$

This gives

$$I = U \frac{2t\sigma}{3\pi} \int_{r_1}^{r_2} \frac{dr}{r} = U \left(\frac{2t\sigma}{3\pi} \ln \frac{r_2}{r_1} \right)$$

Q_2 The total resistance of the potentiometer is obtained from the above result through

$$R = \frac{U}{I} = \frac{3\pi}{2t\sigma \ln(r_2/r_1)} \tag{3.13}$$

Numerically, we obtain $R = 116.2\ \Omega$.

Q_3 The U_{CB} voltage is determined as in (3.12) by substituting C for A,

$$U_{CB} = \int_{\overrightarrow{CB}} \mathbf{E} \cdot d\mathbf{s} = \int_{\phi=\alpha}^{\phi=\frac{3}{2}\pi} E(r) r\ d\phi = \left(\frac{3\pi}{2} - \alpha \right) \frac{2U}{3\pi} = U \left(1 - \frac{\alpha}{3\pi/2} \right)$$

Q_4 Define the average length of the resistor as $l = r_{av} 3\pi/2$, where r_{av} is the average radius of the potentiometer, $r_{av} = (r_2 + r_1)/2$.
Take $r_2 = r_{av}(1+\delta)$ and $r_1 = r_{av}(1-\delta)$. By using (3.13) you find

$$R = \frac{3\pi}{2t\sigma \ln\left(\frac{1+\delta}{1-\delta}\right)}$$

Taking into account that, for small δ,

$$\ln\left(\frac{1+\delta}{1-\delta}\right) \approx 2\delta = \frac{r_2 - r_1}{r_{av}}$$

you can obtain the approximation

$$R_{\text{approx}} = \frac{r_{av} 3\pi/2}{\sigma\,(t(r_2 - r_1))} = \frac{l}{\sigma S}$$

thus $R_{\text{approx}} = 117.8\ \Omega$, giving an excess error of 1.4 %.

3.6 Application Example (The Wheatstone Bridge)

The Wheatstone bridge is a very simple circuit network which finds application in instrumentation and measurement. The circuit, represented in Figure 3.10 permits the experimental determination of an unknown resistance R based on previous knowledge of R_1, R_2 and R_3 (one of them being a variable resistor).

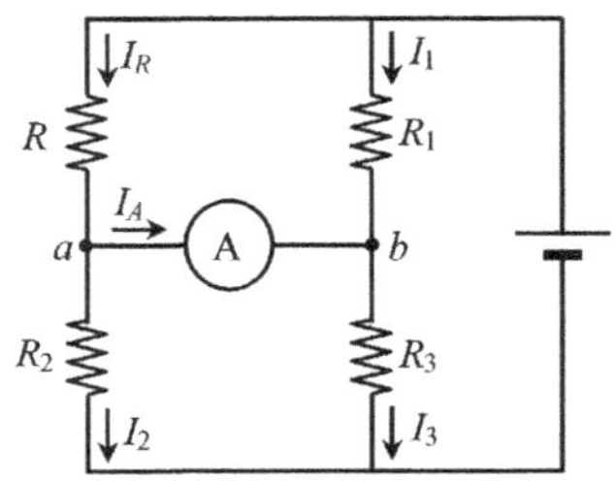

Figure 3.10 The Wheatstone bridge

Assume that R_3 has been adjusted so as to ensure that the ammeter placed between a and b measures zero current – a balanced bridge ($I_A = 0$ and $U_{ab} = 0$).

Questions

Q_1 By using KVL and KCL, write the equations governing the circuit.

Q_2 Determine the relationship among R, R_1, R_2 and R_3 when the bridge is balanced.

Solutions

Q_1 $RI_R + U_{ab} - R_1 I_1 = 0$; $R_3 I_3 + U_{ab} - R_2 I_2 = 0$.
$I_A + I_2 - I_R = 0$; $I_3 - I_A - I_1 = 0$.

Q_2 By making $U_{ab} = 0$ and $I_A = 0$, you find

$$R = \frac{R_1 R_2}{R_3}$$

3.7 Joule Losses, Generator Applied Field

You should know that resistors get hot when submitted to currents. We have already mentioned that this effect – the Joule effect – is associated with electron random collisions inside the atomic lattice of the resistor medium. The energy dissipated by this process may vary from point to point inside the resistor. We will now see that dissipation is proportional to E^2, meaning that hot spots in a resistor are regions where E has attained increased values.

Consider, as shown in Figure 3.11, that an infinitesimal volume dV of the conductor contains an infinitesimal amount of free charge $dq = \rho_f dV$.

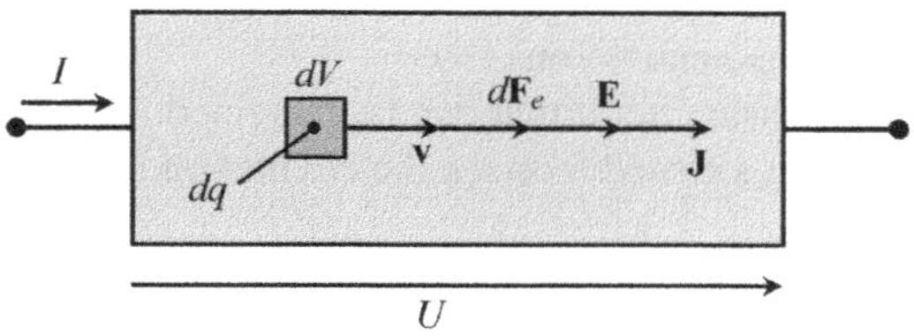

Figure 3.11 Vectors involved in the analysis of Joule losses in a resistor

Under the influence of the impressed **E** field, the free charge dq drifts with a velocity **v** driven by an elemental electric force $d\mathbf{F}_e = dq\mathbf{E}$.

The activity of the latter force produces an elemental power dp which is dissipated in dV

$$dp = \mathbf{v} \cdot d\mathbf{F}_e = \rho_f \mathbf{v} \cdot \mathbf{E}\, dV = \hat{p}_J\, dV$$

Taking into account that $\mathbf{J} = \rho_f \mathbf{v}$ from (3.3), integration of the above result over the resistor's whole volume V yields the total power losses (Joule losses), that is

$$P_J = \int_V \hat{p}_J\, dV; \quad \hat{p}_J = \mathbf{J} \cdot \mathbf{E} = \sigma E^2 \tag{3.14}$$

The local power losses density $\hat{p}_J$ (W/m^3) is thus shown to increase with E^2.

The preceding formulation for the power losses in a conductor not only permits the evaluation of the total losses in a resistor, but, further, has the additional advantage of allowing for a detailed perception of what is happening locally inside the resistor; for instance, where the hot spots are localized.

Sometimes this detailed knowledge is unimportant. Quite often you may only want to evaluate the resistor's total losses based on its voltage and current.

For illustrative purposes, let us revisit the resistor geometry previously shown in Figure 3.8(b). By making $J = I/S$ and $E = U/\delta$, we obtain from (3.14)

$$P_J = \int_V \frac{UI}{S\delta} dV = \frac{UI}{S\delta} \overbrace{\int_V dV}^{S\delta} = UI = GU^2 = RI^2 \qquad (3.15)$$

Not surprisingly, we found $P_J = RI^2$, a result that you are certainly familiar with.

Let us pause for a moment to consider an apparently puzzling and paradoxical question. From the key equations (3.1), you have learnt that, on the one hand, the field lines of **E** are open (curl $\mathbf{E} = 0$) and, on the other hand, the field lines of **J** are closed (div $\mathbf{J} = 0$); in addition, from $\mathbf{J} = \sigma\mathbf{E}$, you can see that field lines of **E** and **J** should run parallel.

But how can all this happen? What is missing?

Clearly there is more to stationary currents than could be accounted for by (3.1) and (3.2).

In the framework of stationary phenomena, where do you think the energy necessary for driving the free charges in motion comes from? Where does the heat transferred to the conductor lattice by electron collision processes (Joule losses) come from?

The answer, as you might have already guessed, is: from generators (batteries, photovoltaic cells, electromechanical devices, and so on).

Therefore, the simplest electric circuit that can be imagined must include a generator and an external conductor, forming a closed loop for the circulation of currents – see Figure 3.12.

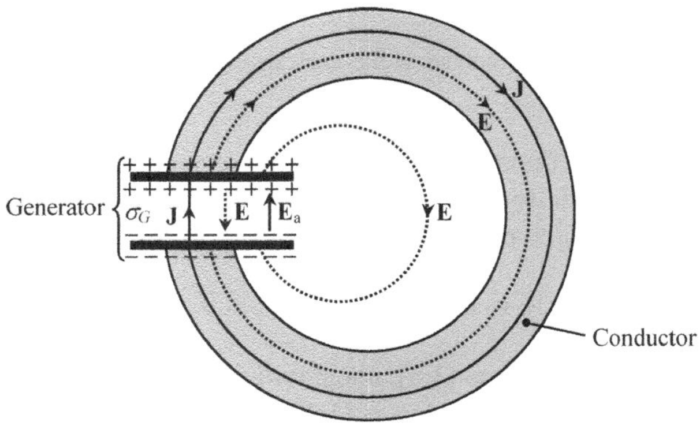

Figure 3.12 A trivial electric circuit made of a generator and an external conductor loop. While the **J**-field lines are closed, the **E**-field lines are open

As shown in Figure 3.12, the relationship between **J** and **E** inside the generator cannot be the same as in (3.2); there, instead, you have to employ

$$\mathbf{J} = \sigma_G(\mathbf{E} + \mathbf{E}_a) \qquad (3.16)$$

where σ_G denotes the conductivity of the internal medium of the generator, and $\mathbf{E}_a$, the so-called applied electric field, represents from a macroscopic point of view the per-unit-charge internal force responsible for maintaining the separation of the electric charges residing at the positive and negative generator terminals (that would otherwise collapse together).

From (3.16) and Figure 3.12 you can easily observe that, when the generator is disconnected ($\mathbf{J} = 0$), the opposite fields $\mathbf{E}$ and $\mathbf{E}_a$ have the same magnitude; however, when the external conductor is connected you will get $|\mathbf{E}| < |\mathbf{E}_a|$.

3.8 Generator Electromotive Force, Power Balance

The actual electric circuit in Figure 3.12, containing a generator and an external load (a resistor), is symbolically represented in Figure 3.13.

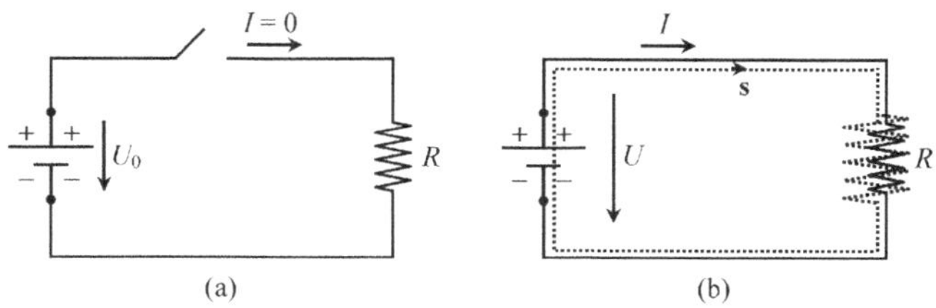

Figure 3.13 Symbolic description of the circuit depicted in Figure 3.12. (a) With the switch open the generator voltage is given by its electromotive force. (b) With the switch closed the generator voltage is smaller than its electromotive force

When the load is disconnected ($\mathbf{J} = 0$) a voltage U_0 appears between the positive and negative terminals of the generator

$$U_0 = \int_{\overrightarrow{+-}} \mathbf{E} \cdot d\mathbf{s}$$

According to (3.16), when $\mathbf{J} = 0$, $\mathbf{E} = -\mathbf{E}_a$ inside the generator. Therefore, the above result can be rewritten as

$$U_0 = \int_{\overrightarrow{+-}} \mathbf{E} \cdot d\mathbf{s} = \int_{\substack{\overrightarrow{-+} \\ \text{generator}}} \mathbf{E}_a \cdot d\mathbf{s} \tag{3.17}$$

An intrinsic characteristic of the generator, since it only depends on the internal applied field $\mathbf{E}_a$, voltage U_0 is commonly known by the name of electromotive force (emf).

When the external resistor R is connected across the generator terminals its voltage U decreases compared to U_0.

In order to determine the resulting voltage U let us evaluate the line integral of $(\mathbf{E}+\mathbf{E}_a)$ along the closed path s inside the circuit (Figure 3.13(b)). The line integration is performed by using two alternative processes:

$$\oint_{s}(\mathbf{E}+\mathbf{E}_a)\cdot d\mathbf{s} = \oint_{s}\mathbf{E}\cdot d\mathbf{s} + \oint_{s}\mathbf{E}_a\cdot d\mathbf{s} = 0 + \int_{\substack{\overrightarrow{-+}\\ \text{generator}}}\mathbf{E}_a\cdot d\mathbf{s} = U_0$$

and

$$\oint_{s}(\mathbf{E}+\mathbf{E}_a)\cdot d\mathbf{s} = \int_{\substack{\overrightarrow{-+}\\ \text{generator}}}(\mathbf{E}+\mathbf{E}_a)\cdot d\mathbf{s} + \int_{\substack{\overrightarrow{+-}\\ \text{resistor}}}(\mathbf{E}+\mathbf{E}_a)\cdot d\mathbf{s}$$

$$= \int_{\substack{\overrightarrow{-+}\\ \text{generator}}}\frac{1}{\sigma_G}\mathbf{J}\cdot d\mathbf{s} + \int_{\substack{\overrightarrow{+-}\\ \text{resistor}}}\mathbf{E}\cdot d\mathbf{s} = r_G I + U$$

from which we conclude $U_0 = r_G I + U$ or, which is the same,

$$U = U_0 - r_G I \tag{3.18}$$

The term $r_G I$ represents the internal voltage drop of the generator, where r_G is its internal resistance (both are zero when ideal generators are considered, that is when $\sigma_G \to \infty$).

The relationship $U(I)$ in (3.18) describes the generator's behavior in terms of its intrinsic parameters U_0 and r_G. In the diagram shown in Figure 3.14 this relationship is represented by the straight line with negative slope.

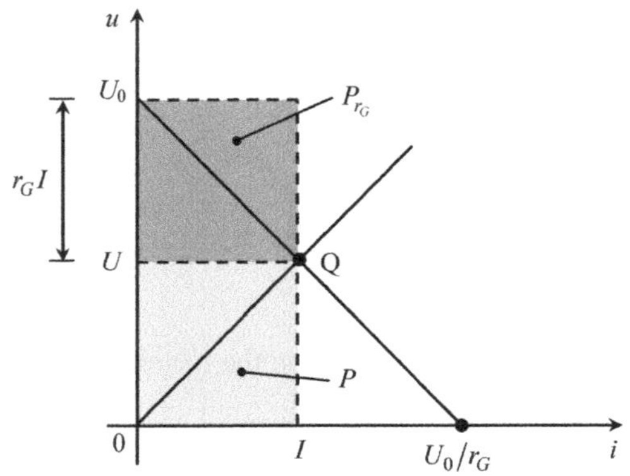

Figure 3.14 Diagram for power balance analysis. The straight line with negative slope describes the generator's features. The straight line with positive slope characterizes the external resistor. Q is the operating point

On the other hand, the resistor characteristic is described by Ohm's law (3.10), $U = RI$, which in Figure 3.14 is represented by the straight line with positive slope.

The intersection of the two lines permits the identification of the circuit's operating point (point Q), from where U and I can simultaneously be obtained.

At last we turn our attention to power balance analysis. From $U_0 = r_G I + U$ we readily get

$$U_0 I = r_G I^2 + UI$$

The left-hand side of this equation represents the total power produced by the generator's applied field $P_G = U_0 I$. On the right-hand side, the first term represents dissipation losses internal to the generator, $Pr_G = r_G I^2$, whereas the second term represents the available power delivered to the load $P = UI$; hence

$$P_G = Pr_G + P \tag{3.19}$$

This power balance equation can be graphically interpreted by using Figure 3.14. While the area of the upper rectangle (with sides I and $r_G I$) corresponds to Pr_G, the area of the lower rectangle (with sides I and U) corresponds to P. Summing the two areas, we obtain P_G (a rectangle whose sides are I and U_0) as in (3.19).

3.9 Proposed Homework Problems

Problem 3.9.1

In order to monitor and control the unavoidable stress phenomena occurring in some mechanical structures, they are usually provided with embedded resistor-type strain gauges. The simplest scheme used to detect resistance changes due to gauge deformation (tension or compression) utilizes the Wheatstone bridge already analyzed in Application Example 3.6. Assume, as shown in Figure 3.15, that the bridge is initially balanced (fixed resistors R_1, R_2 and R_3 are equal to R). Next, allow the embedded strain gauge resistor on the bridge's upper left arm to be subjected to a small variation ΔR on its resistance, ($\Delta R \ll R$).

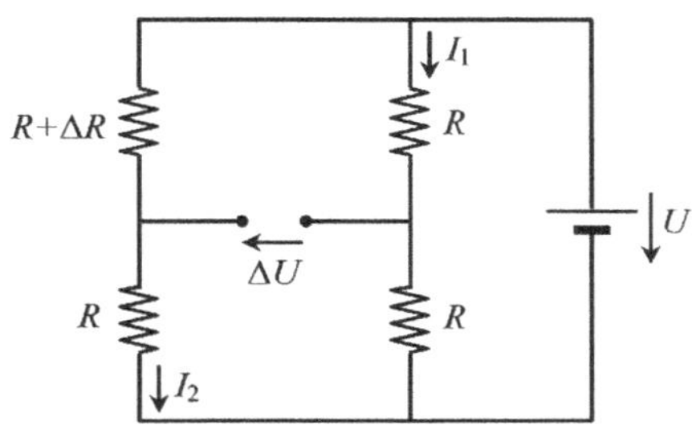

Figure 3.15 Application of the Wheatstone bridge for the detection of strain gauge deformations

Q_1 Write the KVL equations governing the circuit.

Q_2 Find the relationship between the monitored voltage ΔU and ΔR.

Answers

Q$_1$ $I_1 = \dfrac{U}{2R}$; $I_2 = \dfrac{U}{2R + \Delta R}$; $\Delta U = R(I_1 - I_2)$

Q$_2$ $\Delta U \approx \Delta R \dfrac{U}{4R}$

Problem 3.9.2

For security reasons many types of electrical equipment ought to have a connection to ground. Ground electrodes are buried in the soil to provide a means for the flow of undesirable currents. Take the situation depicted in Figure 3.16 where a ground electrode of hemispherical shape is considered. Assume the metallic electrode is a perfect conductor. Choose for the potential $V(\infty) = 0$. Consider the following data: $I = 100\,\text{A}$, $a = 10\,\text{cm}$, $r_1 = 1\,\text{m}$, $r_2 = 2\,\text{m}$, $\sigma_{\text{soil}} = 3.18 \times 10^{-2}\,\text{S/m}$.

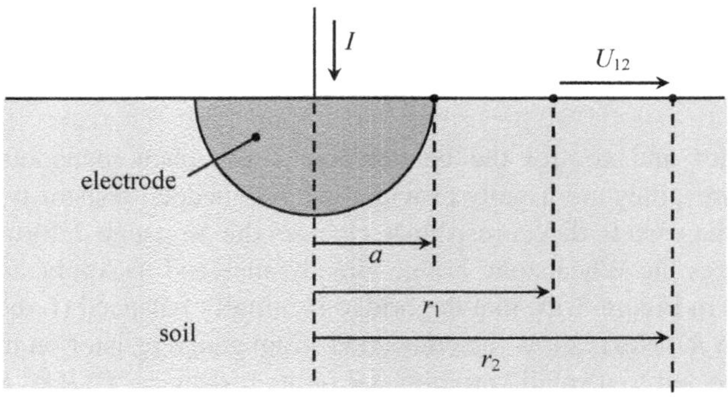

Figure 3.16 Hemispherical ground electrode

Q$_1$ Obtain the equation for the radial dependence of the current density field **J** and potential function V inside the soil.

Q$_2$ Compute the electrode potential V_E. Determine the boundary of the region around the electrode outside of which V becomes smaller than $V_E/10$.

Q$_3$ Compute the electrode resistance

$$R_E = \frac{V_E - V(\infty)}{I}$$

Q$_4$ Evaluate the step voltage U_{12}.

Q$_5$ Find the power P corresponding to the energy dissipated in the soil.

Answers

Q$_1$ $\mathbf{J}(r) = \dfrac{I}{2\pi r^2}\vec{e}_r$; $V(r) = \dfrac{I}{2\pi r \sigma_{\text{soil}}}$ (for $r \geq a$)

Q$_2$ $V_E = 5\,\text{kV}$; $V \leq V_E/10$ for $r \geq 1\,\text{m}$.

Q$_3$ $R_E = 50\ \Omega$.

Q$_4$ $U_{12} = 250\,\text{V}$.

Q$_5$ $P = 500\,\text{kW}$.

Problem 3.9.3

Consider a two-wire transmission line like the one we dealt with in Chapter 2 (Application Example 2.8), where two identical cylindrical conductors, of length $l = 50\,\text{m}$, radius $r = 1\,\text{mm}$, run parallel separated by 4 mm. Assume that the surrounding dielectric medium is a perfect insulator. Line conductors, made of copper, have a conductivity $\sigma = 5.6 \times 10^7\ \text{S/m}$. The line is excited at the sending end by a voltage generator characterized by an electromotive force $U_0 = 50\,\text{V}$ and internal resistance $r_G = 1\ \Omega$. At the receiving end a resistor load $R_L = 50\ \Omega$ is placed (Figure 3.17).

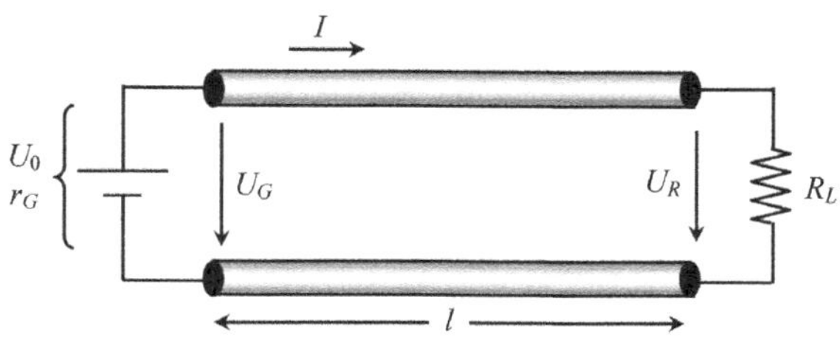

Figure 3.17 A DC link employing a lossy two-conductor line

Q$_1$ In electrostatics we showed that proximity effects would give rise to non-uniform charge distributions over the conductor surfaces. Show, however, that as far as stationary currents are concerned, **J**-field lines are uniformly distributed inside the line conductors. (Note: Only for hight-frequency regimes the distribution of currents in the conductor cross-section becomes non-uniform due to skin effect phenomena – see Chapter 8.)

Q$_2$ Evaluate the resistance R_{cond} of each line conductor.

Q$_3$ Evaluate I, U_G and U_R.

Q$_4$ Determine the generator internal power losses Pr_G, as well as the transmission power losses P_{trans}. Compare the power produced by the generator applied field P_G to the power delivered to the load P.

Answers

Q_1 Consider two neighboring parallel field lines $\mathbf{J}_1$ and $\mathbf{J}_2$ inside the cylindrical conductor. By applying the property

$$\oint_S \mathbf{E}\cdot d\mathbf{s} = \frac{1}{\sigma}\oint_S \mathbf{J}\cdot d\mathbf{s} = 0$$

you get $J_1 = J_2$.

Q_2

$$R_{\text{cond}} = \frac{l}{\sigma \pi r^2} = 284.2\ \text{m}\Omega$$

Q_3 $I = U_0/(r_G + 2R_{cond} + R_L) = 969.6\,\text{mA}$; $U_G = 49.03\,\text{V}$; $U_R = 48.48\,\text{V}$.

Q_4 $Pr_G = 0.94\,\text{W}$; $P_{\text{trans}} = 0.53\,\text{W}$; $P_G = U_0 I = 48.48\,\text{W}$; $P = 47.01\,\text{W} = 97\% P_G$.

Problem 3.9.4

Consider a coaxial cable which is terminated at its receiving end ($y = 0$) by a resistor load $R_L = 1\,\text{k}\Omega$ whose power consumption is kept at $P_L = 250\,\text{W}$. The cable's longitudinal view and respective cross-section are shown in Figure 3.18. The length of the cable is $l = 10\,\text{km}$, and the remaining geometrical parameters are $r_1 = 1\,\text{mm}$, $r_2 = 5\,\text{mm}$, $r_3 = 5.1\,\text{mm}$.

The conductivity of the cable's internal and external conductors is $\sigma_{\text{cond}} = 31.67 \times 10^6\,\text{S/m}$. The dielectric medium is an imperfect insulator with conductivity $\sigma_{\text{insul}} = 5.123 \times 10^{-9}\,\text{S/m}$.

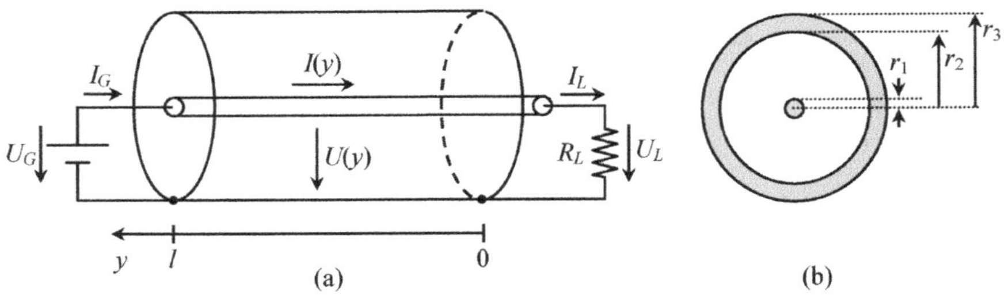

Figure 3.18 A DC link employing a lossy coaxial cable. (a) General view. (b) Cable cross-section

Q_1 Determine the per-unit-length longitudinal cable resistance R (including the internal and external conductors).

Q_2 Determine the per-unit-length transverse conductance G of the dielectric medium.

Q_3 Consider the approximation that cable conductors are perfect (that is, cable voltage U is constant along the longitudinal coordinate y).

Determine the generator voltage U_G at the cable sending end ($y = l$).
Determine the evolution of the cable current intensity along y, $I(y)$, a consequence of the leakage currents crossing the imperfect dielectric.
(Hint: From div $\mathbf{J} = 0$, find dI/dy).
Obtain I_G and determine the power dissipation in the insulating medium P_{insul}.

Q_4 Consider the approximation in which the dielectric medium is a perfect insulator (that is, cable current I is constant along the longitudinal coordinate y).

Determine the generator current I_G at the cable sending end ($y = l$).
Determine the evolution of the cable voltage along y, $U(y)$, a consequence of the voltage drop along the cable's imperfect conductors.
(Hint: From curl $\mathbf{E} = 0$, find dU/dy).
Obtain U_G and determine the power dissipation in the cable conductors P_{cond}.

Answers

Q_1

$$R = \frac{1}{\pi\sigma_{\text{cond}}}\left(\frac{1}{r_1^2} + \frac{1}{r_3^2 - r_2^2}\right) = 20.0\,\text{m}\Omega/\text{m}$$

Q_2

$$G = \frac{2\pi\sigma_{\text{insul}}}{\ln(r_2/r_1)} = 20.0\,\text{nS/m}$$

Q_3

$$U_G = U = U_L = \sqrt{P_L R_L} = 500\,\text{V}$$

$$\frac{d}{dy}I(y) = GU \rightarrow I(y) = I_L + GUy; \quad I_L = U_L/R_L = 500\,\text{mA}$$

$$I_G = I_L + GUl = 600\,\text{mA}$$

$$P_{\text{insul}} = P_G - P_L = 300 - 250 = 50\,\text{W}$$

Q_4

$$I_G = I = I_L = \sqrt{P_L/R_L} = 500\,\text{mA}$$

$$\frac{d}{dy}U(y) = RI \rightarrow U(y) = U_L + RIy; \quad U_L = R_L I_L = 500\,\text{V}$$

$$U_G = U_L + RIl = 600\,\text{V}$$

$$P_{\text{cond}} = P_G - P_L = 300 - 250 = 50\,\text{W}$$

Problem 3.9.5

Consider again the problem discussed in Problem 3.9.4. The approaches referred to in Q_3 and Q_4 are approximations because conductor voltage drops and dielectric leakage currents influence each other; in fact, here, we are dealing with a distributed coupled-phenomena problem. (Note: Distributed coupled phenomena will come to your attention in Part IV where the topic of electromagnetic wave propagation will be handled.)

Q_1 Determine the coupled differential equations describing the evolution of $U(y)$ and $I(y)$.

Q_2 Solve the equations.

Q_3 Make use of the boundary conditions at the receiving end of the cable to obtain the unknown integration constants.

Q_4 Obtain U_G and I_G at the generator terminals.

Q_5 Determine the total power losses and show how they break into conductor and insulator losses.

Answers

Q_1

$$\begin{cases} \dfrac{d}{dy} I(y) = GU(y) \\ \dfrac{d}{dy} U(y) = RI(y) \end{cases} \rightarrow \frac{d^2}{dy^2} \begin{Bmatrix} I(y) \\ U(y) \end{Bmatrix} - RG \begin{Bmatrix} I(y) \\ U(y) \end{Bmatrix} = 0$$

Q_2

$$\begin{cases} U(y) = U_1 \, e^{+y/D} + U_2 \, e^{-y/D} \\ I(y) = \dfrac{1}{R_0} \left(U_1 \, e^{+y/D} - U_2 \, e^{-y/D} \right) \end{cases}$$

where D is the attenuation distance $D = 1/\sqrt{RG} = 50\,\text{km}$ and R_0 is the characteristic resistance $R_0 = \sqrt{R/G} = 1\,\text{k}\Omega$.

Q_3 Boundary conditions:

$$\begin{cases} U_{(y=0)} = U_L = 500\,\text{V} \\ I_{(y=0)} = I_L = 500\,\text{mA} \end{cases} \rightarrow U_1 = U_L = 500\,\text{V}; \;\; U_2 = 0$$

Q_4

$$U_G = U_L \, e^{+l/D} = 610.7\,\text{V}; \;\; I_G = \frac{U_L}{R_0} e^{+l/D} = 610.7\,\text{mA}$$

Q_5 $$P_{\text{losses}} = P_G - P_L = 373 - 250 = 123\,\text{W}.$$

$$P_{\text{cond}} = \int_{y=0}^{y=l} R\, I^2(y)\; dy = 61.5\,\text{W}; \quad P_{\text{insul}} = \int_{y=0}^{y=l} GU^2(y)\; dy = 61.5\,\text{W}$$

Problem 3.9.6

In electrostatics, the experimental determination of the partial capacitances among the conductors of a multiconductor system can be a very delicate subject. However, for homogeneous systems, an experimental procedure using stationary currents can be used indirectly to determine those capacitances. The electrolytic tank technique explores the existing analogy between capacitances and conductances as suggested by (3.11).

Consider a three-dimensional arrangement consisting of three metallic conductors in air (ε_0) as depicted in Figure 3.19(a). Next, assume that the same set of conductors is immersed into a tank filled with an electrolytic liquid of conductivity σ (Figure 3.19(b)).The walls of the tank are made of an insulating material, and the size of the tank is very large compared to the overall conductor system dimensions.

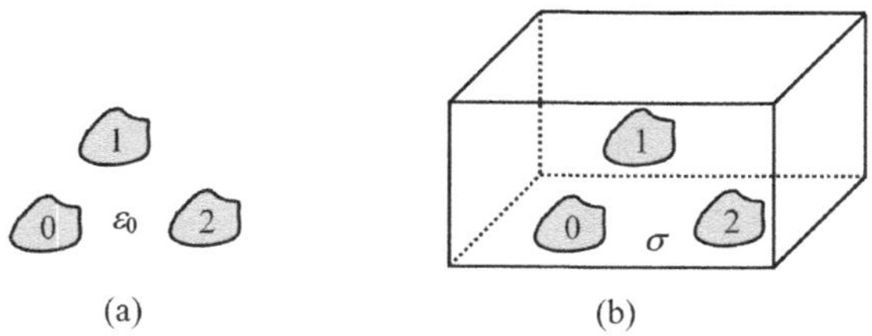

Figure 3.19 A multiconductor system with three metallic bodies. (a) Immersed in air, ε_0. (b) Placed inside a tank filled with an electrolytic liquid of conductivity σ

From the point of view of stationary currents, the structure in Figure 3.19(b) can be replaced by the equivalent circuit in Figure 3.20, where $\hat{G}_{12}$, $\hat{G}_{10}$ and $\hat{G}_{20}$ represent the partial conductances of the system corresponding to the flow of currents through the electrolytic medium.

The following two experiments were conducted:

- Voltages $U_1 = 10\,\text{V}$ and $U_2 = 0$ were applied between accessible terminals. Ammeters used for current measurement gave the following readings: $I_1 = 0.6\,\text{A}$ and $I_2 = -0.2\,\text{A}$.
- Voltages $U_2 = 10\,\text{V}$ and $U_1 = 0$ were applied between accessible terminals. Ammeters used for current measurement produced the following readings: $I_1 = -0.2\,\text{A}$ and $I_2 = 0.5\,\text{A}$.

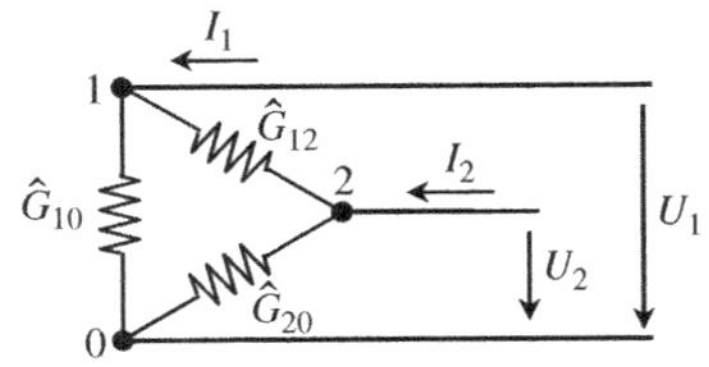

Figure 3.20 Equivalent electric circuit made of partial conductances representing the arrangement depicted in Figure 3.19(b)

Q_1 Why must the size of the electrolytic tank be much larger than the conductor system dimensions?

Q_2 Write the equations governing the equivalent circuit in Figure 3.20.

Q_3 Determine the system partial conductances $\hat{G}_{12}$, $\hat{G}_{10}$ and $\hat{G}_{20}$.

Q_4 Knowing that the conductivity of the electrolyte is $\sigma = 17.68\,\text{mS/m}$, find the partial capacitances and the capacitance matrix $[C]$ that characterize the multiconductor electrostatic configuration in Figure 3.19(a).

Answers

Q_1 The configuration in Figure 3.19(a) is an unbounded system, so should be the one in Figure 3.19(b). This can be achieved (approximately) by placing the tank walls far away from the system conductors.

Q_2 $I_1 = \hat{G}_{12}(U_1 - U_2) + \hat{G}_{10}U_1$; $I_2 = \hat{G}_{12}(U_2 - U_1) + \hat{G}_{20}U_2$.

Q_3 First experiment:

$$\hat{G}_{12} = -\frac{I_2}{U_1} = 20\,\text{mS};\ \hat{G}_{10} = \frac{I_1 + I_2}{U_1} = 40\,\text{mS}$$

Second experiment:

$$\hat{G}_{12} = -\frac{I_1}{U_2} = 20\,\text{mS};\ \hat{G}_{20} = \frac{I_1 + I_2}{U_2} = 30\,\text{mS}$$

Q_4 From

$$\hat{C}_{jk} = \frac{\varepsilon_0}{\sigma}\hat{G}_{jk}, \text{ with } \frac{\varepsilon_0}{\sigma} = 0.5\,\text{ns}$$

we find $\hat{C}_{12} = 10\,\text{pF}$, $\hat{C}_{10} = 20\,\text{pF}$ and $\hat{C}_{20} = 15\,\text{pF}$.
From (2.53) we get the capacitance matrix

$$[C] = \begin{bmatrix} 30 & -10 \\ -10 & 25 \end{bmatrix}\text{pF}$$

4

Magnetic Field of Stationary Currents

4.1 Fundamental Equations

The intrinsic properties of stationary currents were the object of analysis in the preceding chapter. Now you are going to learn about the magnetic fields originated by such currents.

The fundamental laws governing magnetic field stationary problems are those in (PII. 4) together with a constitutive relation concerning the magnetic behavior of material media. That is,

$$\begin{cases} \text{curl } \mathbf{H} = \mathbf{J} \\ \text{div } \mathbf{B} = 0 \end{cases} \tag{4.1}$$

and

$$\mathbf{B} = \mu \mathbf{H} \tag{4.2}$$

where μ is the permeability of the medium (units: H/m, henry per meter).

4.2 Ampère's Law, Magnetomotive Force, Magnetic Voltage

From curl $\mathbf{H} = \mathbf{J}$ in (4.1), you can see that magnetic fields $\mathbf{H}$ can be produced by electric currents.

Let us concentrate on the integral version of the *curl* equation. For that purpose we need to make use of the Stokes theorem – as we did in Chapter 2, (2.4):

$$\int_{S_s} \text{curl } \mathbf{H} \cdot \mathbf{n}_S \, dS = \oint_{\mathbf{s}} \mathbf{H} \cdot d\mathbf{s}$$

where $\mathbf{s}$ is a simply closed oriented path, S_s is an open surface having the path $\mathbf{s}$ as its bounding edge, and $\mathbf{n}_S$ is the Stokes unit normal (the direction of the unit normal is related to the path orientation according to the right-hand screw rule). Since curl $\mathbf{H} = \mathbf{J}$ we obtain

$$\oint_{s} \mathbf{H} \cdot d\mathbf{s} = \int_{S_s} \mathbf{J} \cdot \mathbf{n}_S \, dS \tag{4.3}$$

This very important result, illustrated in Figure 4.1, is Ampère's law. The cyclic integral on the left-hand side of (4.3) is the so-called magnetomotive force (mmf). Note that, like the electromotive force in Chapter 3, this traditional designation is also a misnomer, for mmf is obviously not a force – its units are ampere.

The right-hand side of (4.3) represents the flux of $\mathbf{J}$ across S_s; put another way, it is the current intensity I through S_s.

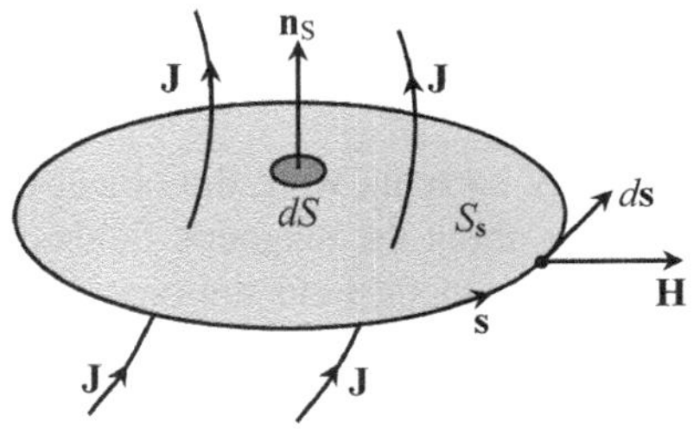

Figure 4.1 Geometrical elements involved in the application of Ampère's law

To help you understand the usefulness of Ampère's law, consider the simple application example depicted in Figure 4.2, where a rectilinear cylindrical conductor carries an electric current of intensity I.

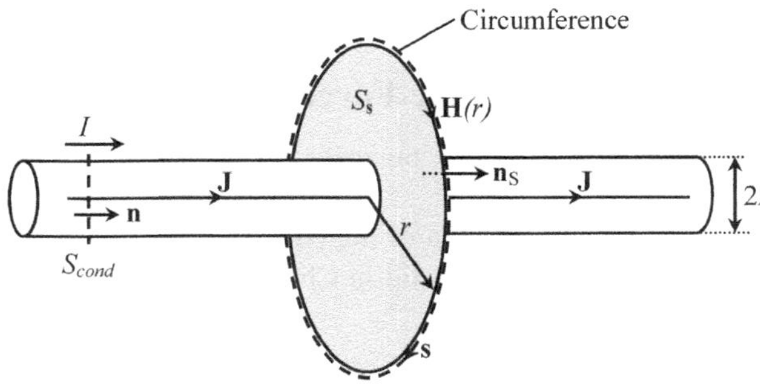

Figure 4.2 Application of Ampère's law to the evaluation of the magnetic field $\mathbf{H}(r)\,\vec{e}_\phi$ encircling a current-carrying cylindrical conductor

Taking into account the particular geometry of the problem (note that the configuration is rotation invariant), it can be ensured that the magnetic field lines around the conductor are coaxial circumferences $\mathbf{H} = H\vec{e}_\phi$, and, in addition, the intensity of $\mathbf{H}$ remains constant along each field line. Therefore, using (4.3), and choosing a closed path **s** coincident with the generic field line **H** of radius r, you have on the one hand

$$\oint_{\mathbf{s}} \mathbf{H} \cdot d\mathbf{s} = \oint_{\mathbf{s}} H\, ds = H \oint_{\mathbf{s}} ds = H \times 2\pi r$$

and on the other hand

$$\int_{S_s} \mathbf{J} \cdot \mathbf{n}_S\, dS = 0 + \int_{S_{cond}} \mathbf{J} \cdot \mathbf{n}\, dS = I$$

In conclusion, outside the conductor, $r > R$, the magnetic field is evaluated as

$$H(r) = \frac{I}{2\pi r} \tag{4.4}$$

However, in other applications where H does not remain constant along the closed integration path **s**, you will eventually need to break **s** into a convenient number of segments, for example

$$\oint_{\mathbf{s}} \mathbf{H} \cdot d\mathbf{s} = \int_{\overrightarrow{ab}} \mathbf{H} \cdot d\mathbf{s} + \int_{\overrightarrow{bc}} \mathbf{H} \cdot d\mathbf{s} + \int_{\overrightarrow{ca}} \mathbf{H} \cdot d\mathbf{s} \tag{4.5a}$$

$$\oint_{\mathbf{s}} \mathbf{H} \cdot d\mathbf{s} = U_{m_{ab}} + U_{m_{bc}} + U_{m_{ca}} \tag{4.5b}$$

The terms on the right-hand side of (4.5) have the common designation of magnetic voltage U_m (units: A, ampere), a concept which is introduced in analogy with the familiar concept of electric voltage, U, in (2.6):

$$\text{Magnetic voltage} : U_{m_{ab}} = \int_{\overrightarrow{ab}} \mathbf{H} \cdot d\mathbf{s} \leftrightarrow \text{Electric voltage} : U_{ab} = \int_{\overrightarrow{ab}} \mathbf{E} \cdot d\mathbf{s} \tag{4.6}$$

It is worth pausing for a moment to ponder the extent of this analogy.

To start with, you may recall that, so far, the electric field **E** is a gradient field (curl $\mathbf{E} = 0$), and as we have already seen in (2.7), the electric voltage between two points is independent of the integration path connecting them.

However, things can be rather different as far as the magnetic field is concerned.

In fact, **H** is not a gradient field (curl $\mathbf{H} \neq 0$) and, because of that, the evaluation of the magnetic voltage between two points depends explicitly on the integration path. You can surely get different results for U_m depending on the chosen paths. To see that this is true, consider the situation in Figure 4.3 where currents are present.

If you apply Ampère's law to the closed path **s**, which you can break into $\widehat{ab}$ plus $\widehat{ba}$, you will get

$$\oint_{\mathbf{s}} \mathbf{H} \cdot d\mathbf{s} = \int_{\substack{\overrightarrow{ab}\\(1)}} \mathbf{H} \cdot d\mathbf{s} + \int_{\substack{\overrightarrow{ba}\\(2)}} \mathbf{H} \cdot d\mathbf{s} = \overbrace{\int_{\substack{\overrightarrow{ab}\\(1)}} \mathbf{H} \cdot d\mathbf{s}}^{U_{m_1}} - \overbrace{\int_{\substack{\overrightarrow{ab}\\(2)}} \mathbf{H} \cdot d\mathbf{s}}^{U_{m_2}} = \overbrace{\int_{S_s} \mathbf{J} \cdot \mathbf{n}_S\, dS}^{I_S}$$

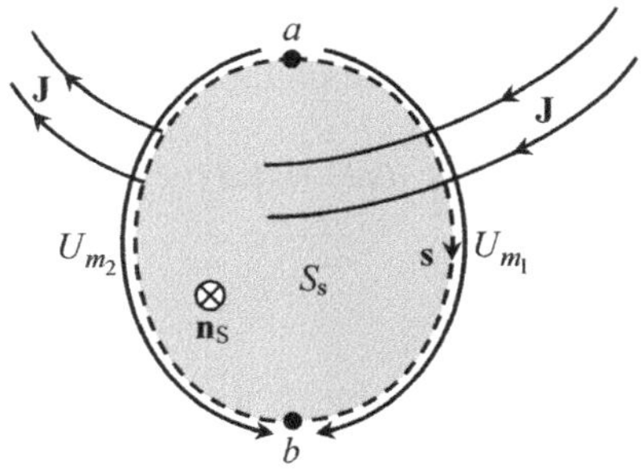

Figure 4.3 Magnetic voltages are not uniquely defined. U_{m_1} and U_{m_2} will coincide if and only if $\mathbf{J} = 0$

from which you can conclude $U_{m_1} - U_{m_2} = I_S$, clearly showing that the magnetic voltages between the same pair of points are not the same, $U_{m_1} \neq U_{m_2}$. If and only if currents are absent, $I_S = 0$, will you get the same result for both magnetic voltages.

You also learnt in Chapters 2 and 3 that the electric voltages along a closed loop sum to zero (KVL), but make no mistake: KVL does not apply to magnetic voltages; the sum of magnetic voltages along a closed path is, in general, not zero.

4.3 Magnetic Induction Field, Magnetic Induction Flux

We saw in Chapter 3 that the current density vector **J** obeys the equation div $\mathbf{J} = 0$, which allowed the derivation of a set of important properties.

From (4.1), you can see that the magnetic induction field **B** also obeys div $\mathbf{B} = 0$, and consequently, by analogy, **B** and **J** must share the same properties. That is, **B**-field lines must be closed, and their flux across a closed surface bounding a given volume must be zero,

$$\int_{S_V} \mathbf{B} \cdot \mathbf{n}_o \, dS = 0 \tag{4.7}$$

However, there is a subtle difference we would like to emphasize. While **J**-field properties are particular to stationary regimes (if $\partial/\partial t \neq 0$ then div $\mathbf{J} \neq 0$), **B**-field properties are absolutely general, and are valid for either stationary or time-varying regimes. In fact, div $\mathbf{B} = 0$ contains no approximations whatsoever.

By following the analogy trend, the same way the current intensity I was introduced in (3.6), here we define the magnetic flux ϕ across an open surface S (units: Wb, weber)

$$\text{Current intensity: } I = \int_S \mathbf{J} \cdot \mathbf{n} \, dS \;\leftrightarrow\; \text{Magnetic flux: } \phi = \int_S \mathbf{B} \cdot \mathbf{n} \, dS \tag{4.8}$$

Although the magnetic flux ϕ is a scalar quantity, a reference arrow is usually associated with it; such an arrow is just a reminder for the arbitrary unit normal **n** in (4.8) – see Figure 4.4.

At this point, and in connection with the equivalence in (4.8), an important remark is in order. While currents are physically associated with free electric charges in motion, (3.3),

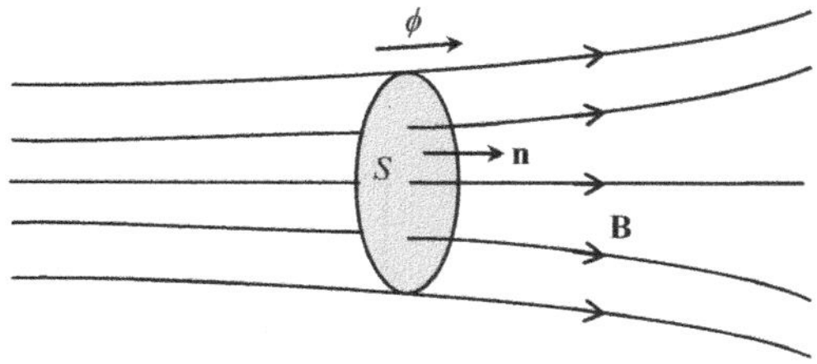

Figure 4.4 Definition of the magnetic flux ϕ

magnetic fluxes cannot be interpreted in identical terms. The reason is very simple: magnetic charges do not exist in nature!

In Section 4.2, we drew your attention to the fact that KVL is of no use in the analysis of magnetic field problems. However, an analogous law to KCL can be utilized here. In reality, from (4.7), you can immediately recognize that the sum of outward-going magnetic fluxes across a closed surface must yield zero as a result (see the example in Figure 4.5)

$$0 = \int_{S_V} \mathbf{B} \cdot \mathbf{n}_o \, dS = \int_{S_1} \mathbf{B} \cdot \mathbf{n}_1 \, dS + \int_{S_2} \mathbf{B} \cdot \mathbf{n}_2 \, dS + \int_{S_3} \mathbf{B} \cdot \mathbf{n}_3 \, dS = \phi_1 + \phi_2 + \phi_3$$

Or, in general,

$$\sum_k \phi_k = 0 \tag{4.9}$$

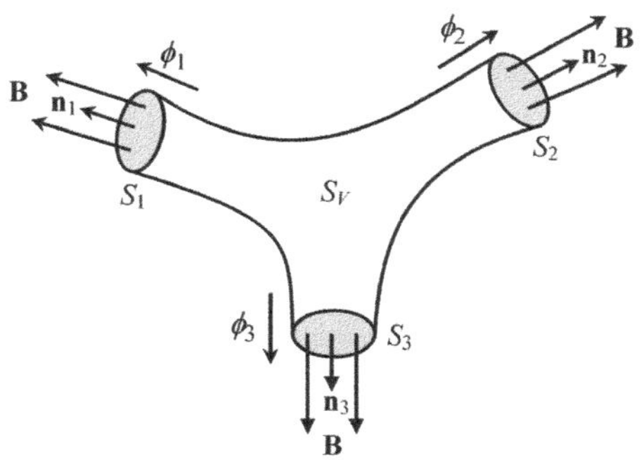

Figure 4.5 The magnetic fluxes sum to zero, $\sum \phi_k = 0$, as a consequence of div $\mathbf{B} = 0$

4.4 Application Example (Power Line Magnetic Fields)

Magnetic field effects produced by overhead power lines on human health are currently a matter of serious public concern. Within the European Union, as far as power–frequency magnetic induction fields are concerned, a safety limit has been set at 100 μT.

Figure 4.6 shows a cross-sectional view of a flat three-phase overhead line, made of cylindrical conductors, running parallel to a poorly conducting soil.

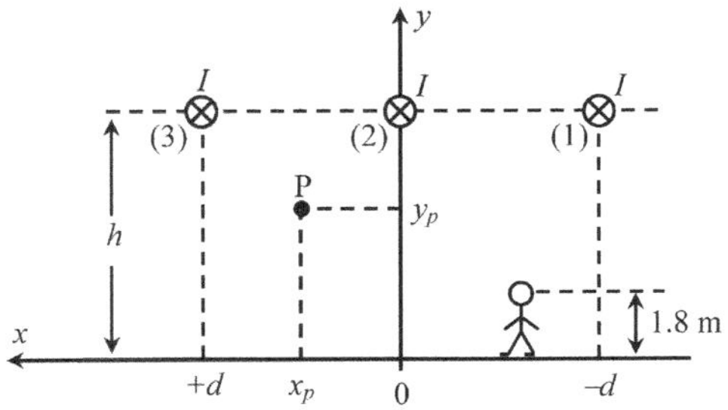

Figure 4.6 Cross-sectional view of a three-phase overhead line used for exemplifying the computation of the magnetic induction field intensity surrounding the structure

Assume that all the line conductors carry the same current I. Assume also that the currents flowing deep in the soil make a negligible contribution to the evaluation of the **B** field in air.

Data: $h = 12\,\text{m}$, $d = 7\,\text{m}$.

Questions

Q_1 Determine the B_x and B_y components of the magnetic induction field at a generic point P in air with coordinates x_P and y_P.

Q_1 Consider a person 1.8 m tall walking perpendicularly to the line, moving along the x axis. Determine analytically the intensity of the magnetic induction field $B(x)$ in the region about head height ($y_H = 1.8\,\text{m}$).

Q_1 Find an expression for B_{max} and determine the permissible line current intensity I such that the 100 μT safety limit is not exceeded.

Solutions

Q_1 Let x_k and y_k ($k = 1, 2, 3$) be the coordinates of the overhead line conductors, and let x_P and y_P be the coordinates of the observation point P.

The distance between conductor k and point P can be written as

$$r_{kP} = \sqrt{(x_k - x_P)^2 + (y_k - y_P)^2}$$

According to the result in (4.4), the magnetic field at P produced by the current I in the kth conductor is

$$\mathbf{H}_k = \frac{I}{2\pi r_{kP}} \vec{e}_{\phi_k}$$

On the other hand, the azimuthal unit vector $\vec{e}_{\phi_k}$ can be broken down into its $\vec{e}_x$ and $\vec{e}_y$ components as

$$\vec{e}_{\phi_k} = \frac{y_k - y_P}{r_{kP}} \vec{e}_x + \frac{x_P - x_k}{r_{kP}} \vec{e}_y$$

Summing the contributions from the three conductors, and noting that $\mathbf{B} = \mu_0 \mathbf{H}$, you obtain

$$\mathbf{B} = \underbrace{\frac{\mu_0 I}{2\pi} \sum_{k=1}^{3} \frac{y_k - y_P}{r_{kP}^2}}_{B_x} \vec{e}_x + \underbrace{\frac{\mu_0 I}{2\pi} \sum_{k=1}^{3} \frac{x_P - x_k}{r_{kP}^2}}_{B_y} \vec{e}_y$$

Q_2

$$B(x) = \sqrt{B_x^2 + B_y^2} = \frac{\mu_0 I}{2\pi} \sqrt{\left(\sum_{k=1}^{3} \frac{h - y_H}{r_{kP}^2(x)} \right)^2 + \left(\sum_{k=1}^{3} \frac{x - x_k}{r_{kP}^2(x)} \right)^2}$$

where $r_{kP}^2(x) = (x_k - x)^2 + (h - y_H)^2$, with $x_3 = -x_1 = d$ and $x_2 = 0$.

$B(x)$ is an even function of x with a maximum occurring at $x = 0$ (see sketch in Figure 4.7)

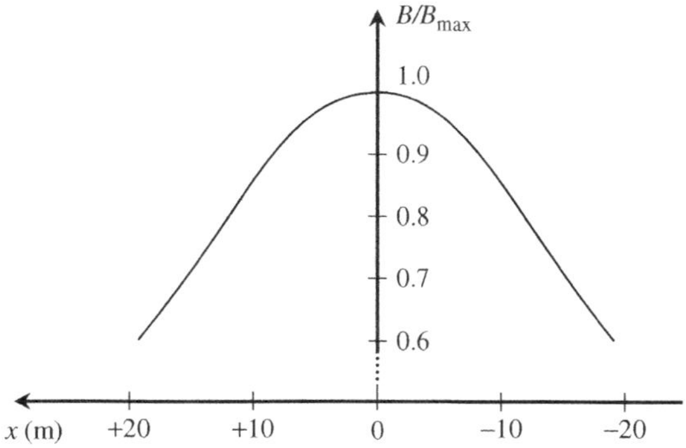

Figure 4.7 Graphical plot of the normalized magnetic induction field intensity sensed by a person moving along x. B_{max} should not exceed 100 μT

Q_3 At $x = 0$ and $y = y_H$ we find $\mathbf{B} = B_{max} \vec{e}_x$ where

$$B_{max} = \frac{\mu_0 I}{2\pi} \times \underbrace{\left(\frac{1}{(h - y_H)} + \frac{2(h - y_H)}{(h - y_H)^2 + d^2} \right)}_{K}$$

where, taking into account the problem data, $K = 0.231\,34\ \text{m}^{-1}$.

Putting $B_{max} = 100\,\mu\text{T}$, the permissible current can be evaluated as

$$I = \frac{2\pi B_{max}}{\mu_0 K} = 2.16\,\text{kA}$$

4.5 Magnetic Materials, Ferromagnetic Media, Saturation and Hysteresis

In empty space (vacuum), **B** and **H** are related through

$$\mathbf{B} = \mu_0 \mathbf{H} \tag{4.10}$$

In a material medium such a relationship must be modified in order to take into account electronic orbital perturbations as well as electron spin motions in the atomic structure of the medium. From a formal point of view, in a way similar to (2.12), we can write

$$\mathbf{B} = \mu_0 (\mathbf{H} + \mathbf{M}(\mathbf{H})) \tag{4.11}$$

where **M** is the so-called magnetization vector, or magnetic dipole moment per unit volume.

For linear media **M** and **H** are proportional to each other:

$$\mathbf{M} = \chi_m \mathbf{H} \tag{4.12}$$

where the dimensionless constant χ_m is the magnetic susceptibility of the medium.

By combining (4.11) and (4.12) we obtain

$$\mathbf{B} = \mu_0(1 + \chi_m)\, \mathbf{H} = \mu_0 \mu_r\, \mathbf{H} = \mu\, \mathbf{H} \tag{4.13}$$

where μ_r is the relative permeability of the medium.

For the vast majority of materials you will find $\mu_r \approx 1$ and, because of that, they are commonly classified as non-magnetic materials.

However, there is a special group of materials – with important applications in electrical engineering – where things are rather different.

Ferromagnetic media such as cobalt, nickel and iron, as well as several alloys whose composition includes them (ferroxcube, permalloy, mumetal, alnico, etc.), can have very large relative permeabilities, typically in the range 10^2 to 10^5.

Unfortunately, these materials display a nonlinear behavior. In Figure 4.8(a) we present a sketch of a typical first magnetization curve $B(H)$ showing that for weak intensities (linear zone) B and H are proportional, $B = \mu H$, but as the **H** field keeps increasing, the corresponding **B** field grows at a slower pace (saturation zone), tending to a limiting saturation value B_S.

Note that the above description of the first magnetization curve is not strictly accurate for two reasons. Firstly, for weak fields the linear zone is not exactly linear as $d^2B/dH^2 \neq 0$. Secondly, for very intense fields, the curve $B(H)$ tends to a straight line with positive slope μ_0; however, in most graphical representations this straight line appears as a quasi-horizontal line.

But there is more to ferromagnetic media than the saturation feature.

If, after reaching saturation, you try to demagnetize the material by progressively decreasing the **H** field to zero, you will see that the operating point does not descend to the origin – see Figure 4.8(b); when $H = 0$ you will get a positive remanence induction $B = B_R$.

In order to make $B = 0$ you will need to magnetize the material in the reverse direction with a coercive field $H = -H_C$ – see Figure 4.8(b).

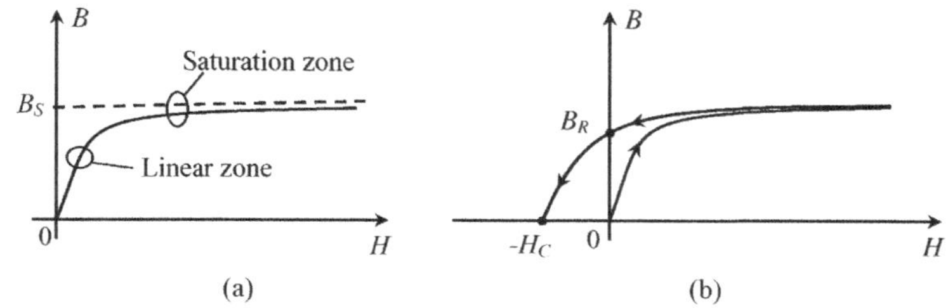

Figure 4.8 Saturation and hysteresis in ferromagnetic materials. (a) The ascending first magnetization curve includes a linear zone and a saturation zone. (b) The descending demagnetization curve does not coincide with the magnetization curve

If you let the magnetization–demagnetization process repeat itself slowly and alternately, swinging from positive saturation to negative saturation, you will obtain a closed curve in the *BH* plane, the so-called hysteresis loop – Figure 4.9(a).

Depending on the type of material being utilized, the hysteresis loop can be very thin or very fat. In the first case we say we are dealing with a soft magnetic material, in the second case with a hard magnetic material (Figure 4.9(b)). While soft magnetic materials are usually employed in electric machines and transformers, hard magnetic materials find their main application in permanent magnets and in magnetic recording.

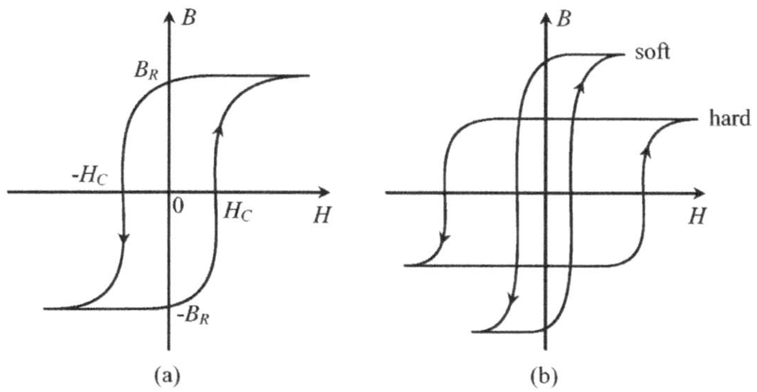

Figure 4.9 Hysteresis loop *B* against *H*. (a) Typical shape of a hysteresis loop, where H_C and B_R respectively denote the coercive field and the remanence induction. (b) Soft magnetic materials exhibit a thin hysteresis loop and, conversely, hard magnetic materials exhibit a very thick hysteresis loop

4.6 Magnetic Circuits

The electric circuits introduced in Section 3.2 were conceptually defined as perfect tubes for the circulation of **J**-field closed lines.

By analogy, magnetic circuits (ordinarily made of ferromagnetic media) are defined as tubes for the circulation of **B**-field closed lines.

There is, however, a distinctive point that we must address at once. In electric circuits the conductivity of the current-carrying conductors and the conductivity of the external medium (insulation) are related by a factor typically in excess of 10^{16}.

In the case of linear magnetic circuits, the permeability of the ferromagnetic media rarely exceeds the one for the external medium by a factor of 10^5. This circumstance reveals that magnetic circuits are not perfect tubes for the circulation of **B**-field lines which, to some extent, can escape from the main circuit – a phenomenon called dispersion (Figure 4.10).

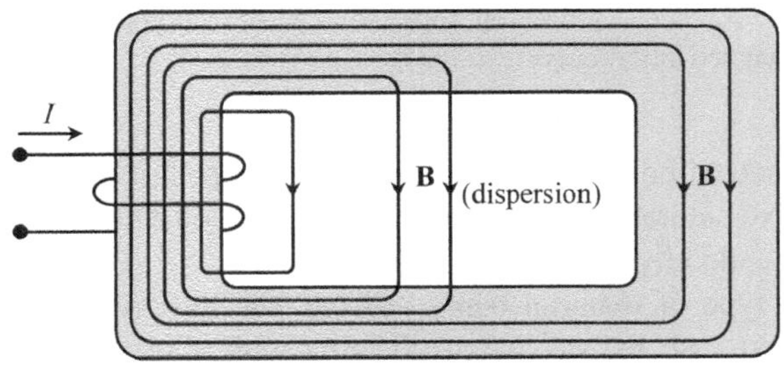

Figure 4.10 Dispersion of magnetic field lines

A frequently used approximation, the neglecting of dispersion phenomena, prevents you from obtaining very accurate results; however, on the other hand, it not only facilitates your task, but also provides you with good result estimates, which, often, are sufficient from an engineering standpoint. Moreover, those estimates can help you exert some type of control over the numerical results produced by dedicated software packages available for magnetic circuit analysis.

4.7 Application Example (Three-Legged Transformer)

To illustrate the application of the magnetic-field laws to the analysis of magnetic circuits, consider the example depicted in Figure 4.11 – a three-legged transformer with two horizontal yokes. For the purposes of generalization, we assume that the assembly may contain three thin protective pads made of a non-magnetic material, which are inserted between the transformer legs and the upper yoke. Also for the sake of generality, we will assume that the transformer legs are not identical.

The magnetic circuit is driven by currents I_a and I_b which flow in two separate coils with N_a and N_b turns respectively.

Simplifying assumptions:

- The yokes are made of exceptionally good magnetic material ($\mu_Y \to \infty$).
- Dispersion is neglected.
- Fields B and H are uniform inside each of the small protective pads.
- Fields B and H are uniform inside each of the transformer legs.

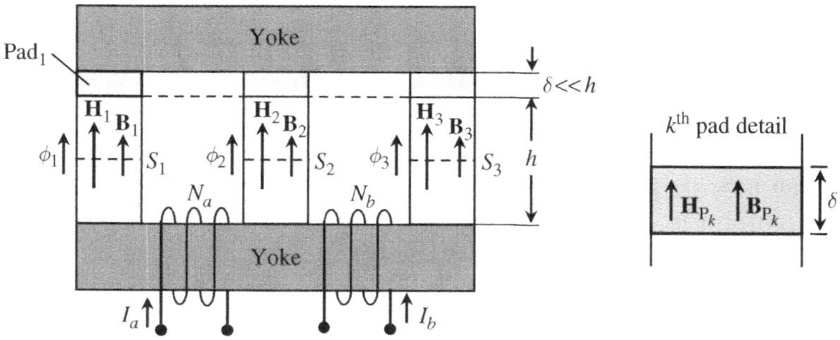

Figure 4.11 Geometry of a three-legged transformer

Assume that geometrical data of the configuration are known, and that the magnetic characteristics $B_k(H_k)$ are also known for legs 1, 2 and 3.

Establish a complete set of equations allowing for the determination of B_1, B_2, B_3, B_{P1}, B_{P2}, B_{P3} and H_1, H_2, H_3, H_{P1}, H_{P2}, H_{P3} as a function of the driving currents.

Solution

From div $\mathbf{B} = 0$ or, which is the same, from $\int_{S_V} \mathbf{B} \cdot \mathbf{n}_o dS = 0$ we immediately obtain

$$\phi_1 = B_1 S_1 = B_{P1} S_1 \rightarrow B_1 = B_{P1} \tag{4.14a}$$

$$\phi_2 = B_2 S_2 = B_{P2} S_2 \rightarrow B_2 = B_{P2} \tag{4.14b}$$

$$\phi_3 = B_3 S_3 = B_{P3} S_3 \rightarrow B_3 = B_{P3} \tag{4.14c}$$

$$\phi_1 + \phi_2 + \phi_3 = 0 \rightarrow B_1 S_1 + B_2 S_2 + B_3 S_3 = 0 \tag{4.14d}$$

Next, we apply Ampère's law (4.3) to the rectangular closed path $\mathbf{s}_a$ drawn clockwise inside and along the transformer legs and yokes of the circuit's left window. For that purpose, the integration path needs to be broken down into several segments, each one corresponding to a different part of the circuit (see details in Figure 4.12).

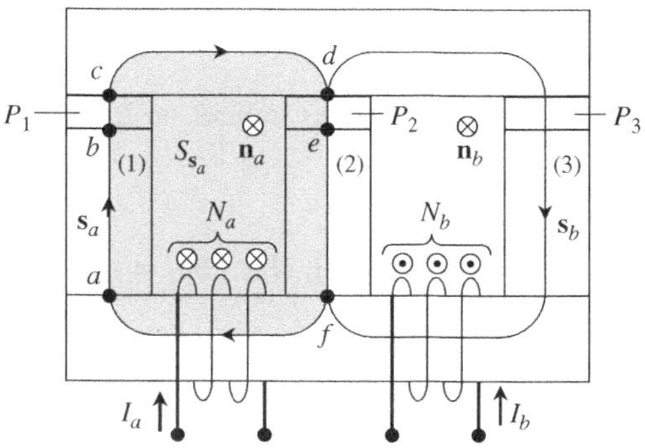

Figure 4.12 Application of Ampère's law to the closed circulation path $\mathbf{s}_a$

Then we find

$$\oint_{\mathbf{s}_a} \mathbf{H} \cdot d\mathbf{s} = \int_{\overrightarrow{ab}} \mathbf{H}_1 \cdot d\mathbf{s} + \int_{\overrightarrow{bc}} \mathbf{H}_{P1} \cdot d\mathbf{s} + \int_{\overrightarrow{cd}} \mathbf{H}_Y \cdot d\mathbf{s} + \int_{\overrightarrow{de}} \mathbf{H}_{P2} \cdot d\mathbf{s} + \int_{\overrightarrow{ef}} \mathbf{H}_2 \cdot d\mathbf{s} + \int_{\overrightarrow{fa}} \mathbf{H}_Y \cdot d\mathbf{s}$$

This identity can now be greatly simplified by taking into account the fact that fields are assumed to be uniform in each circuit part, and noting also that $H_Y = B_Y/\mu_Y = 0$. Hence, the result for the left-hand side of Ampère's law is

$$\oint_{\mathbf{s}_a} \mathbf{H} \cdot d\mathbf{s} = H_1 h + H_{P1}\delta - H_{P2}\delta - H_2 h \tag{4.15}$$

As for the right-hand side of Ampère's law, taking into account the orientation of the Stokes unit normal $\mathbf{n}_a$, we find

$$\int_{S_{s_a}} \mathbf{J} \cdot \mathbf{n}_a \, dS = \underbrace{I_a + I_a + \cdots + I_a + I_a}_{N_a \text{ times}} = +N_a I_a \tag{4.16}$$

By equating (4.15) and (4.16) we get

$$H_1 h + H_{P1}\delta - H_{P2}\delta - H_2 h = N_a I_a \tag{4.17}$$

Next, we need to apply Ampère's law again to a new closed path $\mathbf{s}_b$ (clockwise oriented) corresponding to the circulation along the transformer's right window. Following exactly the same steps as above we obtain

$$H_2 h + H_{P2}\delta - H_{P3}\delta - H_3 h = -N_b I_b \tag{4.18}$$

You should note that $N_b I_b$ has a minus sign. This happens because the reference direction of current I_b through S_{s_b} is antiparallel to the Stokes unit normal $\mathbf{n}_b$.

Counting the equations contained in (4.14), (4.17) and (4.18), we reach a total of six. Since the number of unknown fields is 12, you can see that we still lack six more equations. At this point you may be tempted to reapply Ampère's law to the closed circulation path including the yokes and legs 1 and 3 of the transformer. Do not do that! The equation you would obtain would not be independent of (4.17) and (4.18). In fact, the six equations that are missing must come from the material constitutive relations.

As for the transformer legs, we use

$$B_1 = B_1(H_1); \;\; B_2 = B_2(H_2); \;\; B_3 = B_3(H_3) \tag{4.19}$$

And for the non-magnetic pads, we use

$$B_{P1} = \mu_0 H_{P1}; \;\; B_{P2} = \mu_0 H_{P2}; \;\; B_{P3} = \mu_0 H_{P3} \tag{4.20}$$

From a physical point of view, the problem is completely posed and solved; the only thing that remains to be done lies on the mathematical side, that is the evaluation of a numerical solution of a complete set of equations. The only troubling aspect that you might face is related to the nonlinear relationships in (4.19), which may require nonlinear equation solver software.

4.8 Magnetic Reluctance

The foregoing application example can be simplified enormously if nonlinear phenomena are ruled out. In fact, if the transformer legs are assumed to be operating in their linear zones then, instead of (4.19), you may write

$$B_1 = \mu_1 H_1; \; B_2 = \mu_2 H_2; \; B_3 = \mu_3 H_3 \tag{4.21}$$

For the particular case of linear media, the analysis of the magnetic circuit in Figure 4.11 collapses into a much simpler problem – a problem that, as we will show, can be dealt with using only three basic equations.

We will turn to these in a moment, but first we must make a slight detour to introduce the new concept of magnetic reluctance R_m (units: H^{-1}), something which is analogous to the resistance in Ohm's law for electric circuits:

$$U = RI \;\leftrightarrow\; U_m = R_m \phi \tag{4.22}$$

As shown in Figure 4.13, consider a kth component part of a linear magnetic circuit where a magnetic flux ϕ_k flows. Due to linearity, the magnetic voltage U_{m_k} between the ends a and b of the component ought to be proportional to ϕ_k; the proportionality constant is the magnetic reluctance of the component part under analysis:

$$R_{m_k} = \frac{U_{m_k}}{\phi_k} = \frac{\int_{\overrightarrow{ab}} \mathbf{H}_k \cdot d\mathbf{s}}{\int_{S_k} \mathbf{B}_k \cdot \mathbf{n}\, dS} \tag{4.23}$$

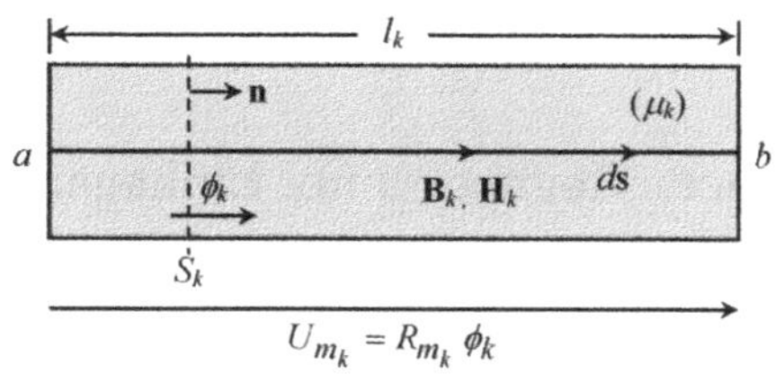

Figure 4.13 Magnetic reluctance of a section of a linear magnetic circuit

If, in addition, the magnetic circuit part is homogeneous, with uniform fields inside, then, from (4.23), we will obtain the approximation

$$R_{m_k} = \frac{H_k l_k}{B_k S_k} = \frac{l_k}{\mu_k S_k} \tag{4.24}$$

where l_k is the length of the magnetic circuit part, S_k its cross-sectional area and μ_k its magnetic permeability.

The analysis of linear magnetic circuits can be made simple by selecting the magnetic fluxes ϕ as their primary unknowns. After that, the evaluation of the B and H field intensities follows immediately.

Let us rework (4.17) and (4.18). From (4.17) and (4.22), we have

$$\underbrace{(H_1 h}_{U_{m_1}} + \underbrace{H_{P1}\delta)}_{U_{mP1}} - \underbrace{(H_2 h}_{U_{m_2}} + \underbrace{H_{P2}\delta)}_{U_{mP2}} = N_a I_a \rightarrow (R_{m_1} + R_{m_{P1}})\phi_1 - (R_{m_2} + R_{m_{P2}})\phi_2 = N_a I_a$$

From (4.18) and (4.22), we have

$$\underbrace{(H_2 h}_{U_{m_2}} + \underbrace{H_{P2}\delta)}_{U_{mP2}} - \underbrace{(H_3 h}_{U_{m_3}} + \underbrace{H_{P3}\delta)}_{U_{mP3}} = -N_b I_b \rightarrow (R_{m_2} + R_{m_{P2}})\phi_2 - (R_{m_3} + R_{m_{P3}})\phi_3 = -N_b I_b$$

The above equations, together with (4.14d), yield finally

$$\begin{cases} (R_{m_1} + R_{m_{P1}})\phi_1 - (R_{m_2} + R_{m_{P2}})\phi_2 = N_a I_a \\ (R_{m_2} + R_{m_{P2}})\phi_2 - (R_{m_3} + R_{m_{P3}})\phi_3 = -N_b I_b \\ \phi_1 + \phi_2 + \phi_3 = 0 \end{cases} \tag{4.25}$$

where the several intervening magnetic reluctances can be evaluated using (4.24).

The solution of the three-equation set in (4.25) is absolutely trivial, and, from there, by using (4.14), the magnetic induction field intensities can be readily obtained:

$$B_1 = B_{P1} = \frac{\phi_1}{S_1}; \ B_2 = B_{P2} = \frac{\phi_2}{S_2}; \ B_3 = B_{P3} = \frac{\phi_3}{S_3} \tag{4.26}$$

The last step, namely the evaluation of the magnetic field intensities along the circuit parts, is now rather obvious. From (4.20) and (4.21), we obtain

$$H_1 = \frac{B_1}{\mu_1}; \ H_2 = \frac{B_2}{\mu_2}; \ H_3 = \frac{B_3}{\mu_3}; \ H_{P1} = \frac{B_1}{\mu_0}; \ H_{P2} = \frac{B_2}{\mu_0}; \ H_{P3} = \frac{B_3}{\mu_0} \tag{4.27}$$

4.9 Inductor, Inductance, Magnetic Flux Linkage, Magnetic Energy

In its simplest form an inductor is a closed conductor, typically with the shape of a single loop or multiple loops (a coil), that is used to create magnetic fields when submitted to currents.

A characteristic parameter of inductors is the inductance L (units: H, henry).

To introduce this new concept, consider the trivial case of a single-conductor loop in air – see Figure 4.14. As soon as a current I is made to flow in the conductor loop (using a generator), a magnetic induction field **B** with closed lines embracing the conductor will be produced.

Let us define a closed integration path **s** coinciding with the conductor loop and oriented according to the reference direction assigned to I. Further, let S_s be an open surface having the path **s** as its bounding edge. The unit normal $\mathbf{n}_S$ to S_s is defined according to the path orientation using Stokes' rule.

When **B** is integrated across S_s we obtain a quantity called the magnetic flux linkage ψ:

$$\psi = \int_{S_s} \mathbf{B} \cdot \mathbf{n}_S \, dS \tag{4.28}$$

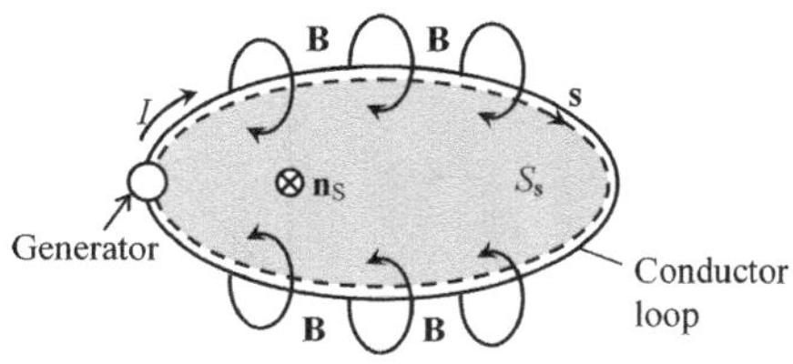

Figure 4.14 Magnetic flux linked with a current-carrying conductor loop

For the case of linear media, the intensity of **B** is proportional to I and, therefore, from (4.28), the same will happen with ψ:

$$\psi = LI \tag{4.29}$$

The positive proportionality constant between ψ and I is the so-called inductance L. This important parameter depends only on the permeability of the material media and on the geometry of the inductor. For nonlinear media the inductance concept is useless, since ψ and I are no longer proportional to each other.

We should emphasize that the magnetic flux linkage does not depend on the particular shape of the surface S_s through which the **B**-field is integrated, (4.28). The flux ψ is linked with the circulation path **s** and depends solely on it. In fact, as a natural consequence of div $\mathbf{B} = 0$, when we consider two distinct open surfaces S_s of different shape but bounded by the same path **s**, the fluxes through the two surfaces will be exactly equal (Figure 4.15).

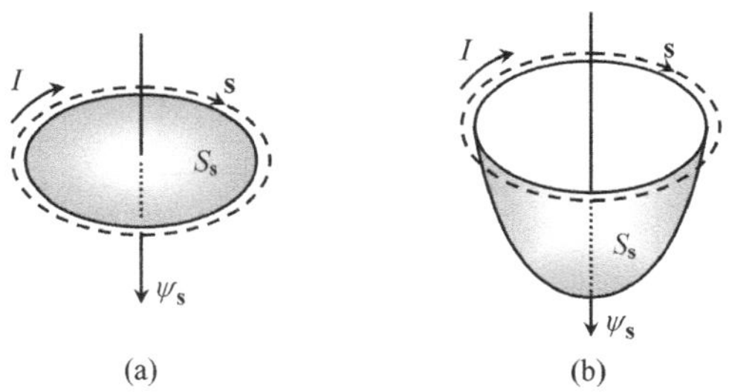

Figure 4.15 The magnetic flux linkage ψ in (a) and (b) is the same. The particular shape of the open surfaces bounded by the path **s** is irrelevant for the evaluation of ψ

At this point it is important to draw your attention to the existing analogy between the capacitance of a capacitor (Chapter 2) and the inductance of an inductor – see Figure 4.16.

$$\mathbf{D} = \varepsilon \mathbf{E} \;\rightarrow\; Q = CU \;\Leftrightarrow\; \mathbf{B} = \mu \mathbf{H} \rightarrow \psi = LI$$

A rather common way to achieve large inductance values consists of utilizing coils of many turns wound around a closed magnetic core of high permeability, as in Figure 4.17.

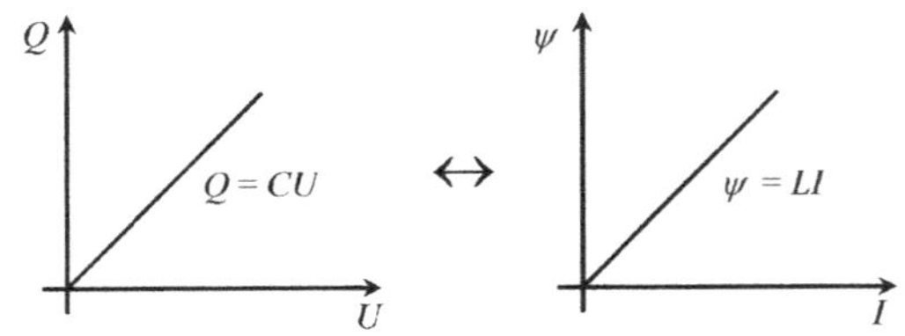

Figure 4.16 Analogy between the definitions of C and L

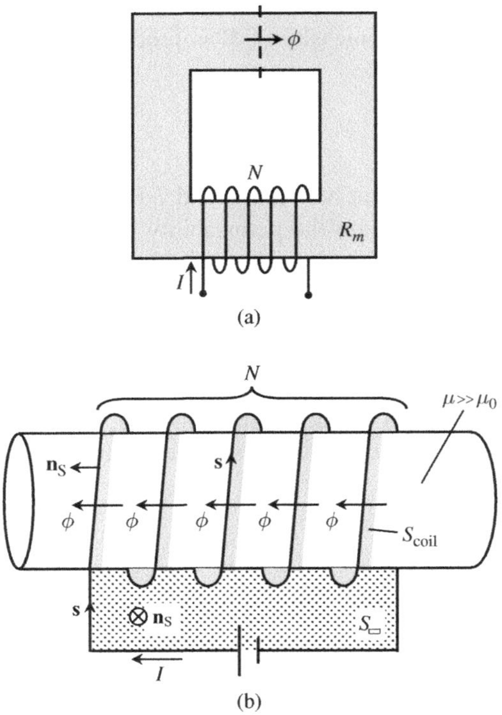

Figure 4.17 (a) A coil of N turns wound around a closed magnetic core of high permeability. (b) Zooming in to the coil region

On the one hand, neglecting dispersion phenomena, we find for the flux linkage

$$\psi_s = \int_{S_\square} \mathbf{B} \cdot \mathbf{n}_S \, dS + \int_{S_{coil}} \mathbf{B} \cdot \mathbf{n}_S \, dS = 0 + \underbrace{(\phi + \phi + \cdots + \phi + \phi)}_{N \text{ times}} = N\phi \qquad (4.30)$$

where S_{coil} is a helical-type surface bounded by the coil contour, which is intersected N times by the magnetic induction flux ϕ, and $S_\square$ is a rectangular planar surface across which the flux is zero (dispersion neglected). On the other hand, from magnetic circuit analysis (using Ampère's law) we have

$$R_m \, \phi = NI \qquad (4.31)$$

where the core magnetic reluctance R_m decreases with increasing core permeability. Hence, from (4.30) and (4.31), we get the usual approximation

$$L = \frac{\psi_s}{I} = \frac{N^2}{R_m} \tag{4.32}$$

When the current of an inductor is increased from zero to its final state, the inductor core becomes magnetized; that is, the magnetic induction field **B** in the inductor core grows in intensity until its stationary final state is reached (the same thing happens to the flux ψ_s).

The magnetization process requires the generator, responsible for driving the inductor current, to expend a certain amount of energy. Part of it is associated with irreversible Joule losses in the inductor resistance. Another part, of a completely different nature, has to do with the fact that the progressive increase of **B**, just by itself, demands additional work from the generator; this work is not lost, but remains available in the form of magnetic energy W_m, stored in the inductor core, and can be retrieved later during the demagnetization process.

If you have an inquisitive mind (and we hope you do), you may now be wondering about the reason why a continuous increase of **B** does indeed require additional work from the generator. To answer this question we cannot avoid a small leap ahead to a topic that actually belongs to Chapter 5, where the time dynamics of **B** fields is analyzed.

From Maxwell's equations in (I.1) we known that

$$\text{curl } \mathbf{E} = -\frac{\partial \mathbf{B}}{\partial t}$$

This shows that electric fields can be originated by time-varying **B** fields. This new type of electric field (so-called electric induction field $\mathbf{E} = \mathbf{E}_i$) has its field lines closed, embracing $\mathbf{B}(t)$.

As shown in Figure 4.18(a), during the magnetization process, that is when dB/dt is positive, the electric induction field $\mathbf{E}_i$ opposes the circulation of currents in the conductor loop (inductor) and, therefore, the generator is forced to provide additional work to sustain the current flow. Conversely (Figure 4.18(b)), during the demagnetization process, when dB/dt is negative, the electric induction field $\mathbf{E}_i$ favors the circulation of currents in the inductor and, as a result, the generator recovers energy.

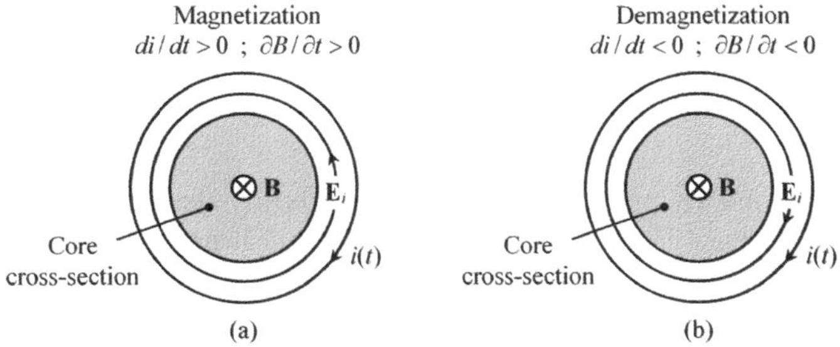

Figure 4.18 Magnetization and demagnetization. (a) During magnetization the electric induction field $\mathbf{E}_i$ opposes the current flow. (a) During demagnetization the electric induction field $\mathbf{E}_i$ favors the current flow

If hysteresis phenomena are absent (as we will assume in the analysis that follows next) then the recovered energy during the demagnetization process will be the same that was spent during magnetization.

The magnetic energy stored in the magnetic field of an inductor, W_m, is evaluated in perfect analogy with the process we used earlier, in Section 2.7, to obtain the electric energy stored in the electric field of a capacitor, W_e:

$$W_e = \int_{\substack{\text{initial}\\\text{state}}}^{\substack{\text{final}\\\text{state}}} u\,dq \quad \Leftrightarrow \quad W_m = \int_{\substack{\text{initial}\\\text{state}}}^{\substack{\text{final}\\\text{state}}} i\,d\psi \tag{4.33}$$

This result for W_m is graphically interpreted in Figure 4.19, for linear and nonlinear magnetic media.

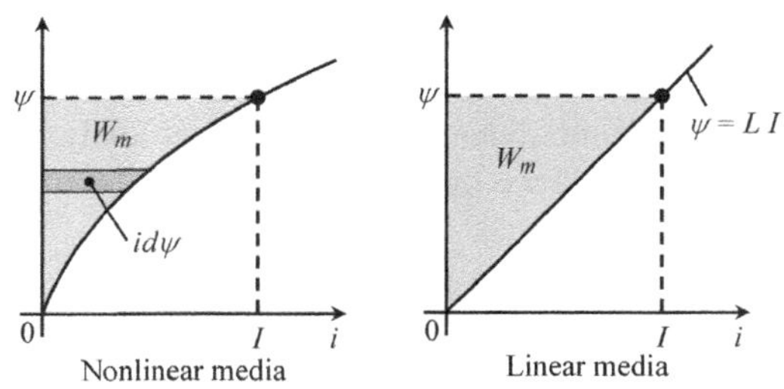

Figure 4.19 Graphical interpretation of the magnetic energy stored in an inductor, for linear and nonlinear magnetic media

For the simplest case of linear magnetic media, by using $\psi = Li$ in (4.33), we find

$$W_m = L\int_0^I i\,di = \frac{1}{2}LI^2 = \frac{1}{2}\frac{\psi^2}{L} = \frac{1}{2}\psi I \tag{4.34}$$

The magnetic energy stored in the inductor core can be determined by volume integration:

$$W_m = \int_{V_{\text{core}}} \hat{w}_m dV \tag{4.35}$$

where $\hat{w}_m$ represents the volumetric magnetic energy density (units: J/m^3).

Analogous to Figure 4.19, you will find in Figure 4.20 a graphical interpretation of the computation of $\hat{w}_m$ for linear and nonlinear media.

For linear media, the magnetic energy density (the area of the shaded triangle in Figure 4.20) is

$$\hat{w}_m = \frac{1}{2}BH = \frac{1}{2}\frac{B^2}{\mu} = \frac{1}{2}\mathbf{B}\cdot\mathbf{H} \tag{4.36}$$

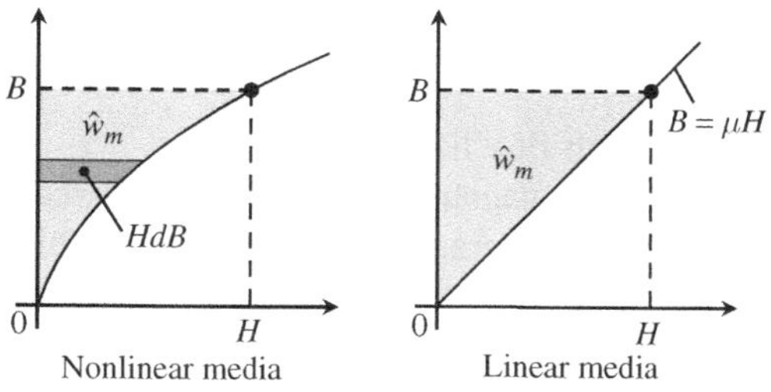

Figure 4.20 Graphical interpretation of the magnetic energy density stored in the magnetic field, for linear and nonlinear magnetic media

4.10 Application Example (Coaxial Cable)

Figure 4.21 shows a cross-sectional view of a coaxial cable of indefinite length l, with transversal dimensions r_1, r_2 and r_3. The insulation medium and the cable conductors are characterized by $\mu = \mu_0$. The cable current is I.

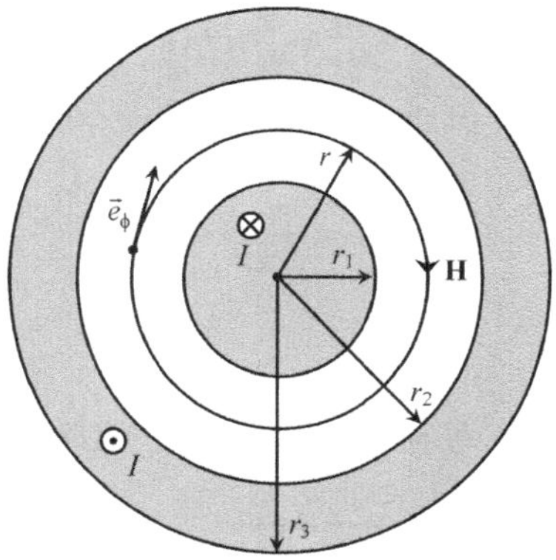

Figure 4.21 Cross-sectional view of a coaxial cable. Conductor currents give rise to a magnetic field $\mathbf{H} = H(r)\vec{e}_\phi$ whose field lines are concentric circumferences

Questions

Q_1 By application of Ampère's law determine the magnetic field intensity $H(r)$ for $0 < r < \infty$.

Q_2 Determine the flux ψ linked with the cable.

Q_3 Based on the above result determine the per-unit-length external inductance of the cable.

Q_4 By volume integration, find the per-unit-length magnetic energy stored in the insulation medium ($r_1 < r < r_2$) and, based on that, confirm the result obtained in question Q_3.

Q_5 Numerical application. Take $I = 100\,\text{mA}$. Consider $r_1 = 1\,\text{mm}$, $r_2 = 5\,\text{mm}$, $r_3 = 5.1\,\text{mm}$.

Compute $H(r_1)$ and $H(r_2)$.
Compute the per-unit-length flux linkage.
Compute the per-unit-length magnetic energy stored between cable conductors.
Compute the per-unit-length external inductance of the cable.

Solutions

Q_1 Due to the cable geometry, the field lines of **H** are closed circumferences coaxial with the cable's longitudinal axis, and the intensity of H remains invariant along any given field line, $\mathbf{H} = H(r)\vec{e}_\phi$. For the determination of H we utilize Ampère's law, employing integration paths **s** that exactly match the circumferential field lines, so that

$$\mathbf{H} \cdot d\mathbf{s} = H\ ds = H\ r\ d\phi$$

For $0 < r < r_1$:

$$\underbrace{\oint_{\mathbf{s}} \mathbf{H} \cdot d\mathbf{s}}_{2\pi r H} = \underbrace{\int_{S_s} \mathbf{J} \cdot \mathbf{n}_S\ dS}_{J_1(\pi r^2)}$$

Since, in the inner conductor of the cable, $J_1 = I/(\pi r_1^2)$, we get

$$H(r) = \frac{I}{2\pi r} \times \left(\frac{r}{r_1}\right)^2$$

For $r_1 < r < r_2$:

$$\underbrace{\oint_{\mathbf{s}} \mathbf{H} \cdot d\mathbf{s}}_{2\pi r H} = \underbrace{\int_{S_s} \mathbf{J} \cdot \mathbf{n}_S\ dS}_{+I} \rightarrow H(r) = \frac{I}{2\pi r}$$

For $r_2 < r < r_3$:

$$\underbrace{\oint_{\mathbf{s}} \mathbf{H} \cdot d\mathbf{s}}_{2\pi r H} = \underbrace{\int_{S_s} \mathbf{J} \cdot \mathbf{n}_S\ dS}_{I - J_2\pi(r^2 - r_2^2)}$$

Since, in the outer conductor of the cable,

$$J_2 = \frac{I}{\pi(r_3^2 - r_2^2)}$$

we get

$$H(r) = \frac{I}{2\pi r} \times \frac{r_3^2 - r^2}{r_3^2 - r_2^2}$$

For $r > r_3$:

$$\underbrace{\oint_{\mathbf{s}} \mathbf{H} \cdot d\mathbf{s}}_{2\pi r H} = \underbrace{\int_{S_s} \mathbf{J} \cdot \mathbf{n}_S \, dS}_{+I-I} \to H = 0$$

Summarizing,

$$\mathbf{H} = H(r)\,\vec{e}_\phi; \begin{cases} H(r) = \dfrac{I}{2\pi r} \times \left(\dfrac{r}{r_1}\right)^2; \text{ for } 0 < r < r_1 \\ H(r) = \dfrac{I}{2\pi r}; \text{ for } r_1 < r < r_2 \\ H(r) = \dfrac{I}{2\pi r} \times \dfrac{r_3^2 - r^2}{r_3^2 - r_2^2}; \text{ for } r_2 < r < r_3 \\ H(r) = 0; \text{ for } r_3 < r \end{cases} \tag{4.37}$$

Q_2 The flux linked with the cable is obtained by integrating the **B**-field lines external to the conductors crossing a rectangular surface of length l that belongs to a radial plane between r_1 and r_2. The rectangular path **s** is oriented according to the reference direction assigned to I – see Figure 4.22:

$$\psi = \int_{S_s} \mathbf{B} \cdot \mathbf{n}_S \, dS = \int_{S_s} B \, dS = \int_{r_1}^{r_2} \mu_0 H(r) \, l \, dr = I \frac{\mu_0 l}{2\pi} \ln \frac{r_2}{r_1}$$

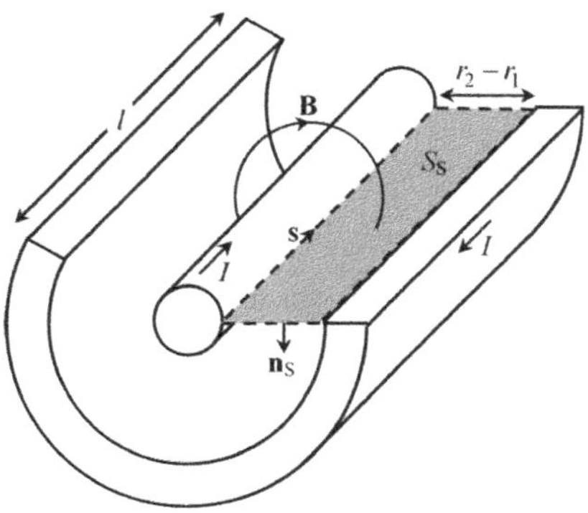

Figure 4.22 Integration surface being used for the evaluation of the cable's external inductance

Q_3 The per-unit-length external inductance L of the cable is obtained from

$$L = \frac{\psi / I}{l} = \frac{\mu_0}{2\pi} \ln \frac{r_2}{r_1} \tag{4.38}$$

Q_4 For a cable section of length l, the magnetic energy stored in the space between cable conductors is

$$W_m = \int_V \frac{\mu_0 H^2}{2} dV = \frac{\mu_0}{2} \int_{r_1}^{r_2} H^2 \underbrace{2\pi l r\, dr}_{dV} = I^2 \frac{\mu_0 l}{4\pi} \ln \frac{r_2}{r_1}$$

Putting $l = 1\,\text{m}$, and using (4.34), the per-unit-length cable inductance is obtained from the magnetic energy stored as

$$L = \frac{2W_m}{I^2} = \frac{\mu_0}{2\pi} \ln \frac{r_2}{r_1}$$

Q_5 $H(r_1) = 15.9\,\text{A/m}$; $H(r_2) = 3.2\,\text{A/m}$;
$\psi = 32.2\,\text{nWb/m}$; $W_m = 1.6\,\text{nJ/m}$; $L = 321.9\,\text{nH/m}$.

4.11 Hysteresis Losses

In Section 4.5, when magnetic media were discussed, we referred to the hysteresis loop as a special feature of ferromagnetic materials. Consider again the graphical representation of the hysteresis loop in Figure 4.9(a), but now utilizing the variables ψ and i – see Figure 4.23.

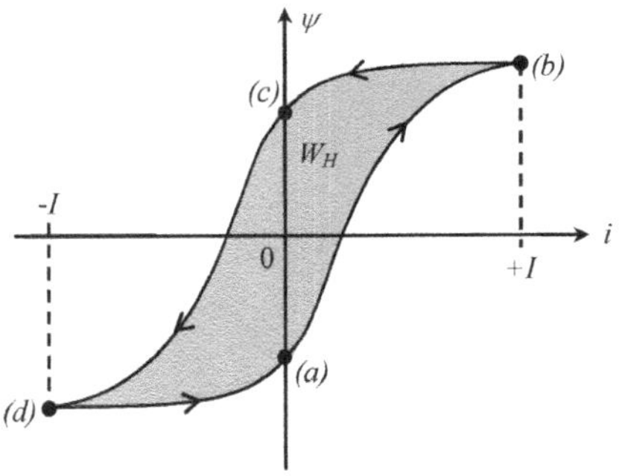

Figure 4.23 Hysteresis loop ψ against i. The shaded area represents the energy lost during one complete period

As in (4.33), during the magnetization process corresponding to moving the operation point from (a) to (b), the work of the generator is determined by the area between the curve $\widehat{ab}$ and the ψ axis:

$$W_{ab} = \int_{\psi_a}^{\psi_b} i\, d\psi > 0$$

This quantity is positive ($i > 0$; $d\psi > 0$) and therefore represents energy being delivered to the inductor core.

During the demagnetization process corresponding to moving the operation point from (b) to (c), the work of the generator is determined by the area between the curve $\widehat{bc}$ and the ψ axis:

$$W_{bc} = \int_{\psi_b}^{\psi_c} i\, d\psi < 0$$

This quantity is negative ($i > 0$; $d\psi < 0$) and represents energy being delivered back to the generator.

During the magnetization process corresponding to moving the operation point from (c) to (d), the work of the generator is determined by the area between the curve $\widehat{cd}$ and the ψ axis:

$$W_{cd} = \int_{\psi_c}^{\psi_d} i\, d\psi > 0$$

This quantity is positive ($i < 0$; $d\psi < 0$), is equal to W_{ab} and again represents energy being delivered to the inductor core.

Finally, closing the hysteresis loop, during the demagnetization process corresponding to moving the operation point from (d) to (a), the work of the generator is determined by the area between the curve $\widehat{ba}$ and the ψ axis:

$$W_{da} = \int_{\psi_d}^{\psi_a} i\, d\psi < 0$$

This quantity is negative ($i < 0$; $d\psi > 0$), is equal to W_{bc} and again represents energy being delivered back to the generator.

The energy spent during the magnetization intervals $(W_{ab} + W_{cd})$ exceeds the energy recovered during the demagnetization ones $(W_{bc} + W_{da})$. The missing energy per cycle (hysteresis losses) is thus given by the internal area of the hysteresis loop – see again Figure 4.23:

$$W_H = \oint i\, d\psi \tag{4.39}$$

The lost energy is irreversibly transformed into heat in the inductor core.

We will see in Chapter 5 that inductor core heating can result from another physical mechanism (eddy currents).

4.12 Multiple Circuit Systems

The analysis of magnetic coupling phenomena is of crucial importance in a variety of electrical engineering problems where multiple circuit systems are in play, such as printed circuit boards, power lines, transformers, multiple inductor circuits, etc. To give you a couple of simple examples, consider the illustrations in Figure 4.24, which describe a two-conductor transmission line above a conducting plane, and a two-coil system.

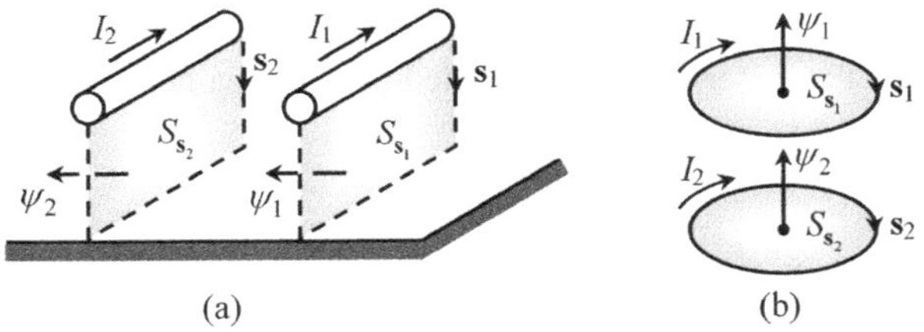

Figure 4.24 Magnetically coupled circuits. (a) Two-conductor transmission line above ground. (b) Two single-turn coil system

Assuming a linear behavior for the systems (neglecting saturation and hysteresis), the magnetic fluxes linked with the circuits 1 and 2 can be expressed as linear combinations of their own currents:

$$\begin{aligned} \psi_1 &= \int_{S_{s_1}} \mathbf{B}(I_1, I_2) \cdot \mathbf{n}_{S_1}\, dS = L_{11} I_1 + L_{12} I_2 \\ \psi_2 &= \int_{S_{s_2}} \mathbf{B}(I_1, I_2) \cdot \mathbf{n}_{S_2}\, dS = L_{21} I_1 + L_{22} I_2 \end{aligned} \tag{4.40}$$

The magnetic coupling between circuits is described by the mutual inductances L_{12} and L_{21}.

In the general case of n-coupled circuits and, in analogy with the electrostatic matrix formulation presented in (2.44), we can write

$$\underbrace{\begin{bmatrix} \psi_1 \\ \vdots \\ \psi_k \\ \vdots \\ \psi_n \end{bmatrix}}_{[\psi]} = \underbrace{\begin{bmatrix} L_{11} & \cdots & L_{1k} & \cdots & L_{1n} \\ \vdots & & \vdots & & \vdots \\ L_{k1} & \cdots & L_{kk} & \cdots & L_{kn} \\ \vdots & & \vdots & & \vdots \\ L_{n1} & \cdots & L_{nk} & \cdots & L_{nn} \end{bmatrix}}_{[L]} \underbrace{\begin{bmatrix} I_1 \\ \vdots \\ I_k \\ \vdots \\ I_n \end{bmatrix}}_{[I]} \tag{4.41}$$

The square $n \times n$ real matrix $[L]$ is called the inductance matrix. The entries of $[L]$ can be determined using various methods. They can be found experimentally by measuring fluxes and currents, they can be found numerically using dedicated computer programs and in some cases, when very simple geometries are considered, they can also be determined analytically.

Whatever the method, the results you will obtain should be checked against a few general properties that matrix $[L]$ must necessarily obey. These properties are almost the same as those that were reported earlier for the capacitance matrix $[C]$ of multiple conductor systems (Chapter 2), namely:

$[L]$ is a symmetric matrix;
$[L]$ is a positive definite matrix.

In addition, if the contour paths $\mathbf{s}_k$ used for the definition of the linkage fluxes ψ_k are oriented according to the reference directions of the I_k currents (Figure 4.24) then the L_{kk} entries (self-inductances) will all be positive.

However, the mutual inductances $L_{ij} = L_{ji}$ can have positive or negative algebraic signs, depending on the arbitrarily chosen reference directions for the I_i and I_j currents; in the first case we say we have concordant coupling, whereas in the second case the coupling is discordant.

A positive dimensionless parameter usually introduced to characterize the extent of the magnetic coupling between the *i*th and *j*th circuits is the so-called magnetic coupling factor k_{ij}. It is defined as the ratio of the absolute value of the circuits' mutual inductance and the geometric mean value of the circuits' self-inductances

$$k_{ij} = \frac{|L_{ij}|}{\sqrt{L_{ii}L_{jj}}}; \quad 0 < k_{ij} < 1 \tag{4.42}$$

If, on the one hand, the two circuits are non-interacting ($L_{ij} = 0$) then you will have $k_{ij} = 0$. On the other hand, if all the magnetic field lines produced by current I_i pass across the integration surface where ψ_j is evaluated (or vice versa), then the magnetic coupling is said to be perfect and you will have $k_{ij} = 1$.

Do you have any questions? Can you see the reason why the magnetic coupling factor cannot exceed 1?

It is just a consequence of the fact that $[L]$ is a positive definite matrix. In fact, such a property requires that the reduced determinant

$$\begin{vmatrix} L_{ii} & L_{ij} \\ L_{ji} & L_{jj} \end{vmatrix}$$

must be positive.

Then, taking into account that $L_{ij} = L_{ji}$, you can obtain $L_{ii}L_{jj} - L_{ij}^2 > 0$, from which you conclude

$$\frac{L_{ij}^2}{L_{ii}L_{jj}} = k_{ij}^2 < 1 \;\rightarrow\; k_{ij} < 1$$

The limiting case $k_{ij} = 1$ is a mathematical abstraction which usually goes hand in hand with the ordinary approximation of neglecting dispersion phenomena.

In order to illustrate the procedure used for determining the $[L]$ matrix, consider the important example of a single-core transformer with two windings – see Figure 4.25.

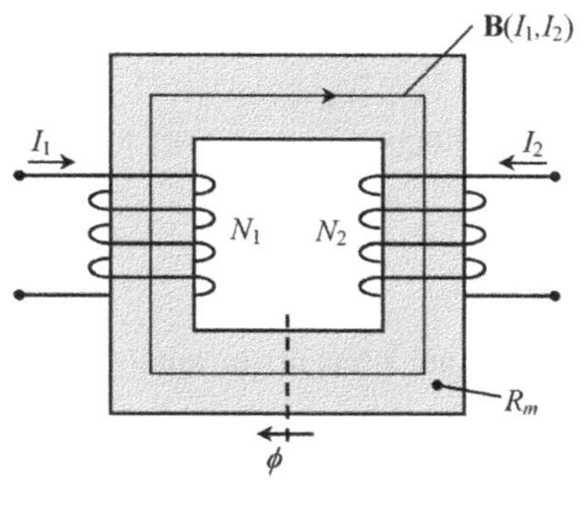

Figure 4.25 Single-core transformer with two windings

Neglecting dispersion, the magnetic flux ϕ circulating in the core is obtained with the help of Ampère's law

$$R_m\phi = N_1 I_1 + N_2 I_2 \tag{4.43}$$

where R_m is the core magnetic reluctance, which depends only on the core magnetic permeability and on the core geometry.

The magnetic fluxes linked with both transformer windings are (neglecting dispersion)

$$\psi_1 = \int_{S_{s_1}} \mathbf{B} \cdot \mathbf{n}_{S_1}\, dS = \underbrace{\phi + \phi + \cdots + \phi + \phi}_{N_1 \text{ times}} = N_1\, \phi \tag{4.44a}$$

$$\psi_2 = \int_{S_{s_2}} \mathbf{B} \cdot \mathbf{n}_{S_2}\, dS = \underbrace{\phi + \phi + \cdots + \phi + \phi}_{N_2 \text{ times}} = N_2\, \phi \tag{4.44b}$$

From (4.44) and (4.43), you can obtain

$$\begin{cases} \psi_1 = \dfrac{N_1^2}{R_m} I_1 + \dfrac{N_1 N_2}{R_m} I_2 \\ \psi_2 = \dfrac{N_1 N_2}{R_m} I_1 + \dfrac{N_2^2}{R_m} I_2 \end{cases} \tag{4.45}$$

By comparing (4.45) to (4.40), you finally find

$$L_{11} = \frac{N_1^2}{R_m};\; L_{22} = \frac{N_2^2}{R_m};\; L_{12} = L_{21} = L_M = \frac{N_1 N_2}{R_m} \tag{4.46}$$

Three remarks are in order:

1. In this case you have $k = 1$ for the magnetic coupling factor; this should be no surprise since dispersion was neglected.
2. A positive value resulted for the mutual inductance in this analysis. However, had you decided to change the reference direction of I_1 or I_2 (but not both simultaneously) then L_M would turn out to be negative.
3. The results in (4.46) pertain to the particular configuration shown in Figure 4.25; do not make the mistake of using them as a general recipe for analyzing magnetic circuits with other configuration types.

To end this section we must address the important issue of the magnetic energy stored in a multiconductor linear system.

You saw in Section 4.6 that the magnetic energy of a single circuit could be evaluated through

$$W_m = \frac{\psi I}{2}$$

When multiple, n-coupled, circuits are present, in analogy with the electrostatic situation (2.47), you can write

$$W_m = \frac{\psi_1 I_1}{2} + \cdots + \frac{\psi_k I_k}{2} + \cdots + \frac{\psi_n I_n}{2} = \sum_{k=1}^{n} \frac{\psi_k I_k}{2} \tag{4.47}$$

Put in matrix form

$$W_m = \frac{1}{2}\left[\psi_1 \cdots \psi_k \cdots \psi_n\right]\begin{bmatrix} I_1 \\ \vdots \\ I_k \\ \vdots \\ I_n \end{bmatrix} = \frac{1}{2}[\psi]^T[I]$$

Using $[\psi] = [L][I]$ from (4.41), and noting that matrix $[L]$ is symmetric, you find

$$W_m = \frac{1}{2}[I]^T[L][I] \tag{4.48}$$

You should bear in mind that the above result is analogous to

$$W_e = \frac{1}{2}[U]^T[C][U]$$

in (2.49).

For a system of two-coupled circuits like those shown in Figure 4.24 and Figure 4.25, from (4.48), you find W_m given by the following quadratic form:

$$W_m = \frac{1}{2}\left(L_{11}I_1^2 + 2L_M I_1 I_2 + L_{22}I_2^2\right) \tag{4.49}$$

4.13 Magnetic Forces and Torques

Magnetic forces and torques can be evaluated using exactly the same type of formulation elaborated in Section 2.11 for the computation of electric forces and torques (Chapter 2).

Here, you just need to change some of the nomenclature. By using the analogies in Table 4.1, you can immediately write, without further discussion,

Table 4.1 Electric and magnetic forces and torques

Electric	Magnetic
$\mathbf{F}_e = -\text{grad } W_e\|_Q$	$\mathbf{F}_m = -\text{grad } W_m\|_\psi$
$T_\alpha = -\left(\frac{\partial W_e}{\partial \alpha}\right)_Q$	$T_\alpha = -\left(\frac{\partial W_m}{\partial \alpha}\right)_\psi$

Magnetic force:
$$\begin{cases} \mathbf{F}_m = f_x\vec{e}_x + f_y\vec{e}_y + f_z\vec{e}_z = -\text{ grad } W_m|_\psi \\ f_x = -\left(\dfrac{\partial W_m}{\partial x}\right)_\psi;\ f_y = -\left(\dfrac{\partial W_m}{\partial y}\right)_\psi;\ f_z = -\left(\dfrac{\partial W_m}{\partial z}\right)_\psi \end{cases} \tag{4.50}$$

Magnetic pressure: $P = \hat{w}_m$ (4.51)

Magnetic torque: $T_\alpha = -\left(\dfrac{\partial W_m}{\partial \alpha}\right)_\psi$ (4.52)

We saw in (2.63) that the maximum pressure achievable by electric forces could be estimated to around $40\,\mathrm{N/m^2}$. However, magnetic forces are much more powerful. Considering the case of a magnetic induction field in an air gap with a typical value of $B = 1\,\mathrm{T}$, you will find from (4.51)

$$P = \hat{w}_m = \frac{B^2}{2\mu_0} \approx 400\,\mathrm{kN/m^2}$$

which is a four-fold more intense pressure than the one found in the electric case.

4.14 Application Example (U-Shaped Electromagnet)

Consider the U-shaped electromagnet shown in Figure 4.26. The ferromagnetic yoke and bar are separated by two small-size air gaps of length x. The magnetic circuit is driven by a current I flowing in a coil with N turns.

Questions

Q_1 Determine analytically the magnetic force f_x exerted on the bar.

Q_2 Particularize the result for the case of a linear system (ignoring saturation).

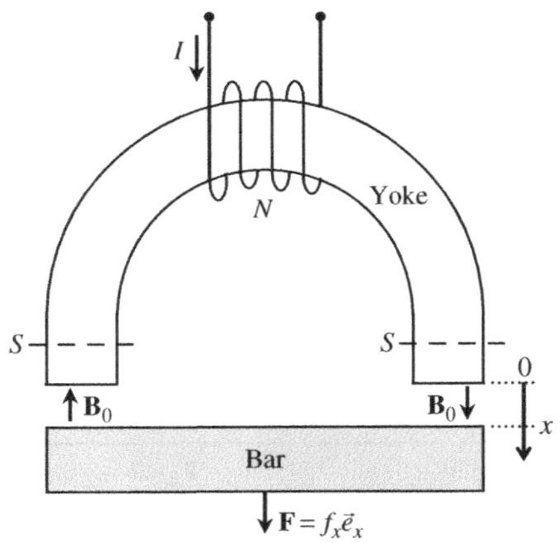

Figure 4.26 U-shaped electromagnet.

Solutions

Q_1 From (4.50) we have

$$f_x = -\left(\frac{\partial W_m}{\partial x}\right)_{\psi} \tag{4.53}$$

Since the computation of the force is made under the assumption that the magnetic flux linked with the coil is invariant with x, the magnetic induction field itself will also be invariant with x. Noting, in addition, that the volume of the ferromagnetic circuit parts is constant, we can write for the global magnetic energy

$$W_m(x) = W_{m_{\text{yoke}}} + W_{m_{\text{bar}}} + W_{m_{\text{air}}}(x)$$

Then, from (4.53), we find

$$f_x = -\left(\frac{\partial W_{m_{\text{air}}}}{\partial x}\right)_{\psi} = -\frac{d}{dx}\left(2 \times \frac{B_0^2}{2\mu_0} \times \underbrace{Sx}_{\substack{\text{air gap}\\ \text{volume}}}\right) = -\frac{S}{\mu_0}B_0^2 \tag{4.54}$$

The magnitude of the air gap field B_0 is evaluated using Ampère's law (Section 4.2). The minus sign in f_x reveals that the force is attractive no matter if B_0 is positive or negative. The above result is valid even if the ferromagnetic circuit parts are operating in the saturation zone.

Q_2 If, however, we are sure that the ferromagnetic parts have their operation points in the linear zone then an alternative option for the calculation of f_x can be presented.

From (4.32), the inductance L of the coil depends on x through

$$L(x) = \frac{N^2}{R_{m_{\text{yoke}}} + R_{m_{\text{bar}}} + \dfrac{2x}{\mu_0 S}} \tag{4.55}$$

From (4.34) the magnetic energy stored in the system can be determined from L by using $W_m = \psi^2/(2L)$, and, consequently, from (4.53) and (4.55), we get

$$f_x = -\left(\frac{\partial W_m}{\partial x}\right)_{\psi} = \frac{1}{2}\left(\frac{\psi}{L}\right)^2 \frac{dL}{dx} = -\frac{(NI)^2}{\mu_0 S}\left(R_{m_{\text{yoke}}} + R_{m_{\text{bar}}} + \frac{2x}{\mu_0 S}\right)^{-2}$$

Therefore we see that the attractive force actuates so as to increase the coil inductance L. The force is independent of the algebraic sign of the driving current. The smaller the air gap, the stronger the force. The force is maximum when the yoke and the bar are in contact ($x = 0$)

$$(f_x)_{\max} = -\frac{1}{\mu_0 S}\left(\frac{NI}{R_{m_{\text{yoke}}} + R_{m_{\text{bar}}}}\right)^2 \tag{4.56}$$

4.15 Proposed Homework Problems

Problem 4.15.1

Consider again the coaxial cable examined in Section 4.10 (Figure 4.21), where the per-unit-length external inductance of the cable was determined. In that problem we also evaluated the **H**-field intensity inside both the inner conductor and the shield. In addition to the predominant magnetic energy stored in the space between cable conductors ($r_1 < r < r_2$), two other contributions should be accounted for: one inside the inner conductor and one inside the shield. These two terms allow for the definition of corresponding cable internal inductances.

Q_1 By volume integration, find the per-unit-length magnetic energy stored inside the inner conductor ($0 < r < r_1$) and, based on that result, show that the inner conductor's contribution to the per-unit-length internal inductance of the cable, L_{inner}, is given by

$$L_{\text{inner}} = \frac{\mu_0}{8\pi}$$

Q_2 By volume integration, find the per-unit-length magnetic energy stored inside the cable shield ($r_2 < r < r_3$) and, based on that result, compute the shield contribution to the per-unit-length internal inductance of the cable, L_{shield}. Check that, for the usual case of thin shields ($r_3 - r_2 \ll r_2$), the following approximation is valid:

$$L_{\text{shield}} = \frac{\mu_0}{6\pi}\left(\frac{r_3}{r_2} - 1\right)$$

Q_3 Evaluate numerically L_{inner} and L_{shield}. Compare to L (external).

Answers

Q_1

$$W_{m_{\text{inner}}} = \int_V \frac{\mu_0 H^2}{2} dV = \frac{\mu_0}{2}\int_0^{r_1} H^2 2\pi r\, dr = I^2 \frac{\mu_0}{16\pi}; \quad L_{\text{inner}} = \frac{2W_{m_{\text{inner}}}}{I^2} = \frac{\mu_0}{8\pi}$$

Q_2

$$L_{\text{shield}} = \frac{2W_{m_{\text{shield}}}}{I^2} = \mu_0 \int_{r_2}^{r_3} \left(\frac{H}{I}\right)^2 2\pi r\, dr = \frac{\mu_0}{2\pi}\frac{1}{(r_3^2 - r_2^2)^2}\left(r_3^4 \ln\frac{r_3}{r_2} + \frac{r_3^4 - r_2^4}{4} - r_3^2(r_3^2 - r_2^2)\right)$$

When $r_3 - r_2 \ll r_2$, the following approximation results:

$$L_{\text{shield}} = \frac{\mu_0}{6\pi}\left(\frac{r_3}{r_2} - 1\right)$$

Q_3 $L_{\text{inner}} = 50\,\text{nH/m}\,(16\,\%L)$; $L_{\text{shield}} = 1.3\,\text{nH/m}(0.4\,\%L)$.

Problem 4.15.2

Figure 4.27 represents an aerial two-wire transmission line of indefinite length l, whose cylindrical conductors of radius r have their axes separated by a distance $2d$. A current I flows in both conductors.

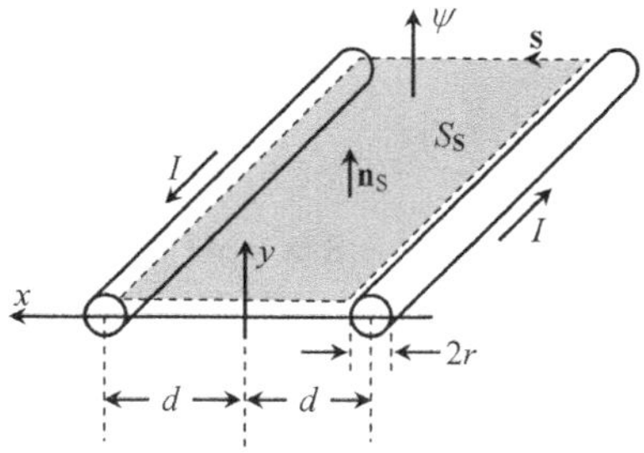

Figure 4.27 Two-wire transmission line. The flux linkage across the shaded rectangle is proportional to the line current, $\psi = LI$.

Q_1 By summing the contributions due to both conductors, find the **H**-field distribution outside the conductors, $H(x)$, in the horizontal plane $y = 0$.

Q_2 Determine the flux ψ linked with the two-wire line.
(Hint: Integrate **B** across a rectangular surface of length l lying on the horizontal plane – see Figure 4.27.)

Q_3 Based on the above result obtain the per-unit-length external inductance of the two-wire line.

Q_4 Numerical application. Take $I = 100\,\text{mA}$. Consider $r = 1\,\text{mm}$, $d = 5\,\text{mm}$.
Compute the **H**-field intensity at $x = 0$ and $x = \pm(d - r)$.
Compute the per-unit-length flux linkage.
Compute the per-unit-length external inductance of the line.
Compute the per-unit-length magnetic energy stored outside the line conductors.

Answers

Q_1

$$\mathbf{H} = H(x)\vec{e}_y;\ H(x) = \frac{I}{2\pi}\left(\frac{1}{d-x} + \frac{1}{d+x}\right)$$

Q_2

$$\psi = \int_{S_s} \mathbf{B}\cdot\mathbf{n}_S\, dS = \int_{S_s} B\, dS = \int_{r-d}^{d-r} \mu_0 H(x)\, l\, dx = 2\mu_0 l \int_0^{d-r} H(x)\, dx = I\left(\frac{\mu_0 l}{\pi}\ln\frac{2d-r}{r}\right)$$

Q_3

$$L = \frac{\psi/I}{l} = \frac{\mu_0}{\pi}\ln\frac{2d-r}{r}$$

This result can be further simplified if the thin-wire approximation ($r \ll d$) is employed, that is,

$$L = \frac{\mu_0}{\pi}\ln\frac{2d}{r}$$

Q_4

$$H(0) = \frac{I}{\pi d} = 6.4\,\text{A/m}; \quad H_{x=\pm(d-r)} = H(0) \times \frac{d^2}{r(2d-r)} = 17.7\,\text{A/m}$$

$$\psi = 87.9\,\text{nWb/m}; \quad L = 878.9\,\text{nH/m}; \quad W_m = 4.4\,\text{nJ/m}$$

Problem 4.15.3

Consider a system of two cylindrical thin wires of length l and radii r, separated by a distance $2d$, running at height h above a perfectly conducting ground plane (conductor 0) – see Figure 4.28. The material media above the conducting plane have non-magnetic characteristics ($\mu = \mu_0$). I_1 and I_2 are the currents in wires 1 and 2.

Data: $r = 1$ mm, $d = 1$ cm, $h = 1$ cm.

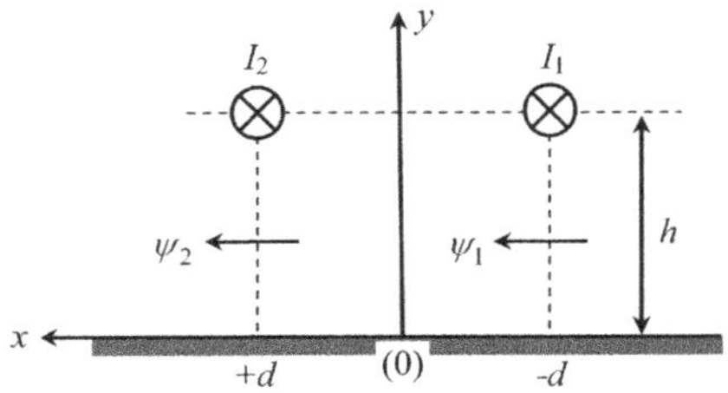

Figure 4.28 Two-wire transmission line above a perfectly conducting ground

Q_1 Take $I_1 \neq 0$ and $I_2 = 0$. Determine the x component of **H** at points belonging to the vertical planes $x = \pm d$ (for $y > 0$).

Q_2 Find the fluxes ψ_1 and ψ_2 linked with the two circuits (circuits 1/0 and 2/0).

Q_3 Determine the per-unit-length inductance matrix $[L]$ of the system.

Q_4 Find the magnetic coupling factor k between the circuits 1/0 and 2/0.

Q_5 Consider the so-called differential-mode or odd-mode operation, where $I_1 = -I_2 = I$.
Find the flux ψ linked with the circuit composed by the wires 1 and 2 and determine the corresponding per-unit-length inductance L_{odd}. Confirm the inductance result based on magnetic energy considerations. Discuss the proximity effect arising from the presence of the ground conductor.

(Hint: The analysis of this problem can be conducted using an equivalent four-conductor system, since the ground current magnetic effects can be accounted for by considering image currents of intensity $-I_1$ and $-I_2$, located at $(-d, -h)$ and $(d, -h)$ respectively.)

Answers

Q_1 At $x=-d$:

$$H_x(y)=\frac{I_1}{2\pi}\left(\frac{1}{h-y}+\frac{1}{h+y}\right)=H_1$$

At $x=+d$:

$$H_x(y)=\frac{I_1}{2\pi}\left(\frac{h-y}{4d^2+(h-y)^2}+\frac{h+y}{4d^2+(h+y)^2}\right)=H_2$$

Q_2

$$\psi_1(I_1)=\int\limits_{S_{s_1}}\mu_0\mathbf{H}\cdot\vec{e}_x dS=\mu_0 l\int\limits_0^{h-r}H_1\,dy=I_1\left(\frac{\mu_0 l}{2\pi}\ln\frac{2h}{r}\right)$$

$$\psi_2(I_1)=\int\limits_{S_{s_2}}\mu_0\mathbf{H}\cdot\vec{e}_x dS=\mu_0 l\int\limits_0^{h-r}H_2\,dy=I_1\left(\frac{\mu_0 l}{2\pi}\ln\sqrt{(h/d)^2+1}\right)$$

Note that the above results are valid provided that $r \ll d$ and $r \ll h$ (thin-wire approximation).

Q_3

$$L_{11}=\frac{1}{l}\left.\frac{\psi_1}{I_1}\right|_{I_2=0}=\frac{\mu_0}{2\pi}\ln\frac{2h}{r}=599.1\,\mathrm{nH/m}$$

$$L_{12}=L_{21}=L_M=\frac{1}{l}\left.\frac{\psi_2}{I_1}\right|_{I_2=0}=\frac{\mu_0}{2\pi}\ln\sqrt{(h/d)^2+1}=69.3\,\mathrm{nH/m}$$

$L_{22}=L_{11}$ (due to the geometrical symmetry of the configuration).

$$[L]=\begin{bmatrix}599.1 & 69.3\\ 69.3 & 599.1\end{bmatrix}\mathrm{nH/m}$$

Q_4 $k=0.12$.

Q_5 $\psi=\psi_1(I_1,I_2)-\psi_2(I_1,I_2)=2(L_{11}-L_M)I$, with $I_1=-I_2=I$ and $L_{11}=L_{22}$.

$$\psi=I\left(\frac{\mu_0 l}{\pi}\ln\frac{2h}{r\sqrt{(h/d)^2+1}}\right)$$

$$L_{\mathrm{odd}}=\frac{1}{l}\frac{\psi}{I}=\frac{\mu_0}{\pi}\ln\left(\frac{2d}{r}\times\frac{1}{\sqrt{1+(d/h)^2}}\right)=1.06\,\mu\mathrm{H/m}$$

$$W_m=\frac{1}{2}L_{11}I_1^2+L_M I_1 I_2+\frac{1}{2}L_{22}I_2^2=\frac{1}{2}\underbrace{(2(L_{11}-L_M))}_{L_{\mathrm{odd}}}I^2$$

$$L_{\text{odd}} = \frac{\mu_0}{\pi}\ln\left(\frac{2d}{r}\times\frac{1}{\sqrt{1+(d/h)^2}}\right) = L|_{h=\infty} - \underbrace{\frac{\mu_0}{2\pi}\ln\left(1+(d/h)^2\right)}_{\Delta L_{\text{proximity effect}}}$$

For $h \to \infty$,

$$L_{\text{odd}} = L|_{h=\infty} = \frac{\mu_0}{\pi}\ln\frac{2d}{r} = 1.20\,\mu\text{H/m}$$

(see Problem 4.15.2)

$$\Delta L_{\text{proximity effect}} = 0.14\,\mu\text{H/m}$$

As you bring the two-wire circuit closer to the ground plane the equivalent self-inductance of the circuit (L_{odd}) progressively diminishes.

Problem 4.15.4

Consider the conductor system analyzed in Problem 4.15.3, and focus your attention on the evaluation of magnetic forces. Assume that $I_1 = -I_2 = I = 1\,\text{A}$.

Q_1 Determine the horizontal component of the per-unit-length force exerted on each wire.

Q_2 Determine the vertical component of the per-unit-length force exerted on each wire.

Q_3 Determine the ratio of the above force components.

Answers

Q_1 Let conductor 2 remain at a fixed position, define $x = 2d$ as the horizontal separation between conductors 1 and 2, and allow conductor 1 to undergo a virtual displacement dx. Then

$$f_x = -\left(\frac{\partial W_m}{\partial x}\right)_\psi = -\frac{\partial}{\partial x}\left(\frac{1}{2}\frac{\psi^2}{L_{\text{odd}}}\right) = +\frac{1}{2}\frac{\psi^2}{L_{\text{odd}}^2}\frac{\partial L_{\text{odd}}}{\partial x} = \frac{I^2}{2}\frac{\partial}{\partial x}\left[\frac{\mu_0}{\pi}\ln\frac{x}{r} - \frac{\mu_0}{2\pi}\ln\left(1+\frac{x^2}{4h^2}\right)\right]$$

$$f_x = I^2\frac{\mu_0}{4\pi d}\times\frac{1}{1+(d/h)^2} = 5\,\mu\text{N/m}$$

The horizontal force is repulsive and actuates so as to increase the self inductance L_{odd} of the circuit made by wires 1 and 2.

Q_2 Define $y = 2h$ as the vertical separation between the wires and their corresponding images below the ground plane, and allow both wires to undergo a virtual displacement dy. Then

$$f_y = -\left(\frac{\partial W_m}{\partial y}\right)_\psi = -\frac{\partial}{\partial y}\left(\frac{1}{2}\frac{\psi^2}{L_{\text{odd}}}\right) = +\frac{1}{2}\frac{\psi^2}{L_{\text{odd}}^2}\frac{\partial L_{\text{odd}}}{\partial y} = \frac{I^2}{2}\frac{\partial}{\partial y}\left[\frac{\mu_0}{\pi}\ln\frac{2d}{r} - \frac{\mu_0}{2\pi}\ln\left(1+\frac{4d^2}{y^2}\right)\right]$$

$$f_y = I^2\frac{\mu_0}{4\pi h}\times\frac{1}{1+(h/d)^2} = 5\,\mu\text{N/m}$$

The vertical force is repulsive and also actuates so as to increase the self-inductance L_{odd} of the circuit made by wires 1 and 2.

Q_3 In the general case you have

$$\frac{f_x}{f_y} = \left(\frac{h}{d}\right)^3$$

In this problem, where $h = d$, you get $f_x = f_y$. Note that the wires and their images define a perfect square in a cross-sectional view, and, therefore, the equality $f_x = f_y$ is just a consequence of the system symmetry.

Problem 4.15.5

Consider a toroidal magnetic circuit with a square cross-section. The inner and outer radius of the toroid are $r_1 = 1$ cm and $r_2 = r_1 + a = 2$ cm (see Figure 4.29). The magnetic circuit is made of a linear homogeneous medium with relative permeability $\mu_r = 1000$. The magnetization current is I.

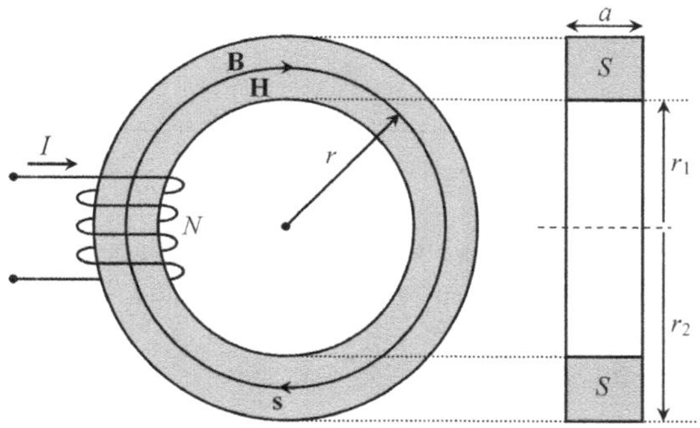

Figure 4.29 Toroidal magnetic circuit excited by a magnetization current I

Q_1 Determine analytically the magnetic reluctance R_m of the toroid.
(Hint: Take into account that, inside the toroid, **H**- and **B**-field lines are circumferences; the intensity of both fields decreases with increasing r.)

Q_2 Compute R_m numerically.

Q_3 Obtain an approximate result for the magnetic reluctance using $R_m = l/(\mu S)$ – see (4.24) – where l is the toroid's perimeter mean value. Compare with the exact result.

Answers

Q_1 From Ampère's law:

$$2\pi r \times H(r) = NI \rightarrow H(r) = \frac{NI}{2\pi r}; \quad B(r) = \mu_0 \mu_r \frac{I}{2\pi r} \text{ for } r_1 < r < r_2$$

The magnetic voltage measured along the toroid, that is along a circular field line of **H**, is

$$U_m = \oint_S \mathbf{H} \cdot d\mathbf{s} = NI$$

The magnetic flux through the toroid's cross-section S is

$$\phi = \int_S \mathbf{B} \cdot \mathbf{n}\, dS = \int_{r_1}^{r_2} B(r) a\, dr = NI \frac{\mu_0 \mu_r a}{2\pi} \ln \frac{r_2}{r_1}$$

Hence,

$$R_m = \frac{U_m}{\phi} = \frac{2\pi}{\mu_0 \mu_r a \ln(r_2/r_1)}$$

Q_2 $R_m = 721.35\,\text{kH}^{-1}$.

Q_3

$$R_m \approx \frac{2\pi(r_2 + r_1)/2}{\mu_0 \mu_r a^2} = 750\,\text{kH}^{-1}$$

(excess error of 4%).

Problem 4.15.6

Consider a magnetic circuit made of two identical toroidal cores whose magnetic reluctance was determined in the preceding problem. As shown in Figure 4.30, three coils with N_1, N_2 and N_3 turns are wound around the cores.

Data: $N_1 = 100$, $N_2 = 50$, $N_3 = 50$.

Q_1 Determine the magnetic fluxes in the cores ϕ_1 and ϕ_2 as a function of I_1, I_2 and I_3.

Q_2 Determine the fluxes linked with the coils ψ_1, ψ_2 and ψ_3.

Q_3 Based on the above results, find the inductance matrix of the system $[L]$.

Q_4 Assume $I_1 = 1\,\text{A}$. Determine I_2 and I_3 such that $\phi_1 = 0.1\,\text{mWb}$ and $\phi_2 = 0$.
Compute ψ_1, ψ_2 and ψ_3, and then obtain the magnetic energy stored, specifying its distribution in space.

Answers

Q_1

$$\begin{cases} R_m \phi_1 = \dfrac{N_1}{2} I_1 + N_2 I_2 \\ R_m \phi_2 = N_3 I_3 - \dfrac{N_1}{2} I_1 \end{cases}$$

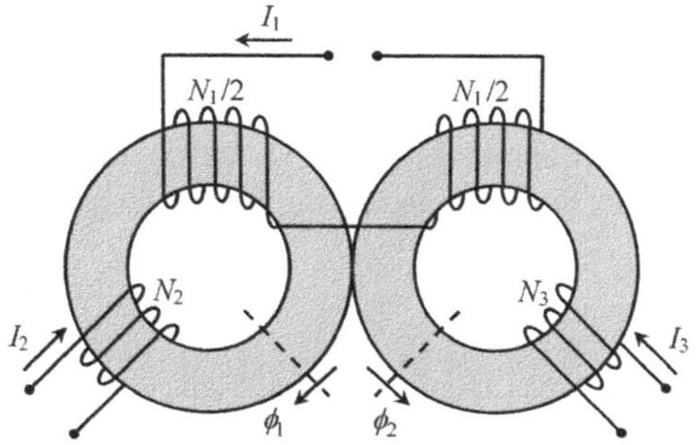

Figure 4.30 Magnetic circuit with two identical toroidal cores driven by a set of three currents, I_1, I_2 and I_3

Q_2

$$\psi_1 = \frac{N_1}{2}(\phi_1 - \phi_2) = \underbrace{\left(\frac{N_1^2}{2R_m}\right)}_{L_{11}} I_1 + \underbrace{\left(\frac{N_1 N_2}{2R_m}\right)}_{L_{12}} I_2 + \underbrace{\left(-\frac{N_1 N_3}{2R_m}\right)}_{L_{13}} I_3$$

$$\psi_2 = N_2\phi_1 = \underbrace{\left(\frac{N_1 N_2}{2R_m}\right)}_{L_{21}} I_1 + \underbrace{\left(\frac{N_2^2}{R_m}\right)}_{L_{22}} I_2 \text{ and } \psi_3 = N_3\phi_2 = \underbrace{\left(-\frac{N_1 N_3}{2R_m}\right)}_{L_{31}} I_1 + \underbrace{\left(\frac{N_3^2}{R_m}\right)}_{L_{33}} I_3$$

Q_3

$$L_{11} = 2L_{12} = -2L_{13} = 2L_{22} = 2L_{33}; \ [L] = \begin{bmatrix} 6.93 & 3.46 & -3.46 \\ 3.46 & 3.46 & 0 \\ -3.46 & 0 & 3.46 \end{bmatrix} \text{mH}$$

Q_4

$$I_2 = \frac{R_m\phi_1 - N_1 I_1/2}{N_2} = 0.443\,\text{A}; \ I_3 = \frac{N_1 I_1}{2N_3} = 1\,\text{A}$$

$$\phi_2 = 0 \rightarrow \begin{cases} \psi_1 = \dfrac{N_1}{2}\phi_1 = \psi_2 = N_2\phi_1 = 5\,\text{mWb} \\ \psi_3 = 0 \end{cases}$$

$$W_m = \sum_{k=1}^{3} \frac{\psi_k I_k}{2} = 3.61\,\text{mJ}$$

All the energy is stored in core 1, since in core 2 the fields **B** and **H** are zero:

$$W_m = \int_{V_{\text{core 1}}} \frac{BH}{2}\,dV = \frac{1}{2}R_m\,\phi_1^2 = 3.61\,\text{mJ}$$

Problem 4.15.7

The two magnetically coupled coils in Figure 4.31 are characterized by known values of their self-inductances L_{11} and L_{22}. The magnetic coupling factor k is unknown.

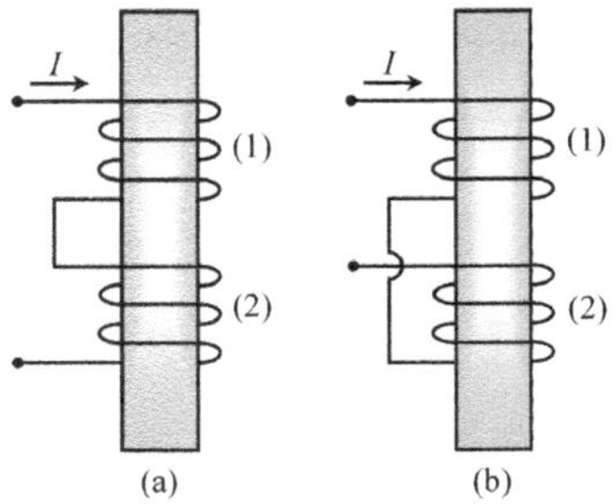

Figure 4.31 Two magnetically coupled coils. (a) Concordant coupling ($L_M > 0$). (b) Discordant coupling ($L_M < 0$)

Q_1 Assume that the coils are connected in series as shown in Figure 4.31(a). Determine the equivalent inductance of the association L_a.

Q_2 Assume that the coils are connected in series as shown in Figure 4.31(b). Determine the equivalent inductance of the association L_b.

Q_3 Show that the coupling factor can be computed as

$$k = \frac{L_a - L_b}{4\sqrt{L_{11}L_{22}}}$$

Answers

Q_1 For the case of concordant coupling ($L_M > 0$):

$$L_a = \frac{\psi_1 + \psi_2}{I} = \frac{(L_{11} + |L_M|)\,I + (L_{22} + |L_M|)\,I}{I} = L_{11} + L_{22} + 2\,|L_M|$$

Q_2 For the case of discordant coupling ($L_M < 0$):

$$L_b = \frac{\psi_1 + \psi_2}{I} = \frac{(L_{11} - |L_M|)\,I + (L_{22} - |L_M|)\,I}{I} = L_{11} + L_{22} - 2\,|L_M|$$

Q_3 $L_a - L_b = 4\,|L_M| = 4k\sqrt{L_{11}L_{22}}$.

Problem 4.15.8

Consider the magnetic circuit represented in Figure 4.32, where, for simplification purposes, the iron parts are assumed to have zero magnetic reluctance ($\mu_{Fe} \rightarrow \infty$). The two air gaps have thickness $\delta = 1\,\text{mm}$. A current I flows in a coil with 1500 turns; this coil is constituted by three series-connected windings as shown in the figure.

Data: $S_3 = 3.183\,\text{cm}^2$; $S_1 = S_2 = S_3/2$; $N_1 = N_2 = N_3 = N = 500$; $I = 1\,\text{A}$. Neglect dispersion phenomena.

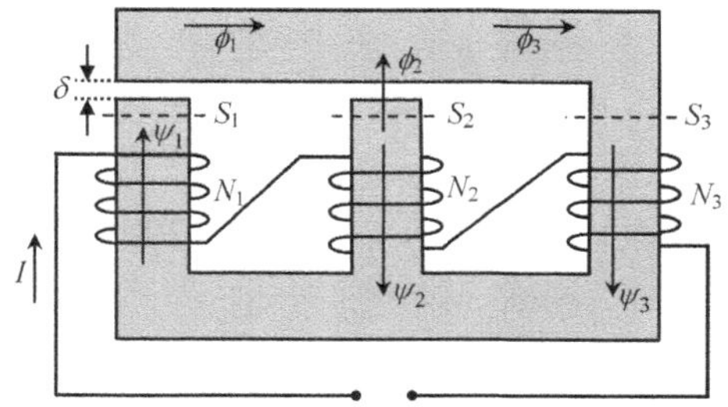

Figure 4.32 Magnetic circuit with a coil made of three series-connected windings

Q_1 Find the magnetic reluctance R_m of each air gap.

Q_2 Obtain the magnetic fluxes ϕ_1, ϕ_2 and ϕ_3.

Q_3 Determine the intensities of the **B** and **H** fields in the air gaps and also in the iron part corresponding to the winding with N_3 turns. Obtain the magnetic energy stored in the circuit.

Q_4 Determine the flux ψ linked to the coil with 1500 turns and find its inductance L.

Q_5 Confirm the above result based on energy considerations.

Answers

Q_1

$$R_{m_1} = R_{m_2} = R_m = \frac{\delta}{\mu_0 S} = 5 \times 10^6\,\text{H}^{-1} \quad (S = S_1 = S_2)$$

Q_2

$$\begin{cases} R_m\phi_1 - R_m\phi_2 = N_1 I + N_2 I \\ R_m\phi_2 = N_3 I - N_2 I \\ \phi_3 - \phi_2 - \phi_1 = 0 \end{cases} \rightarrow \begin{cases} \phi_1 = \phi_3 = 0.2\,\text{mWb} \\ \phi_2 = 0 \end{cases}$$

Q_3 Air gap 1:

$$\begin{cases} B_1 = \phi_1/S_1 = 1.257\,\text{T} \\ H_1 = B_1/\mu_0 = 10^6\,\text{A/m} \end{cases}$$

Air gap 2:

$$\begin{cases} B_2 = \phi_2/S_2 = 0 \\ H_2 = B_2/\mu_0 = 0 \end{cases}$$

Iron (part 3):

$$\begin{cases} B_3 = \phi_3/S_3 = 0.628\,\mathrm{T} \\ H_3 = B_3/\mu_{\mathrm{Fe}} = 0 \end{cases}$$

Fields **B** and **H** are not simultaneously zero only at air gap 1; all the magnetic energy is concentrated there,

$$W_m = \frac{B_1 H_1}{2} \times \underbrace{(S_1 \delta)}_{V_1} = 100\,\mathrm{mJ}$$

Q_4 $\psi = \psi_1 + \psi_2 + \psi_3 = (N_1\phi_1) + (-N_2\phi_2) + (N_3\phi_3) = 200\,\mathrm{mWb}$; $L = \psi/I = 200\,\mathrm{mH}$.

Q_5

$$W_m = \frac{1}{2}LI^2 \rightarrow L = \frac{2W_m}{I^2} = 200\,\mathrm{mH}$$

Problem 4.15.9

A cylindrical solenoid driven by a current I contains in its interior a ferromagnetic cylinder core that can run along the solenoid x axis – see the longitudinal cross-section depicted in Figure 4.33.

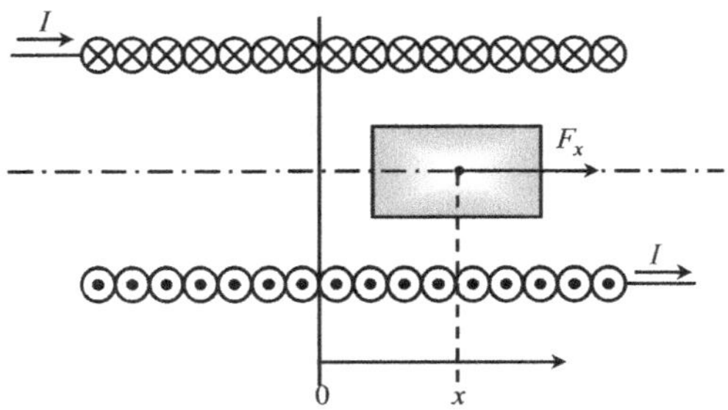

Figure 4.33 Longitudinal cross-section of a cylindrical solenoid, containing a ferromagnetic core (shaded part) that can freely move along the x axis

Q_1 The solenoid's inductance as a function of the position of the ferromagnetic core can be approximated by $L(x) \approx L_0 - \alpha x^2$, for small values of x. Is α positive or negative?

Q_2 Determine the magnetic force F_x exerted on the core. Check that it is positive for $x < 0$ but negative for $x > 0$.

Q_3 Taking into account that the mass of the core is M, determine the mechanical equation that governs the movement of the core along x (neglect friction phenomena). What type of movement will you get? What would it happen if a small friction factor were present?

Answers

Q_1 $\alpha > 0$.

Q_2

$$F_x = -\left(\frac{\partial W_m}{\partial x}\right)_\psi = \frac{1}{2}I^2\frac{dL}{dx} = -\alpha I^2 x$$

Note that the force acts so as to bring the core to its central position inside the solenoid, irrespective of the algebraic sign of I.

Q_3

$$M\frac{d^2x}{dt^2} + \alpha I^2 x = 0$$

The solution to the above equation is an oscillatory movement described by

$$x(t) = X\cos(\omega t + \theta)$$

where the oscillation's angular frequency ω is controlled by the solenoid current I, $\omega = I\sqrt{\alpha/M}$. The particular values of X and θ depend on the particular initial conditions of the problem (initial position and initial velocity of the core).

If friction is present you will get a damped oscillation which ultimately brings the core to rest at its central position $x = 0$.

Problem 4.15.10

The results in (4.50)–(4.52) allow the magnetic forces and torques to be obtained from the negative derivative of the magnetic energy at constant flux.

Q_1 How would you compute F and T under constant current conditions?

Q_2 Particularize the result for linear systems.

Answers

Q_1

$$dW_G = i\,d\psi, \quad dW_{mec} = f_x dx.$$

$$dW_G = dW_m + dW_{mec} \rightarrow dW_m = i\,d\psi - f_x\,dx.$$

As shown in Figure 4.34, the areas above and below the curve $\psi(i)$ are not identical; while the first represents the magnetic energy W_m, the second defines the so-called magnetic co-energy $W'_m = i\psi - W_m$.

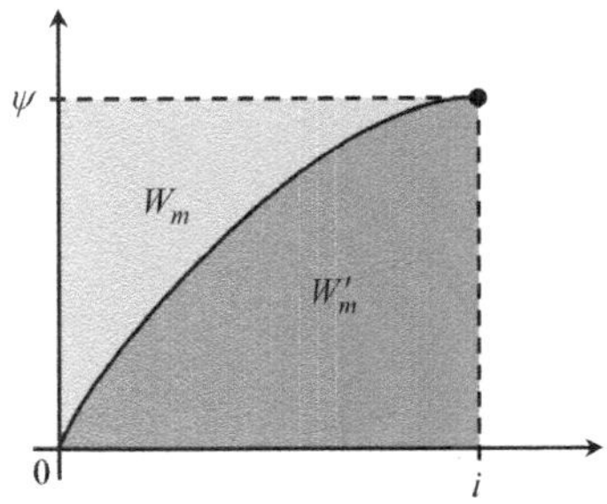

Figure 4.34 Magnetic energy and magnetic co-energy of a nonlinear system

Differentiating W'_m we find

$$dW'_m = i\,d\psi + \psi\,di - dW_m = \psi\,di + f_x\,dx \tag{4.57a}$$

Let W'_m be a function of the independent variables i and x, $W'_m(i, x)$; therefore

$$dW'_m = \left(\frac{\partial W'_m}{\partial i}\right)_x di + \left(\frac{\partial W'_m}{\partial x}\right)_i dx \tag{4.57b}$$

By comparing the results in (4.57) we get

$$f_x = +\left(\frac{\partial W'_m}{\partial x}\right)_i \tag{4.58}$$

Torque evaluation follows the same rationale:

$$T_\alpha = +\left(\frac{\partial W'_m}{\partial \alpha}\right)_i$$

Q_2 For linear systems, where $W'_m = W_m$, we naturally obtain

$$f_x = +\left(\frac{\partial W_m}{\partial x}\right)_i \text{ and } T_\alpha = +\left(\frac{\partial W_m}{\partial \alpha}\right)_i$$

Part III

Slow Time-Varying Fields

Introduction

For time-varying electromagnetic phenomena the time variable t is ubiquitous, hence you need to consider the original Maxwell's equations exactly as they are:

$$\begin{cases} \text{curl } \mathbf{E} = -\dfrac{\partial \mathbf{B}}{\partial t} \\ \text{div } \mathbf{B} = 0 \\ \text{curl } \mathbf{H} = \mathbf{J} + \dfrac{\partial \mathbf{D}}{\partial t} \\ \text{div } \mathbf{D} = \rho \end{cases} \tag{PIII.1}$$

Does this mean that the subjects treated in Part II have the recycle bin as their destiny? Well, it depends on how fast or slow the fields vary with time.

It makes sense that for slow time-varying fields – the so-called quasi-stationary regime – most of the results obtained in Part II should hold true and provide very good approximations. The strategy ordinarily used to deal with quasi-stationary regimes consists of treating the capacitive effects associated with $\partial \mathbf{D}/\partial t$ (electric induction phenomena) separately from the inductive effects associated with $\partial \mathbf{B}/\partial t$ (magnetic induction phenomena).

Before we can continue, a crucial question needs to be clarified. What is a slow time-varying regime? How do you draw the line between slow and fast fields?

Although we are going to prove this in Part IV, you probably have already heard that electromagnetic waves propagate in free space with a velocity $c = 3 \times 10^8$ m/s. In the particular case of space–time sinusoidal waves, the time periodicity and the space periodicity are characterized, respectively, by the period T and the wavelength λ, these two parameters being correlated through $\lambda = cT$.

Consider a sinusoidal voltage generator connected to a load through a pair of perfectly conducting wires of length l. Let u_G and i_G be the generator voltage and current; likewise let u_L and i_L be the load voltage and current. Voltages u_G and u_L share the same shape and the same period T, but, in general, they do not coincide with each other, $u_G(t) \neq u_L(t)$, because, in fact, the electromagnetic wave originating at the generator site takes some time to reach the distant load; the corresponding time delay (or propagation time) is given by $\tau = l/c$.

The same argument applies equally to the generator and load currents, that is $i_G(t) \neq i_L(t)$.

You can say that the regime is a quasi-stationary regime whenever the delay time is negligibly small compared to T, in which case you can employ the very good approximations $u_G = u_L$ and $i_G = i_L$. Note, in addition, that the inequality $\tau \ll T$ can be rewritten as

$$\underbrace{(l/c)}_{\tau} \ll \underbrace{(\lambda/c)}_{T} \rightarrow l \ll \lambda \tag{PIII.2}$$

In short, quasi-stationary regimes are those for which the length of the structure under analysis is much shorter than the lowest wavelength characterizing the field dynamics. If the opposite happens then you will be dealing with a rapid time-varying phenomenon. This is how you draw a line between the problems we are going to analyze in Part III and Part IV.

Part III is subdivided into three chapters. Chapter 5 is concerned with magnetic induction phenomena where electric fields **E** are produced by **B** fields originated by time-varying conduction currents $\mathbf{J}(t)$. In Chapter 5 we assume that the magnitude of $\partial\mathbf{D}/\partial t$ is small compared to the magnitude of **J** (neglecting of capacitive effects). Hence, the key equations for Chapter 5 are

$$\begin{cases} \text{curl } \mathbf{E} = -\dfrac{\partial \mathbf{B}}{\partial t} \\ \text{div } \mathbf{B} = 0 \\ \text{curl } \mathbf{H} \approx \mathbf{J} \end{cases} \tag{PIII.3}$$

Chapter 6 is concerned with electric induction phenomena, where magnetic fields **H** are produced by **E** fields originated by time-varying charge distributions $\rho(t)$. Magnetic induction phenomena from Chapter 5 are assumed to be negligibly small. Key equations for Chapter 6 are

$$\begin{cases} \text{curl } \mathbf{H} = \mathbf{J} + \dfrac{\partial \mathbf{D}}{\partial t} \\ \text{div } \mathbf{D} = \rho \\ \text{curl } \mathbf{E} \approx 0 \end{cases} \tag{PIII.4}$$

Chapter 7 deals, to a great extent, with electrical engineering applications of the theoretical results presented in Chapters 5 and 6. There we will address the steady-state harmonic regime and the transient regime for circuit analysis using the standard lumped parameters approach, which applies to slow time-varying phenomena.

A final word is in order concerning the notation employed in the analysis of time-varying phenomena: While in Part II upper-case italic symbols have been used to denote stationary quantities, from now on we will utilize lower-case italic symbols for time-varying quantities.

5

Magnetic Induction Phenomena

5.1 Fundamental Equations

The fundamental laws governing magnetic induction problems are those in (PIII.3)

$$\begin{cases} \text{curl } \mathbf{E} = -\dfrac{\partial \mathbf{B}}{\partial t} \\ \text{div } \mathbf{B} = 0 \\ \text{curl } \mathbf{H} \approx \mathbf{J} \end{cases} \tag{5.1}$$

together with the constitutive relations concerning $\mathbf{B}(\mathbf{H})$ and $\mathbf{J}(\mathbf{E})$.

In order to spur on your interest in this new topic, we will point out just two very simple situations that cannot be justified in the framework of stationary fields, and that require the above equations for their correct interpretation. Here, you will find cases where closed circuits, containing no generators whatsoever, can have currents flowing in them. Conversely, you also will find cases where open circuits can display voltages across their terminals despite the absence of generators.

5.2 Gradient and Induction Electric Fields, Potential Vector

As you know, for purely stationary phenomena, the electric field vector is a gradient field originated by electric charges. From Part II, we found $\mathbf{E} = \mathbf{E}_g$, with

$$\begin{gathered} \text{curl } \mathbf{E}_g = 0 \rightarrow \mathbf{E}_g = -\text{grad } V \\ \text{div } \mathbf{E}_g = \rho/\varepsilon \end{gathered} \tag{5.2}$$

For time-varying phenomena we necessarily have $\mathbf{E} \neq \mathbf{E}_g$. Yet, it may be useful to consider the electric field vector to be broken down into two different contributions, the first of which is the gradient electric field, and the second is a new contribution, termed the electric induction field $\mathbf{E}_i$:

$$\mathbf{E} = \mathbf{E}_g + \mathbf{E}_i \tag{5.3}$$

Electromagnetic Foundations of Electrical Engineering J. A. Brandão Faria

The properties of $\mathbf{E}_i$ can be obtained simply by substituting (5.3) into Maxwell's equations for homogeneous media:

$$\text{curl } \mathbf{E} = \text{curl } \mathbf{E}_g + \text{curl } \mathbf{E}_i = -\partial \mathbf{B}/\partial t$$
$$\text{div } \mathbf{E} = \text{div } \mathbf{E}_g + \text{div } \mathbf{E}_i = \rho/\varepsilon$$

By taking (5.2) into account we obtain for the electric induction field

$$\begin{cases} \text{curl } \mathbf{E}_i = -\partial \mathbf{B}/\partial t \\ \text{div } \mathbf{E}_i = 0 \end{cases} \tag{5.4}$$

from which we can see that the field lines of $\mathbf{E}_i$ are closed, embracing field lines of time-varying $\mathbf{B}$ – see Figure 5.1. If $\mathbf{B}$ happens to be a stationary field then $\mathbf{E}_i$ no longer exists.

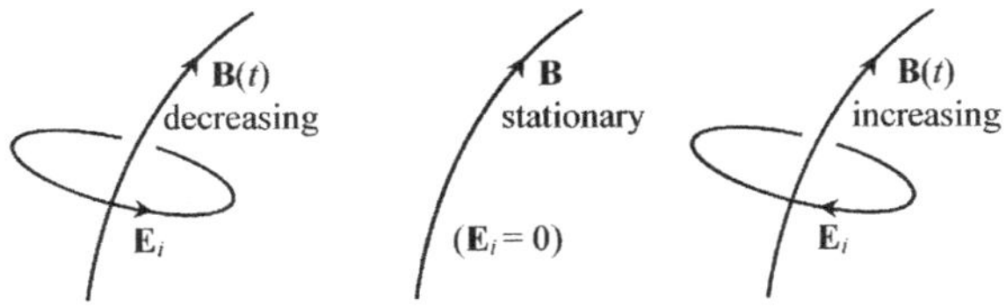

Figure 5.1 Time-varying **B** fields give rise to electric induction $\mathbf{E}_i$ fields

At this point, we recommend that you look again at div $\mathbf{B} = 0$. From vector calculus we know that div curl $\equiv 0$, and therefore we are allowed to define $\mathbf{B}$ as the *curl* of an auxiliary vector function $\mathbf{A}$ (with the same degree of arbitrariness mentioned earlier when the potential function V was introduced in Chapter 2):

$$\mathbf{B} = \text{curl } \mathbf{A} \tag{5.5}$$

Function $\mathbf{A}$, just introduced, is known by the name of potential vector (units: Wb/m, weber per meter). Substituting (5.5) into (5.4), you readily get $\mathbf{E}_i = -\partial \mathbf{A}/\partial t$. This allows you to express the electric field vector as the sum of two contributions where both the old scalar potential and the new potential vector appear:

$$\mathbf{E} = -\left(\text{grad } V + \frac{\partial \mathbf{A}}{\partial t}\right) \tag{5.6}$$

5.3 Revisiting the Voltage Concept

The aspect we are now going to address is new and critical, because it goes against your intuition. So, please pay attention. If you have two ideal voltmeters V_1 and V_2 connected between the same pair of terminals a and b – see Figure 5.2 – and if you are asked about the relationship between their readings, what will your answer be?

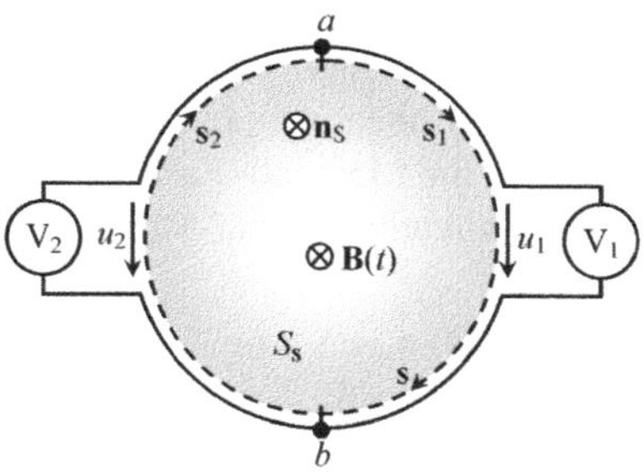

Figure 5.2 For time-varying regimes, $u_1(t) \neq u_2(t) \neq (V_a - V_b)$

Most of you will say that the readings are the same. But that may be well wrong!

Let us go back to the original definition of voltage in (1.7) and replace **E** with its new definition in (5.6):

$$u = \int_{\overrightarrow{ab}} \mathbf{E} \cdot d\mathbf{s} = -\int_{\overrightarrow{ab}} \text{grad } V \cdot d\mathbf{s} - \frac{d}{dt}\int_{\overrightarrow{ab}} \mathbf{A} \cdot d\mathbf{s} = (V_a - V_b) - \frac{d}{dt}\int_{\overrightarrow{ab}} \mathbf{A} \cdot d\mathbf{s} \tag{5.7}$$

Since **A** is not a conservative field (curl $\mathbf{A} \neq 0$), the evaluation of

$$\int_{\overrightarrow{ab}} \mathbf{A} \cdot d\mathbf{s}$$

depends, in general, on the particular path going from a to b. This shows not only that voltage and potential difference are quite different things, $u \neq V_a - V_b$, but also that the evaluation of u requires a clear specification of the integration path between the two points a and b. Then, from (5.7), considering the voltages in Figure 5.2, we have

$$u_1 = (V_a - V_b) - \frac{d}{dt}\int_{\overrightarrow{(ab)}_{s_1}} \mathbf{A} \cdot d\mathbf{s}; \; u_2 = (V_a - V_b) + \frac{d}{dt}\int_{\overrightarrow{(ba)}_{s_2}} \mathbf{A} \cdot d\mathbf{s}$$

By subtracting the preceding results we obtain the difference of the two voltmeters' readings:

$$u_1 - u_2 = -\frac{d}{dt}\left(\int_{\overrightarrow{(ab)}_{s_1}} \mathbf{A} \cdot d\mathbf{s} + \int_{\overrightarrow{(ba)}_{s_2}} \mathbf{A} \cdot d\mathbf{s}\right) = -\frac{d}{dt}\oint_{\mathbf{s}} \mathbf{A} \cdot d\mathbf{s} \tag{5.8}$$

where **s**, the reunion of the subpaths $\mathbf{s}_1$ and $\mathbf{s}_2$, is a closed clockwise-oriented path.

If **A** is null ($\mathbf{B} = 0$) or time invariant then we will get $u_1 = u_2$.

5.4 Induction Law

The cornerstone of magnetic induction phenomena, which allows us to evaluate voltages arising from time-varying **B**-fields, is the Maxwell–Faraday induction law.

From curl $\mathbf{E} = -\partial\mathbf{B}/\partial t$, using the already familiar Stokes theorem, we get

$$\oint_{\mathbf{s}} \mathbf{E} \cdot d\mathbf{s} = -\frac{d}{dt}\int_{S_s} \mathbf{B} \cdot \mathbf{n}_S \, dS \quad \text{or} \quad \oint_{\mathbf{s}} \mathbf{E} \cdot d\mathbf{s} = -\frac{d\psi_S}{dt} \tag{5.9}$$

where, as defined in (4.28), ψ_S is the magnetic flux linked with the circulation path **s**. In (5.9), the unit vector $\mathbf{n}_S$ normal to S_s is oriented according to the Stokes rule (right-hand screw rule).

The induction law in (5.9) simply states that the electromotive force induced along a closed path is equal to the negative derivative of the magnetic flux linked with that path.

The question posed in Figure 5.2, concerning the voltmeters' readings, can be reanswered (avoiding the use of the potential vector **A**) by making use of the induction law:

$$u_1 - u_2 = -\frac{d}{dt}\int_{S_s} \mathbf{B} \cdot \mathbf{n}_S \, dS = -\frac{d\psi_S}{dt}$$

As long as the open surface S_s having the path **s** as its bounding edge is traversed by a time-varying magnetic field, voltages u_1 and u_2 will be different.

5.5 Application Example (Magnetic Noise Effects)

Figure 5.3 shows an indoor electrical socket where a two-wire line is plugged in; the line is left open ($i = 0$) at the opposite end.

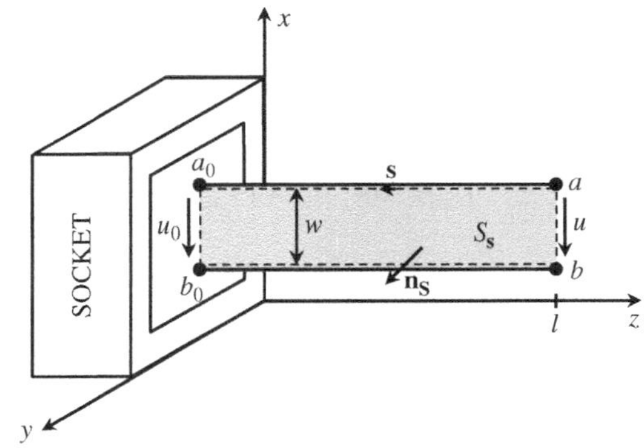

Figure 5.3 The relationship between voltages u_0 and u depends on the magnetic noise interference produced by externally produced **B** fields in the region under analysis

Assume that the two-wire line is longitudinally stretched so as to define a rectangular path of length l and width w, lying in the zx plane. Assume also that the socket voltage is $u_0(t) = U_0 \cos(\omega_0 t)$ and consider electronic equipment placed somewhere in the neighborhood that gives rise, in the region of the line, to a uniform magnetic field noise described by $B(t) = B_M \cos(\omega_B t)$.

Questions

Q_1 Determine $u(t)$ assuming that $\mathbf{B} = B(t)\vec{e}_x$.

Q_2 Determine $u(t)$ assuming that $\mathbf{B} = B(t)\vec{e}_y$.

Q_3 Describe a simple solution to mitigate the magnetic noise effects.

Solutions

Q_1 By applying the induction law to the closed path $\overrightarrow{aa_0b_0ba}$ we find, noting that $\mathbf{B} \perp \mathbf{n}_S$,

$$\underbrace{\oint_S \mathbf{E} \cdot d\mathbf{s}}_{u_0 - u} = -\frac{d}{dt} \underbrace{\int_{S_s} \mathbf{B} \cdot \vec{e}_y dS}_{0}$$

from which we obtain $u(t) = u_0(t) = U_0 \cos(\omega_0 t)$.

Q_2 By applying the induction law to the closed path $\overrightarrow{aa_0b_0ba}$ we find, noting that $\mathbf{B} \parallel \mathbf{n}_S$,

$$\underbrace{\oint_S \mathbf{E} \cdot d\mathbf{s}}_{u_0 - u} = -\frac{d}{dt} \underbrace{\int_{S_s} \mathbf{B} \cdot \vec{e}_y dS}_{Bwl}$$

from which we obtain $u(t) = U_0 \cos(\omega_0 t) - wl\omega_B B_M \sin(\omega_B t)$.

Q_3 If the two-wire line is twisted at regular intervals the Stokes normal $\mathbf{n}_S$ concomitantly switches from $+\vec{e}_y$ to $-\vec{e}_y$. If an odd number of twists are used then we will find $u \approx u_0$ provided that the perturbing magnetic field remains uniform in the region of the line.

5.6 Voltages and Currents in Magnetically Multicoupled Systems

We have already introduced in Section 4.12 the problem of magnetically coupled linear circuits, where self- and mutual inductances were defined. Here we elaborate on that problem considering time-varying currents.

In Figure 5.4 we show a ferromagnetic core provided with two coils. One is connected to a generator and the other is left open. Coil 1, with N_1 turns, has an internal resistance R_1; likewise coil 2, with N_2 turns, has an internal resistance R_2. Magnetic coupling between the coils is ensured through the magnetic field lines in the core that simultaneously embrace both coils. Current $i_1(t)$ flowing in coil 1 is given in the form of a triangular impulse (Figure 5.5(a)) of peak amplitude I_M.

To simplify matters let us assume that dispersion phenomena are negligible; that is, the magnetic flux in the core $\phi(t)$ is the same at every cross-section.

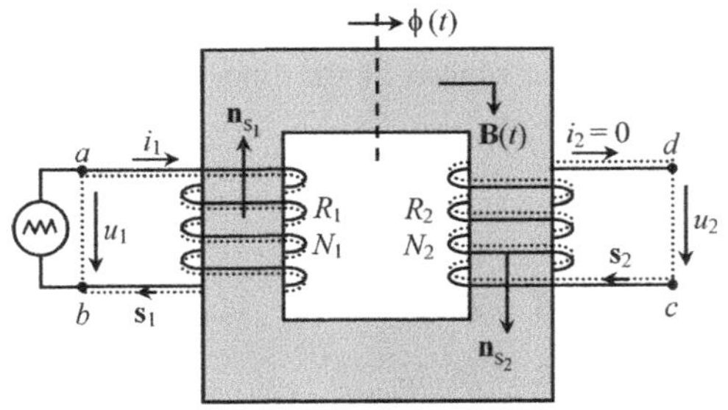

Figure 5.4 Circulation paths $\mathbf{s}_1$ and $\mathbf{s}_2$ used for the application of the induction law in order to evaluate the voltages u_1 and u_2 at the terminals of two magnetically coupled coils

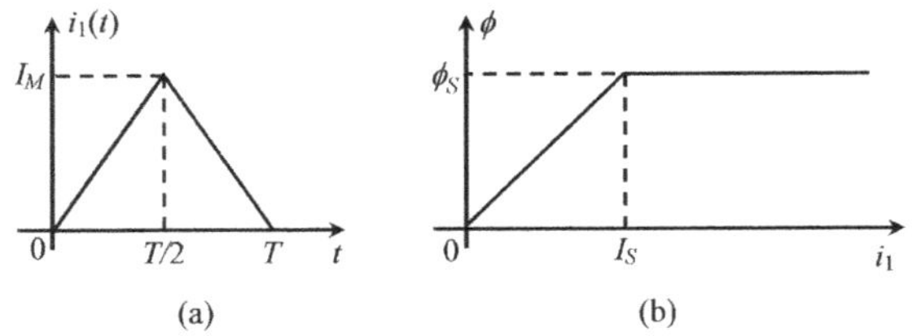

Figure 5.5 (a) Generator current (triangular pulse). (b) Stylized nonlinear magnetic characteristic $\phi(i_1)$ of the ferromagnetic core

Irrespective of the linear or nonlinear character of the core, the determination of ϕ as a function of i_1 is done by application of Ampère's law examined in Chapter 4.

Let us consider that the magnetic characteristic $\phi = \phi(i_1)$ is nonlinear, as stylized in Figure 5.5(b). The knee point $(I_S, \ \phi_S)$ defines an abrupt transition between the linear zone and the saturation zone.

In order to determine the generator voltage u_1 we apply the induction law to the closed circulation path $\mathbf{s}_1$, passing inside the coil conducting turns and through the generator. The path is oriented according to the prescribed reference direction of i_1.

For the sake of clarity, let us determine separately the left- and right-hand sides of the induction law,

$$\oint_{\mathbf{s}_1} \mathbf{E} \cdot d\mathbf{s} = \int_{\substack{\widehat{ab} \\ \text{coil}}} \frac{1}{\sigma_1} \mathbf{J}_1 \cdot d\mathbf{s} + \int_{\substack{\widehat{ba} \\ \text{(G)}}} \mathbf{E} \cdot d\mathbf{s} = +R_1 i_1 - u_1 \tag{5.10a}$$

$$-\frac{d}{dt} \int_{S_{s_1}} \mathbf{B} \cdot \mathbf{n}_{S_1} \, dS = -\frac{d}{dt} \psi_1 = -N_1 \frac{d\phi}{dt} \tag{5.10b}$$

Equating the results in (5.10), we find

$$u_1(t) = R_1 i_1(t) + N_1 \frac{d\phi(t)}{dt} \tag{5.11}$$

Now, let us turn our attention to coil 2 which, remember, carries no current. Nevertheless, as you will see, a voltage u_2 is going to appear across its terminals!

To determine voltage u_2 at the open terminals of coil 2 we reapply the induction law to the closed circulation path $\mathbf{s}_2$, passing inside the coil conducting turns and through the air. The path is oriented according to the prescribed reference direction of i_2. Following the same procedure as before, we find

$$\oint_{\mathbf{s}_2} \mathbf{E} \cdot d\mathbf{s} = \underset{\text{coil}}{\int_{\widehat{cd}}} \frac{1}{\sigma_2} \mathbf{J}_2 \cdot d\mathbf{s} + \underset{\text{air}}{\int_{\widehat{dc}}} \mathbf{E} \cdot d\mathbf{s} = 0 + u_2 \tag{5.12a}$$

$$-\frac{d}{dt} \int_{S_{s_2}} \mathbf{B} \cdot \mathbf{n}_{S_2}\, dS = -\frac{d}{dt} \psi_2 = -N_2 \frac{d\phi}{dt} \tag{5.12b}$$

Note that, in (5.12a), the integration $\widehat{cd}$ along the coil's conductor gives zero because we have assumed $i_2 = 0$, otherwise we would obtain $R_2 i_2$.

Equating the results in (5.12), we find

$$u_2(t) = -N_2 \frac{d\phi(t)}{dt} \tag{5.13}$$

from which you can see that, although coil 2 carries no current, a voltage will appear at its terminals, as a result of induction phenomena.

Next, we analyze the results in (5.11) and (5.13), presenting some pertinent graphics and interpreting from a physical point of view the reason why u_2 exists (despite $i_2 = 0$).

To start with, consider the simple case where the core behaves linearly (that is, $i_1(t) \leq I_S$). Take for instance $I_M = I_S$. Combining the information conveyed in Figure 5.5, you will obtain for $\phi(t)$ a triangular function, with peak value ϕ_S, similar to the one describing $i_1(t)$.

The derivative $d\phi/dt$ appearing in (5.11) and (5.13) is evaluated as

$$\frac{d\phi}{dt} = \begin{cases} +2\phi_S/T; & \text{for } 0 < t < T/2 \\ -2\phi_S/T; & \text{for } T/2 < t < T \end{cases}$$

The evolution of $u_1(t)$ is obtained by summing the resistive voltage component $u_R(t) = R_1 i_1(t)$ with the inductive voltage component $u_L(t) = N_1\, d\phi/dt$. The graphical construction leading to $u_1(t)$ is illustrated in Figure 5.6.

Things are a bit simpler for $u_2(t)$ since the resistive component is absent – see Figure 5.7.

Let us now interpret the results obtained for $u_2(t)$. For this purpose it is helpful to visualize a cross-section of the core leg where coil 2 is wound around (Figure 5.8).

In the time interval 0 to $T/2$ the field $B(t) = \phi(t)/S$ is time increasing. The associated field lines of the electric induction field $\mathbf{E}_i$ are circumferences oriented anticlockwise; this field actuates on the free charged particles of coil 2 ($\mathbf{F}_e = q\mathbf{E}_i$) giving rise to a distribution of positive charges at terminal c and negative charges at terminal d. A gradient electric field $\mathbf{E}_g$ oriented from c to d appears and a negative pulse voltage is revealed in $u_2(t)$.

A similar rationale applies to the time interval from $T/2$ to T during which the field $\mathbf{B}(t)$ is time decreasing. The associated field lines of the electric induction field $\mathbf{E}_i$ are

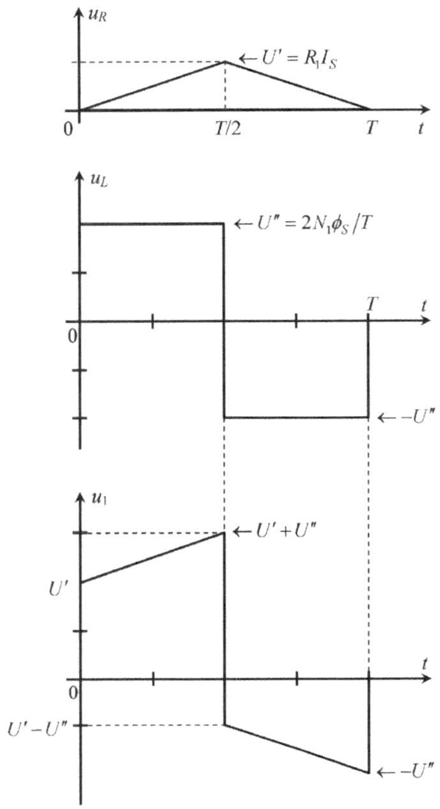

Figure 5.6 Graphical representation of the generator voltage against time, $u_1(t)$, showing its decomposition into resistive and inductive components u_R and u_L, respectively

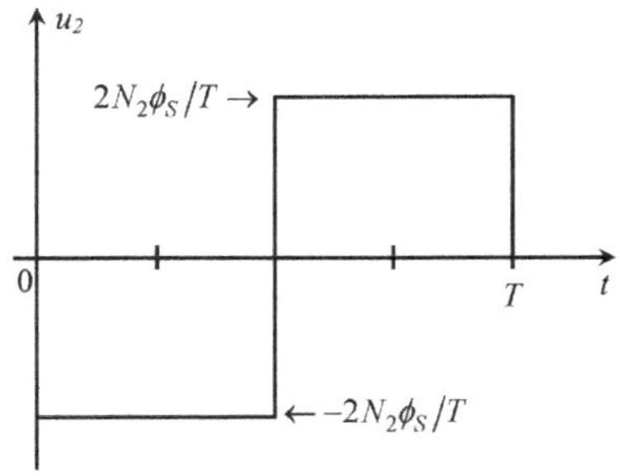

Figure 5.7 Graphical representation of coil 2 voltage against time, $u_2(t)$

circumferences oriented clockwise; this field actuates on the free charged particles of coil 2 giving rise to a distribution of positive charges at terminal d and negative charges at terminal c. A gradient electric field $\mathbf{E}_g$ oriented from d to c appears and a positive pulse voltage is revealed in $u_2(t)$.

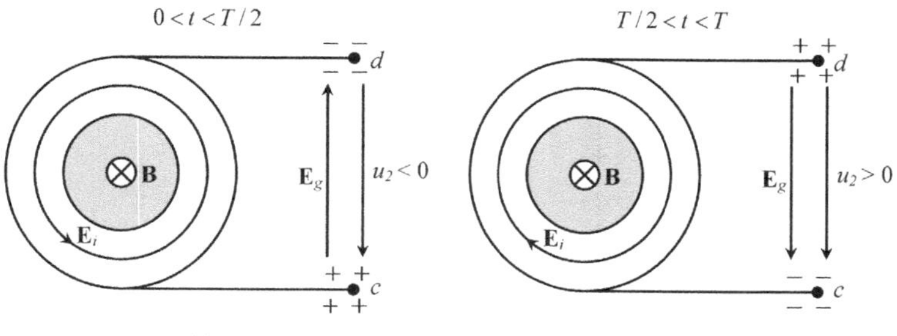

Figure 5.8 Interpretation of the positive and negative pulses appearing in $u_2(t)$. (a) Time-increasing **B** originates an induction electric field $\mathbf{E}_i$ that drives the positive charges to terminal c and negative charges to terminal d. (b) Time-decreasing **B** reverses the orientation of the induction electric field $\mathbf{E}_i$; positive charges are driven to terminal d and negative charges to terminal c

Let us now complicate things a little, allowing the ferromagnetic core to saturate due to imposed higher current intensity.

Make $I_M = 2I_S$. Combining the information conveyed in Figure 5.5, we obtain for $\phi(t)$ a trapezoidal function as shown in Figure 5.9.

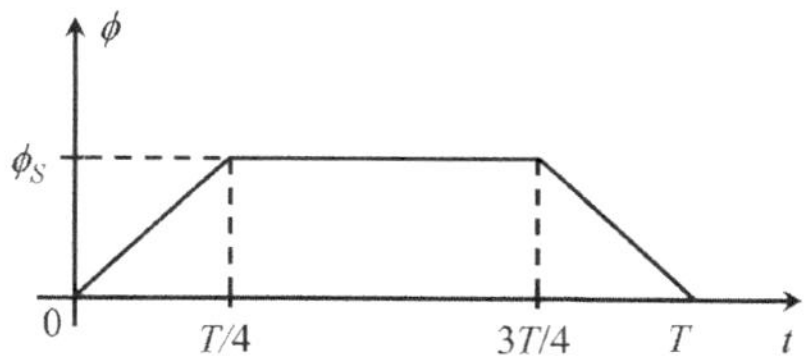

Figure 5.9 Trapezoidal function $\phi(t)$ originated by an intense triangular current pulse $i_1(t)$ that brings the core into saturation

The derivative $d\phi/dt$ shown in (5.11) and (5.13) is evaluated as

$$\frac{d\phi}{dt} = \begin{cases} +4\phi_S/T; & \text{for } 0 < t < T/4 \\ 0; & \text{for } T/4 < t < 3T/4 \\ -4\phi_S/T; & \text{for } 3T/4 < t < T \end{cases}$$

We strongly recommend that you try to redraw new graphs for

$$u_t(t) = R_1 i_1(t) + N_1 \frac{d\phi(t)}{dt} \quad \text{and} \quad u_2(t) = -N_2 \frac{d\phi(t)}{dt}$$

Omitting any details, we present the final results in Figure 5.10. Note that in the time interval from $T/4$ to $3T/4$ the **B** field remains constant with time, induction phenomena are absent, and therefore you get $u_2 = 0$.

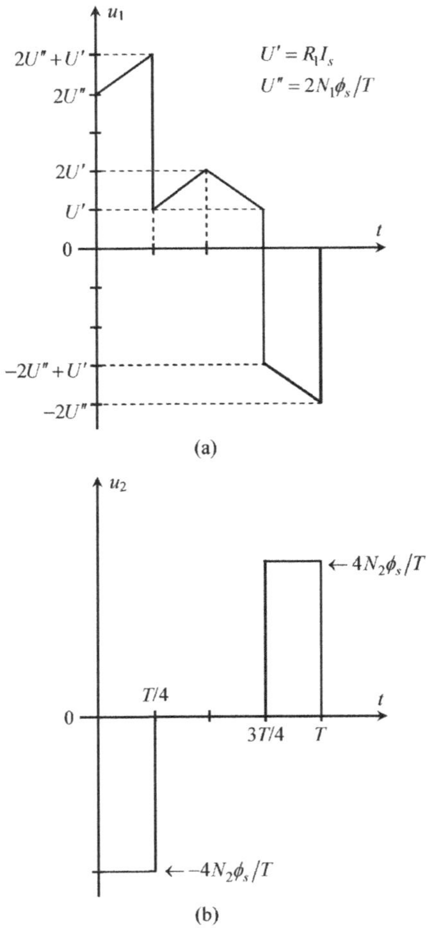

Figure 5.10 Voltage plots against time for the case of a saturated core. (a) Generator voltage (compare Figure 5.6). (b) Voltage of coil 2 (compare Figure 5.7)

The first part of the problem we have just finished solving (linear case) could have been handled more easily if the inductance coefficients characterizing the two-coupled coils were known beforehand. In fact, by using the results of Chapter 4, we could have put

$$\psi_1 = L_{11}i_1 + L_M i_2 \quad \text{and} \quad \psi_2 = L_M i_1 + L_{22}i_2$$

Substituting ψ_1 and ψ_2 above into (5.10b) and (5.12b), and making $i_2 = 0$, we would obtain, as an alternative, the following results:

$$u_1 = R_1 i_1 + L_{11}\frac{di_1}{dt} \quad \text{and} \quad u_2 = -L_M\frac{di_1}{dt} \tag{5.14}$$

Note, however, that you cannot use this alternative when dealing with nonlinear problems (where the concept of inductance does not apply).

In the more general case of a system of N-coupled coils, characterized by their internal resistances R_k (with $k = 1$ to N) and inductance matrix $[L]$, (4.41), the application of the induction law to every coil (using circulation paths coinciding with the reference directions of coil currents) permits coil voltages to be compactly determined from coil currents by making use of the following matrix equation:

$$[u(t)] = [R][i(t)] + [L]\frac{d}{dt}[i(t)] \tag{5.15}$$

where the resistance matrix $[R]$ is diagonal, the column matrix $[i(t)]$ gathers the coil currents, and the column matrix $[u(t)]$ gathers the coil voltages.

A word of caution: depending on how you choose the arbitrary reference directions for the coil voltages, $u_1, \ldots, u_k, \ldots, u_N$, both positive and negative algebraic signs need to be included in the elements of the column matrix $[u(t)]$ in (5.15). If all coil voltages u_k are marked so as to oppose the orientation of the circulation paths $\mathbf{s}_k$ then only positive algebraic signs will appear.

5.7 Application Example (Magnetic Coupling in Printed Circuit Boards)

A printed circuit board consisting of three conducting lands on the surface of a dielectric board above a reference conducting ground plane is shown in Figure 5.11. To simplify things, let's assume that all conductor resistances are negligibly small.

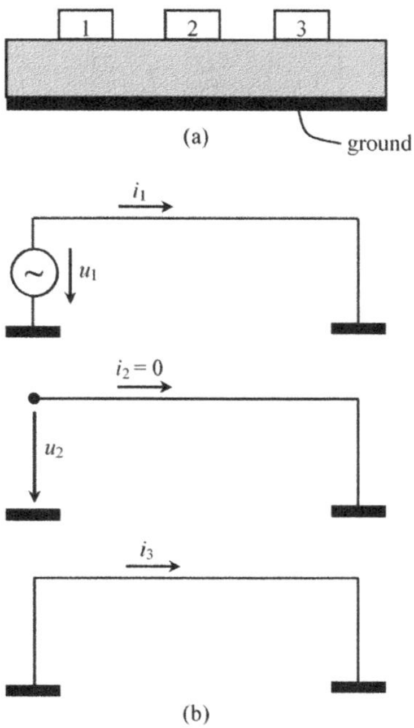

Figure 5.11 Printed circuit board (PCB) with three lands and a ground plane. (a) Cross-sectional view. (b) Enforced boundary conditions at the near and far ends of the PCB lands

All lands are short-circuited to ground at the far end. As for the near end, the situation is described as follows: land 1 is driven by a voltage u_1, land 2 is left open ($i_2 = 0$) and land 3 is short-circuited to ground ($u_3 = 0$). The conductor system is characterized by a symmetric inductance matrix

$$[L] = \begin{bmatrix} L_{11} & L_{12} & L_{13} \\ L_{21} & L_{22} & L_{23} \\ L_{31} & L_{32} & L_{33} \end{bmatrix}$$

Voltage u_1 is a low-frequency sinusoidal function, $u_1(t) = U_1 \cos(\omega t)$.

Questions

Q$_1$ By application of the induction law obtain the governing equations of the system.

Q$_2$ Determine $i_1(t)$, $i_3(t)$ and $u_2(t)$.

Solutions

Q$_1$ Taking into account that $i_2 = 0$, from (5.15), we write

$$\begin{cases} u_1 = L_{11}\dfrac{di_1}{dt} + L_{13}\dfrac{di_3}{dt} \\ u_2 = L_{21}\dfrac{di_1}{dt} + L_{23}\dfrac{di_3}{dt} \\ 0 = L_{31}\dfrac{di_1}{dt} + L_{33}\dfrac{di_3}{dt} \end{cases}$$

Q$_2$ From the third equation we get

$$i_3 = -\frac{L_{31}}{L_{33}} i_1$$

Substituting this information into the first equation we find

$$u_1 = \underbrace{\left(L_{11} - \frac{L_{13}^2}{L_{33}}\right)}_{L_1} \frac{di_1}{dt} \rightarrow i_1(t) = \frac{1}{L_1}\int u_1(t)dt = \frac{U_1}{\omega L_1}\cos(\omega t - \pi/2)$$

Using the already established relationship between i_3 and i_1 we find

$$i_3(t) = \frac{L_{31}U_1}{\omega L_{33}L_1}\cos(\omega t + \pi/2)$$

From the second equation in Q$_1$ we obtain

$$u_2(t) = U_1\left(\frac{L_{21}}{L_1} - \frac{L_{23}L_{31}}{L_1 L_{33}}\right)\cos(\omega t)$$

5.8 Eddy Currents

We have already mentioned in Section 4.11 that ferromagnetic cores, subjected to periodic magnetization–demagnetization processes, dissipate energy because of hysteresis phenomena. There is another physical mechanism that can also cause additional core losses (heating).

You may still remember that in the preceding section (Figure 5.8) voltage u_2 was created by the action of an electric induction field $\mathbf{E}_i$ with closed field lines embracing lines of $\mathbf{B}(t)$. Such a field $\mathbf{E}_i$ exists not only outside the core, but also in its interior. Since ferromagnetic materials are also conductors (conductivity σ_{Fe}) this implies necessarily that electric currents with density $\mathbf{J}_i = \sigma_{\text{Fe}}\mathbf{E}_i$ are allowed to circulate in planes transversal to $\mathbf{B}$, that is in the core cross-sections (see Figure 5.12). These currents are called eddy currents, or Foucault currents.

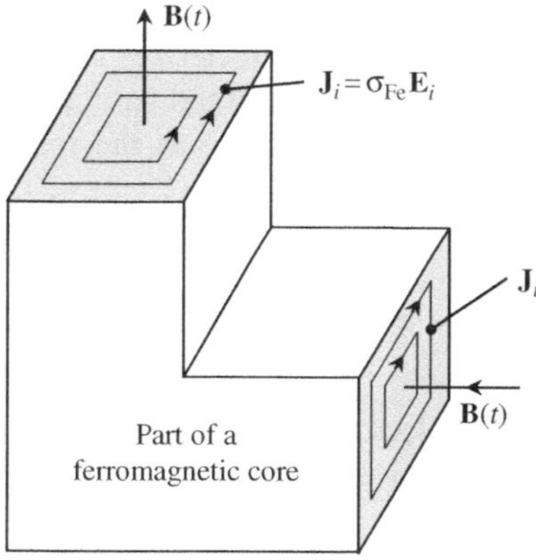

Figure 5.12 Eddy currents in conducting ferromagnetic materials originated by time-varying **B** fields

From (3.14) in Chapter 3, the power losses (Joule effect) associated with these currents are evaluated through

$$P_{\text{Foucault}} = \int_{\text{Core}} \hat{p}_{\text{J}}\, dV; \;\; \hat{p}_{\text{J}} = \mathbf{J}_i \cdot \mathbf{E}_i = \sigma_{\text{Fe}} E_i^2 \propto \left(\frac{dB}{dt}\right)^2 \tag{5.16}$$

This shows that Foucault losses depend on the squared intensity of the electric induction field, which, in turn, depends on the time derivative of $B(t)$. In conclusion, the faster the variation of **B**, the more important the losses.

Another problem created by eddy currents is their own magnetic field which can significantly perturb the original field existing in the ferromagnetic core.

There are two known techniques that can be employed to mitigate eddy current effects. One consists of using laminated cores where each ferromagnetic sheet is electrically insulated from the others using a non-conducting varnish. This technique is relatively effective for time-varying **B** fields up to 20 kHz. For higher frequencies, ferrites should rather be used. Ferrites are electrically non-conductive ceramic compound materials consisting of a mix of iron, zinc and manganese oxides.

5.9 Generalization of the Induction Law to Moving Circuit Systems

This important topic deals with induction phenomena not originated by time-varying **B** fields, but by circuits (subjected to stationary **B** fields) whose geometrical configuration is time dependent.

Even if the velocity of the moving circuits is much smaller than the speed of light, for a rigorous and sound interpretation of the problem a contribution from the theory of relativity would be required. The introductory nature of this textbook prevents us from following such an approach here.

In any case, imagine the following *Gedanken* (thought) experiment. In a certain region of space an electromagnetic field exists. An observer O′ at rest characterizes the electromagnetic field in the region by a pair of vectors $\mathbf{E}'$ and $\mathbf{B}'$. O′ also observes that a particle with electric charge Q moves across the region with velocity $\mathbf{v}$, its trajectory changing according to the exerted Lorentz force $\mathbf{F} = Q\mathbf{E}' + Q\mathbf{v} \times \mathbf{B}'$. A second observer O seated on the particle looks at the region where the particle travels and characterizes the *same* electromagnetic field by a different pair of vectors **E** and **B**. Since the charged particle is at rest ($\mathbf{v} = 0$) with respect to O, this observer interprets its trajectory change as the result of a purely electric force $\mathbf{F} = Q\mathbf{E}$.

In order to make both observations agree we have to conclude that

$$\mathbf{E} = \mathbf{E}' + \underbrace{\mathbf{v} \times \mathbf{B}'}_{\mathbf{E}_\text{v}} \tag{5.17}$$

where $\mathbf{E}_\text{v}$ is the so-called dynamic electric field.

Although we are not going to prove it, **B** could be obtained as $\mathbf{B} = \mathbf{B}' - \mathbf{v} \times \mathbf{E}'/c^2$. Therefore, for typical applications $(v \ll c)$, $\mathbf{B} = \mathbf{B}'$.

The induction law, for bodies at rest subjected to time-varying magnetic fields, has been formulated in (5.9) as

$$\oint_{\mathbf{S}} (\mathbf{E}_g + \mathbf{E}_i) \cdot d\mathbf{s} = -\frac{d}{dt} \int_{S_\text{s}} \mathbf{B}(t) \cdot \mathbf{n}_\text{S} \, dS$$

For moving bodies subjected to stationary magnetic fields, the above equation should be modified to

$$\oint_{\mathbf{S}(t)} (\mathbf{E}_g + \mathbf{E}_\text{v}) \cdot d\mathbf{s} = -\frac{d}{dt} \int_{S_{\text{s}(t)}} \mathbf{B} \cdot \mathbf{n}_\text{S}(t) \, dS \tag{5.18}$$

where the circulation path moves with the moving parts of the circuit, $\mathbf{s} = \mathbf{s}(t)$.

In the most general case of moving bodies subjected to time-varying magnetic fields, the generalization of the Maxwell–Faraday induction law takes the form

$$\oint_{\mathbf{S}(t)} \underbrace{(\mathbf{E}_g + \mathbf{E}_i + \mathbf{E}_\text{v})}_{\mathbf{E}} \cdot d\mathbf{s} = -\frac{d}{dt} \underbrace{\int_{S_{\text{s}(t)}} \mathbf{B}(t) \cdot \mathbf{n}_\text{S}(t) \, dS}_{\psi_{\text{s}(t)}} \tag{5.19}$$

where, it should be stressed, the **E** field on the left-hand side refers to the electric field as observed in the moving reference frame.

5.10 Application Example (Electromechanical Energy Conversion)

We now present an example that illustrates the principle of conversion of mechanical energy into electric energy and the conversion of electric energy into mechanical energy.

Take the situation depicted in Figure 5.13 where a moving bar, of mass M and internal resistance R, slides (frictionless) with velocity $\mathbf{v} = v(t)\vec{e}_x$ over two conducting rails. Perpendicular to the plane defined by the two rails, a uniform time-invariant **B** field is enforced. Neglect the internal resistances of the rails as well as the magnetic field produced by the circulating current.

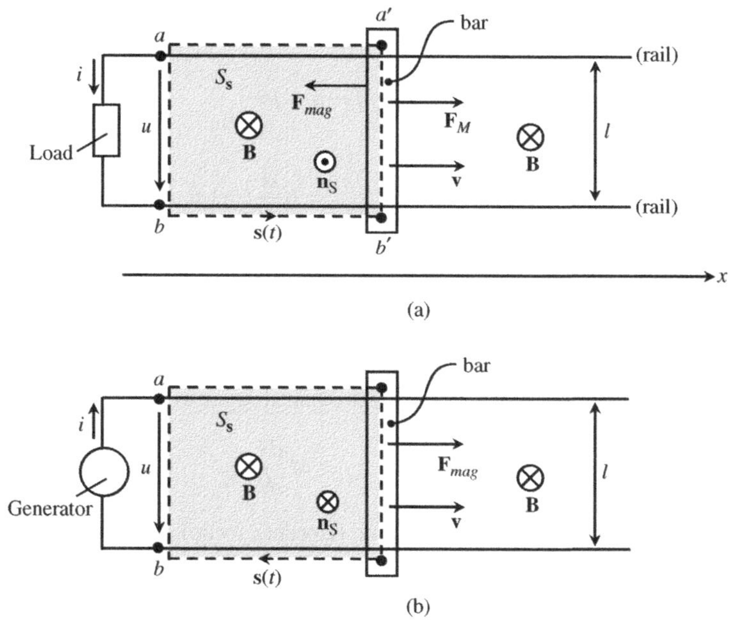

Figure 5.13 Illustration of the principles of electromechanical energy conversion using, as an example, a bar–rail system. (a) An external mechanical force $\mathbf{F}_M$ drives the bar into movement giving rise to an induced emf, electric energy being delivered to the load. (b) An electric power supply (generator) produces a current flow that in conjunction with the **B** field originates a magnetic force $\mathbf{F}_{mag}$, the latter driving the bar into movement (production of mechanical kinetic energy)

Questions

Q_1 In Figure 5.13(a) a passive electric load is connected between terminals a and b where a voltage u appears as a consequence of induction phenomena. The bar is driven by an externally applied mechanical force $\mathbf{F}_M$ which causes the bar to move with velocity $\mathbf{v}$. Determine the voltage u and establish the energy balance equation of the system.

Q_2 Now consider the reverse problem in Figure 5.13(b), where an electric generator provides a voltage u between terminals a and b. The bar is free from any mechanical external force; however, due to the interaction between the **B** field and the current i in the bar, the latter undergoes the action of a magnetic force $\mathbf{F}_{mag}$ which drives the bar into movement with velocity $\mathbf{v}$. Determine the voltage u and establish the energy balance equation of the system.

Solutions

Q_1 In addition to the external mechanical force $\mathbf{F}_M$ the bar is also actuated by an opposite magnetic force $\mathbf{F}_{mag}$ originating from the interaction of the $\mathbf{B}$ field with the current i flowing in the bar.

Let us first apply the induction law to the closed path $\mathbf{s}(t)$ passing by the load, rails and moving bar. The path $\mathbf{s}$ is oriented according to the reference direction assigned to i:

$$\oint_{\mathbf{s}(t)} \mathbf{E} \cdot d\mathbf{s} = u + Ri \tag{5.20a}$$

$$-\frac{d}{dt}\int_{S_{s(t)}} \mathbf{B} \cdot \mathbf{n}_S \, dS = +B\frac{d}{dt}S_{s(t)} = B\frac{d}{dt}(lx) = Bl\frac{dx}{dt} = Blv \tag{5.20b}$$

Equating the results in (5.20) we obtain

$$u = Blv - Ri \tag{5.21}$$

The equation describing the movement of the bar along x is

$$\mathbf{F}_M + \mathbf{F}_{mag} = M\frac{d\mathbf{v}}{dt} \tag{5.22}$$

Taking into account that the magnetic force (Lorentz force) exerted along the length of the bar is

$$\mathbf{F}_{mag} = \int_{\overrightarrow{b'a'}} i \, d\mathbf{s} \times \mathbf{B} = -ilB\vec{e}_x$$

we can write from (5.22)

$$\mathbf{F}_M = M\frac{d\mathbf{v}}{dt} + iBl\vec{e}_x$$

The mechanical power $P_M = \mathbf{F}_M \cdot \mathbf{v}$ associated with the external driving force $\mathbf{F}_M$ is next determined as

$$P_M = M\mathbf{v} \cdot \frac{d\mathbf{v}}{dt} + iBlv = \frac{d}{dt}\left(\frac{1}{2}Mv^2\right) + i(u + Ri) = \frac{d}{dt}\left(\frac{1}{2}Mv^2\right) + ui + Ri^2 \tag{5.23}$$

The term $Mv^2/2$ is the kinetic energy of the bar W_K, the term ui is the electric power available to the load P_L, and Ri^2 is the power losses (Joule effect) in the bar P_J. Thus, we conclude for the power balance

$$P_M = dW_K/dt + P_L + P_J$$

Integration over time gives $W_M = W_K + W_L + W_J$.

This result illustrates the conversion of external mechanical energy into electric energy.

Q_2 Reapplying the induction law to path $\mathbf{s}(t)$ oriented according to the reference direction assigned to i, we obtain (see Figure 5.13(b))

$$\oint_{\mathbf{s}(t)} \mathbf{E} \cdot d\mathbf{s} = -u + Ri \tag{5.24a}$$

$$-\frac{d}{dt} \int_{S_{s(t)}} \mathbf{B} \cdot \mathbf{n}_S \, dS = -B \frac{d}{dt} S_{s(t)} = -Blv \tag{5.24b}$$

Equating the results in (5.24) we obtain $u = Blv + Ri$.

The equation describing the movement of the bar along x is

$$\mathbf{F}_{\text{mag}} = M \frac{d\mathbf{v}}{dt} \quad \text{or} \quad iBl\vec{e}_x = M \frac{d\mathbf{v}}{dt}$$

The inner product with $\mathbf{v}$ on both sides of the above equation yields

$$iBlv = \frac{d}{dt}\left(\frac{1}{2}Mv^2\right) \quad \text{or} \quad i(u - Ri) = \frac{d}{dt}\left(\frac{1}{2}Mv^2\right)$$

Rearranging terms

$$ui = \frac{d}{dt}\left(\frac{1}{2}Mv^2\right) + Ri^2 \quad \text{or} \quad P_G = dW_K/dt + P_J$$

where P_G is the electric power delivered by the generator.

Integrating over time we get the energy balance, $W_G = W_K + W_J$.

This result illustrates the conversion of electric energy into mechanical kinetic energy.

5.11 DC Voltage Generation

This section deals only with the functioning principle of the simplest DC generator.

Consider, as shown in Figure 5.14, a rotating disk (Faraday's disk) of radius R illuminated perpendicularly by a uniform time-invariant $\mathbf{B}$ field. Brush-type conducting contacts are established with the disk shaft and with the disk periphery, allowing a voltage U to be collected at the accessible open terminals a and b. The angular velocity of the rotating disk is ω.

Consider the application of the induction law to the closed path $\mathbf{s}$ (accompanying the disk movement). Since currents are absent we find

$$\oint_{\mathbf{s}} \mathbf{E} \cdot d\mathbf{s} = \int_{\substack{\overrightarrow{ab} \\ \text{air}}} \mathbf{E} \cdot d\mathbf{s} = U \tag{5.25a}$$

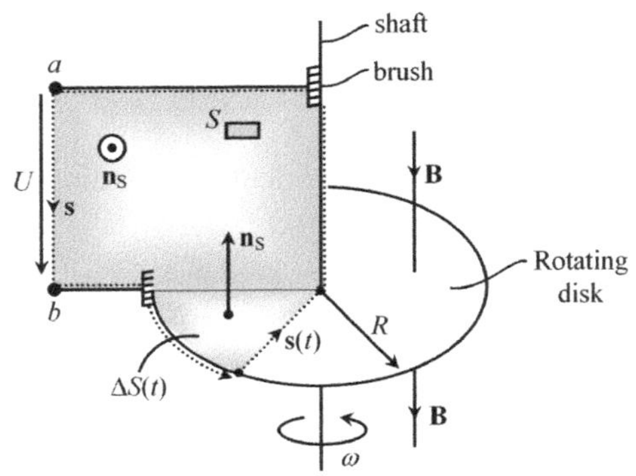

Figure 5.14 Illustration of DC voltage generation principles using, as an example, Faraday's rotating disk. Voltage U is proportional to both B and ω

As for the right-hand side of the induction law equation, concerning the time derivative of the linkage flux, you may evaluate it as

$$-\frac{d\psi_{s(t)}}{dt} = -\frac{d}{dt}\left(\int_{S_{\square}} \mathbf{B}\cdot\mathbf{n}_S\, dS + \int_{\Delta S(t)} \mathbf{B}\cdot\mathbf{n}_S\, dS\right) \tag{5.25b}$$

Noting that the flux across the rectangular surface $S_{\square}$ is null and time invariant, the first contribution on the right-hand side of (5.25b) is zero. As for the second contribution, noting that $\mathbf{B} \uparrow\downarrow \mathbf{n}_S$ and that the area $\Delta S(t)$ steadily increases with time, $\Delta S(t) = \frac{1}{2}R^2\omega t$, we get

$$\psi_{s(t)} = -B\Delta S = -\tfrac{1}{2}BR^2\omega t \quad \text{and} \quad -d\psi_s/dt = \tfrac{1}{2}BR^2\omega$$

Finally, equating the results from (5.25), we obtain a DC voltage given by

$$U = \frac{BR^2\omega}{2} \tag{5.26}$$

Taking into account that $\omega = 2\pi N_{rps}$ (where N_{rps} is the number of rotations per second), and that the magnetic flux through the whole disk is $\phi = B\pi R^2$, the result in (5.26) can be rewritten in the more insightful form $U = N_{rps}\phi$.

5.12 AC Voltage Generation

Similar to the above, this section deals only with the functioning principle of the simplest AC generator.

Consider, as shown in Figure 5.15, a rectangular single-turn coil of area A rotating with angular velocity ω around its own axis in a region where a uniform time-invariant $\mathbf{B}$ field

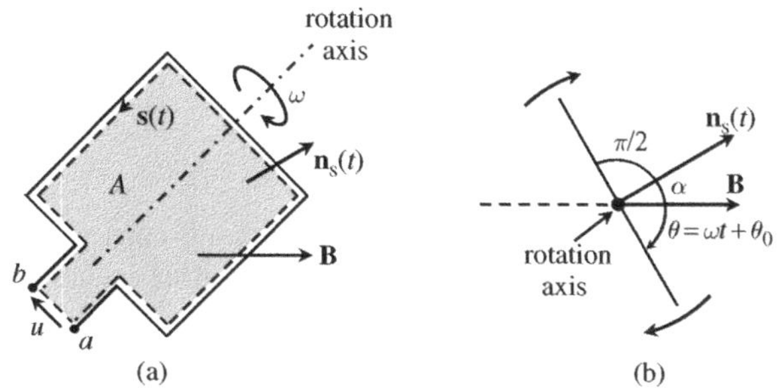

Figure 5.15 Illustration of AC voltage generation principles using, as an example, a rotating rectangular coil immersed in a uniform **B** field. (a) Perspective view and circulation path for the application of the induction law. (b) Side view, showing that the angle between **B** and the Stokes normal changes with time

exists perpendicular to the axis of rotation. A voltage $u(t)$ is collected at the coil's accessible terminals a and b which are left open.

Consider the application of the induction law to the closed path $\mathbf{s}(t)$ (accompanying the coil rotation). Since currents are absent we find

$$\oint_{S} \mathbf{E}\cdot d\mathbf{s} = \int_{\substack{\overrightarrow{ba}\\ \text{air}}} \mathbf{E}\cdot d\mathbf{s} = -u \tag{5.27a}$$

As for the right-hand side of the induction law equation, concerning the time derivative of the linkage flux, we may evaluate it as

$$-\frac{d\psi_s}{dt} = -\frac{d}{dt}\int_A \mathbf{B}\cdot\mathbf{n}_S(t)\, dS$$

Noting that the angle α between $\mathbf{B}$ and $\mathbf{n}_S$ changes with time, $\alpha = \pi/2 - \theta(t)$, where $\theta(t) = \omega t + \theta_0$ (with arbitrary θ_0), we have

$$\mathbf{B}\cdot\mathbf{n}_S(t) = B\cos\alpha = B\sin(\omega t + \theta_0)$$

and therefore

$$-\frac{d\psi_s}{dt} = -\frac{d}{dt}\int_A \mathbf{B}\cdot\mathbf{n}_S(t) dS = -AB\omega\cos(\omega t + \theta_0) \tag{5.27b}$$

Equating the results from (5.27), we obtain an AC voltage given by $u(t) = AB\omega\cos(\omega t + \theta_0)$.

If the coil contains N turns tightly packed, the above result is modified to

$$u(t) = U_M\cos(\omega t + \theta_0), \text{ with } U_M = AB\omega N \tag{5.28}$$

5.13 Proposed Homework Problems

Problem 5.13.1

An overhead line conductor at height h_L above ground carries a sinusoidal current $i_L(t) = I\sin(\omega t)$. As shown in Figure 5.16, a fence of length l is situated just beneath the line and parallel to it. The fence has two horizontal supporting conducting wires, one at the soil level and the other at height h_F. Both wires are connected at one end of the fence by a conducting post, whereas at the opposite end a wooden post is used.

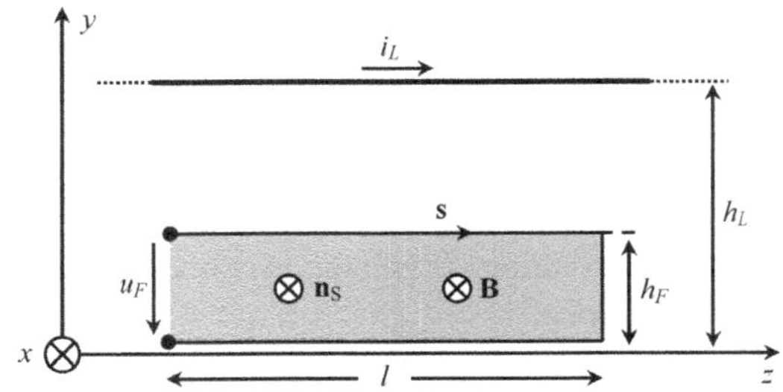

Figure 5.16 Induced voltage in a wire fence placed near to an overhead line

Assume that the return current in the poorly conducting soil contributes negligibly to the evaluation of the **B** field in air.

Data: $I = 5\,\text{kA},\ \omega = 2\pi f,\ f = 50\,\text{Hz},\ h_L = 12\,\text{m},\ h_F = 3\,\text{m},\ l = 200\,\text{m}.$

Q_1 From Chapter 4, determine the **B** field originated by the overhead line conductor in the fence region ($x = 0$).

Q_2 Determine the fence voltage $u_F(t)$ between the two supporting wires at the end where the wooden post has been placed.

Answers

Q_1

$$\mathbf{B}(y,t) = \mu_0 \frac{i_L(t)}{2\pi(h_L - y)} \vec{e}_x, \quad \text{for } 0 < y < h_F$$

Q_2 Application of the induction law to the closed path **s** along the fence wires and wooden post gives

$$\underbrace{\oint_{\mathbf{s}} \mathbf{E} \cdot d\mathbf{s}}_{-u_F} = -\frac{d\psi_F}{dt}, \quad \text{with } \psi_F = \int_{S_s} \mathbf{B} \cdot \mathbf{n}_S \, dS = \int_{y=0}^{y=h_F} B(y,t)\, l \, dy$$

$$\psi_F(t) = \underbrace{\left(\frac{\mu_0 l}{2\pi}\ln\frac{h_L}{h_L - h_F}\right)}_{L_M} i_L(t); \quad L_M = 11.5\,\mu\text{H}$$

$$u_F(t) = L_M \frac{di_L(t)}{dt} = \underbrace{L_M I\omega}_{U}\cos(\omega t); \quad U = 18.1\,\text{V}$$

Problem 5.13.2

A coil with N_1 turns and a conducting ring are placed in a transformer core – see Figure 5.17. The coil, with resistance R_1, is driven by a voltage generator. The ring can be left open or short-circuited depending on the switch position. The transformer core has a uniform cross-section S. The self-inductance of the coil is L_{11}. Neglect dispersion phenomena and the resistance of the ring.

Data: $N_1 = 10$, $R_1 = 10\,\Omega$, $S = 1\,\text{cm}^2$, $L_{11} = 10\,\text{mH}$.

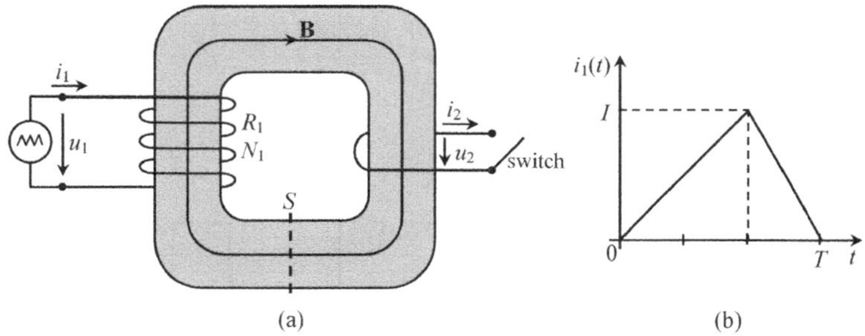

Figure 5.17 A transformer core with a coil of N_1 turns and one ring. (a) General view. (b) Generator current

Q_1 Using your knowledge from Chapter 4, determine the self-inductance of the ring L_{22} as well as the mutual inductance L_M.

Q_2 With the switch open ($i_2 = 0$) the generator current $i_1(t)$ is described by an asymmetrical triangular pulse of duration $T = 3\,\text{ms}$ and peak value $I = 0.1\,\text{A}$.

Using the induction law, determine the voltages $u_1(t)$ and $u_2(t)$.

Using your knowledge from Chapter 4, determine the time evolution of $B(t)$ in the core.

Q_3 Now consider that the switch is closed ($u_2 = 0$) and that the generator voltage u_1 is the same as you determined in Q_2.

Evaluate $i_1(t)$ and $i_2(t)$.

Determine the magnetic induction field in the transformer core. Comment on the result.

Answers

Q_1

$$L_{11} = \frac{N_1^2}{R_m} \rightarrow R_m = 10^4\,\mathrm{H}^{-1}, \quad L_{22} = \frac{N_2^2}{R_m} = 0.1\,\mathrm{mH}, \quad L_M = \frac{N_1 N_2}{R_m} = 1.0\,\mathrm{mH}$$

Q_2

$$u_1 = R_1 i_1 + L_{11}\frac{di_1}{dt}; \quad u_2 = -L_M\frac{di_1}{dt}$$

(see the illustrative plots in Figure 5.18)

$$\psi_1 = \begin{cases} L_{11} i_1 \\ N_1 \phi = N_1 BS \end{cases} \quad \rightarrow B(t) = \frac{L_{11}}{N_1 S} i_1(t)$$

$B(t)$ is a triangular pulse as $i_1(t)$, with peak value of 1 T.

Q_3

$$u_2 = 0 = -\frac{d\psi_2}{dt} \rightarrow \psi_2(t) = 0 \rightarrow B(t) = 0 \rightarrow \psi_1(t) = 0$$

$$u_1 = R_1 i_1 + \frac{d\psi_1}{dt} = R_1 i_1 \rightarrow i_1(t) = \frac{u_1(t)}{R_1}$$

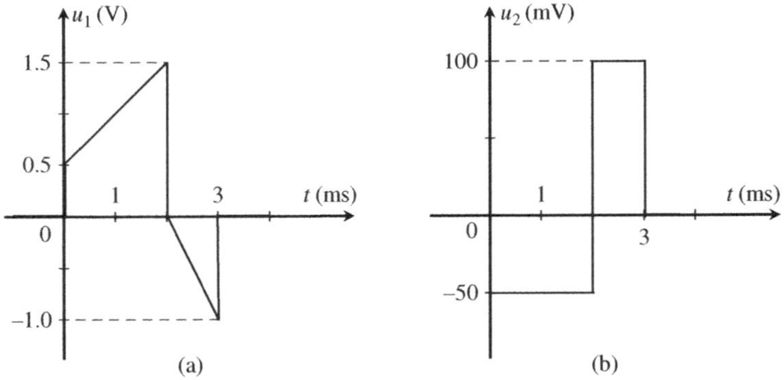

Figure 5.18 Voltage plots against time when the switch is open. (a) Coil voltage. (b) Ring voltage

Current $i_1(t)$ has the shape of the voltage $u_1(t)$ established in Figure 5.18 apart from a scale factor determined by the value of R_1.

Since the magnetic flux in the core is zero, then, from Ampère's law (Chapter 4), you will have $R_m \phi = 0 = N_1 i_1 + N_2 i_2$ (with $N_2 = 1$), and, consequently,

$$i_2(t) = -N_1 i_1(t) = -N_1 \frac{u_1(t)}{R_1}$$

Current $i_2(t)$ has the shape of $i_1(t)$, but with opposite sign, and scaled by a factor of ten ($N_1 = 10$).

When the switch is closed, the new current i_2 creates its own magnetic field which opposes the original one existing in the core (this is known in the literature as Lenz's 'law'). In our case, because the ring is a perfect conductor, the ring's magnetic field has exactly the same magnitude as the one previously existing in the core due to the excitation coil. Therefore, their sum is zero for the resultant field, $\mathbf{B}_1 + \mathbf{B}_2 = 0$.

Firstly, and most important to note, the idea that current $i_2(t)$ always reacts against the induction action of $i_1(t)$ – something equivalent to the action/reaction principle you have heard about in physics – is a false concept.

The reaction of $i_2(t)$ is critically dependent on the type of load connected to the second coil (ring). If, for instance, a capacitor is connected to the terminals of the second coil it may well happen that the total **B** field in the core may actually increase as compared to $\mathbf{B}_1$ (we will come back to this topic in Section 7.3.2).

Secondly, the problem treated here exemplifies a technological application of induction phenomena, namely induction heating. You saw that the current in the ring is N_1 times bigger than the one in the excitation coil. In reality, the ring has an internal resistance and because of that, energy dissipation due to the Joule effect will take place there:

$$W_J = \int_0^T R_{\text{ring}}\, i_2^2(t)\, dt$$

Depending on the system's design parameters, the structure in Figure 5.17 may be engineered so as to ensure that the heat generated in the metallic ring melts it down.

Problem 5.13.3

A ferromagnetic core (see Figure 5.19) is excited by a sinusoidal current flowing in an inductor (not shown) which gives rise to a magnetic flux $\phi(t) = \phi_M \sin(\omega t)$ circulating in the core. As shown in the figure, a single-turn coil is wound around the core. The coil is left open ($i = 0$), but due to induction phenomena a voltage $u(t)$ appears across its terminals a and b. In order to visualize the coil voltage, two oscilloscopes O_1 and O_2 are connected between a and b.

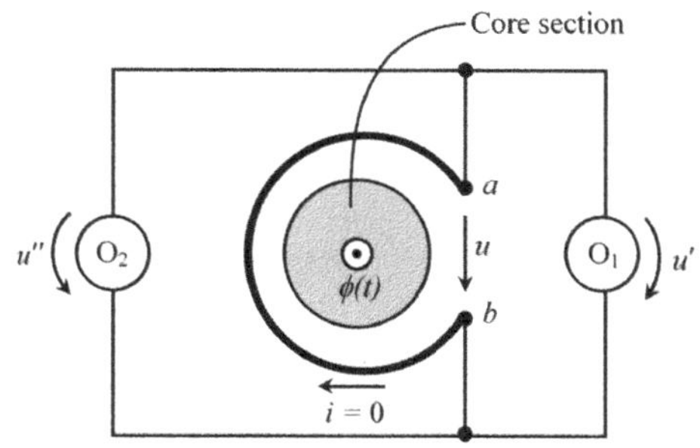

Figure 5.19 The single-turn coil voltage $u(t)$ is read differently by the oscilloscopes O_1 and O_2, both connected between a and b

Q_1 Determine $u(t)$.

Q_2 Determine the voltages $u'(t)$ and $u''(t)$ retrieved by both oscilloscopes.

Q_3 Repeat the problem for the case of a two-turn coil.

Q_4 Generalize the above result for an N-turn coil.

Answers

Q_1

$$u(t) = \frac{d\phi}{dt} = U\cos(\omega t), \text{ with } U = \omega\phi_M$$

Q_2

$$u'(t) = u(t); \; u''(t) = 0$$

Q_3

$$u'(t) = u(t) = 2\frac{d\phi}{dt} = 2U\cos(\omega t); \quad u''(t) = \frac{d\phi}{dt} = U\cos(\omega t)$$

Q_4

$$u'(t) = u(t) = N\frac{d\phi}{dt} = NU\cos(\omega t); \quad u''(t) = (N-1)\frac{d\phi}{dt} = (N-1)\; U\cos(\omega t)$$

Only for $N \gg 1$ will you have similar oscilloscope readings, $u' \approx u''$.

Problem 5.13.4

Two inductors are connected in series as shown in Figure 5.20. Inductors are characterized by internal resistances R_1 and R_2 and self-inductances L_{11} and L_{22}. The magnetic coupling factor between the inductors is $k = 0.75$.

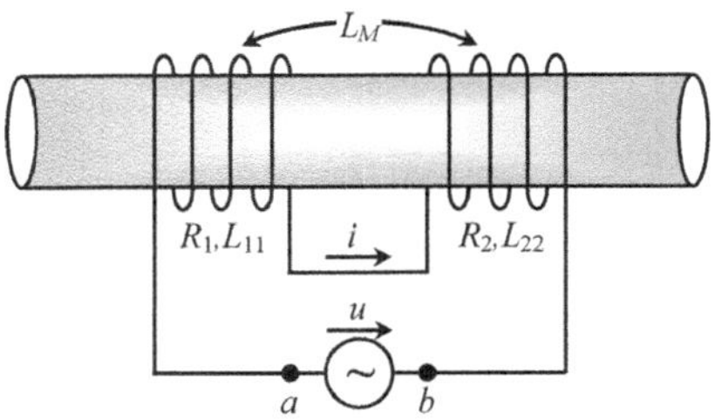

Figure 5.20 Two series-connected magnetically coupled inductors

The current common to both inductors is $i(t) = I\cos(\omega t)$.

Q_1 Determine the mutual inductance L_M between the inductors.

Q_2 Write an analytical expression for the applied voltage $u(t)$ between terminals a and b.

Q_3 Considering that $R_1 = 60\ \Omega,\ R_2 = 140\ \Omega,\ L_{11} = 0.1\,\text{H},\ L_{22} = 0.4\,\text{H},\ I = 0.25\,\text{A}$, and $\omega = 1\,\text{krad/s}$, determine $u(t)$ numerically.

Answers

Q_1 $L_M = -150\,\text{mH}$

Q_2

$$\begin{cases} \oint_S \mathbf{E}\cdot d\mathbf{s} = -u + R_1 i + R_2 i \\ -\dfrac{d\psi_S}{dt} = -\dfrac{d}{dt}(\psi_1 + \psi_2) = -\dfrac{d}{dt}\left((L_{11} + L_M)i + (L_M + L_{22})i\right) \end{cases}$$

$$u = Ri + L\frac{di}{dt}, \quad \text{where } R = R_1 + R_2 \text{ and } L = L_{11} + L_{22} + 2L_M$$

$$u(t) = RI\cos(\omega t) - \omega LI\sin(\omega t)$$

Q_3 $R = 200\ \Omega,\ L = 200\,\text{mH},\ \omega L = 200\ \Omega$:

$$u(t) = 50\cos(\omega t) - 50\sin(\omega t)\ \text{V}$$

$$u(t) = 50\sqrt{2}\ \cos(\omega t + \pi/4)\ \text{V}$$

Problem 5.13.5

Consider the three-legged transformer shown in Figure 5.21, where the three vertical legs share the same geometrical and magnetic properties. For simplification purposes assume that the reluctances of the upper and lower yokes are negligibly small. The inductor placed on the left leg, with N_1 turns and self-inductance L_{11}, is driven by a sinusoidal voltage generator. The inductor placed on the right leg, with N_2 turns, is left open. Around the central leg a perfectly conducting ring can be switched from open state to short-circuit state depending on the switch position.

Assume that the internal resistance of inductor 1 is negligible, and that dispersion phenomena are absent.

Data: $L_{11} = 1\,\text{H},\ u_1(t) = U_1\cos(\omega t + \pi/2),\ U_1 = 325\,\text{V},\ \omega = 100\pi\,\text{rad/s},\ N_1 = 100,\ N_2 = 400.$

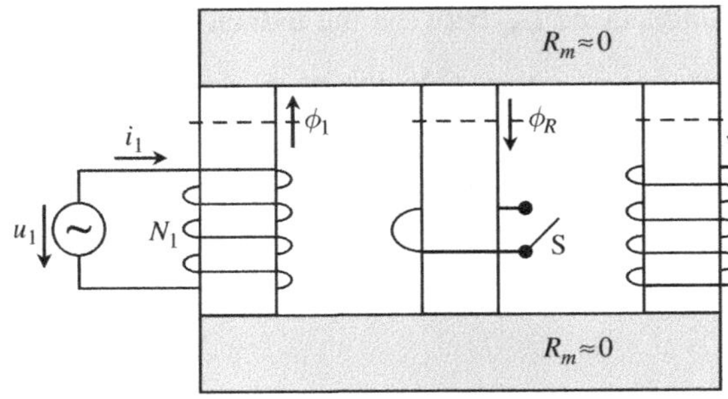

Figure 5.21 A three-legged magnetic circuit containing a ring in the center leg which can be switched on or off

Q_1 Find the magnetic flux $\phi_1(t)$ in the transformer's left leg.

Q_2 Consider that the ring is in the open state. Determine $\phi_2(t)$ and $u_2(t)$.

Q_3 Consider that the ring is in the short-circuit state. Determine $\phi_2(t)$ and $u_2(t)$.

Answers

Q_1

$$u_1(t) = N_1 \frac{d\phi_1(t)}{dt} \rightarrow \phi_1(t) = \frac{1}{N_1} \int u_1(t)\, dt = \Phi_1 \cos(\omega t), \quad \Phi_1 = \frac{U_1}{\omega N_1} = 10.35\,\text{mWb}$$

Q_2 $\phi_1 = \phi_R + \phi_2$; $\phi_R = \phi_2$ (due to symmetry reasons); $\phi_2 = \phi_1/2$.

$$u_2(t) = N_2 \frac{d\phi_2(t)}{dt} = \frac{N_2}{2} \frac{d\phi_1(t)}{dt} = \frac{N_2}{2N_1} u_1(t) = U_2 \cos(\omega t + \pi/2), \text{ with } U_2 = 650\text{ V}$$

Q_3 Because the ring is short-circuited, $\phi_R = 0 \rightarrow \phi_2 = \phi_1$:

$$u_2(t) = N_2 \frac{d\phi_2(t)}{dt} = N_2 \frac{d\phi_1(t)}{dt} = \frac{N_2}{N_1} u_1(t) = U_2 \cos(\omega t + \pi/2), \text{ with } U_2 = 1.3\,\text{kV}$$

Problem 5.13.6

Consider the situation analyzed in Section 5.12 (AC voltage generation) but where, instead of a single rotating rectangular coil, you have three rotating rectangular coils making angles of $2\pi/3$ to each other – see Figure 5.22 (three-phase AC generator).

Q_1 Write the equations for the voltages $u_1(t)$, $u_2(t)$ and $u_3(t)$ at the coil terminals.

Q_2 Determine $u_1(t) + u_2(t) + u_3(t)$.

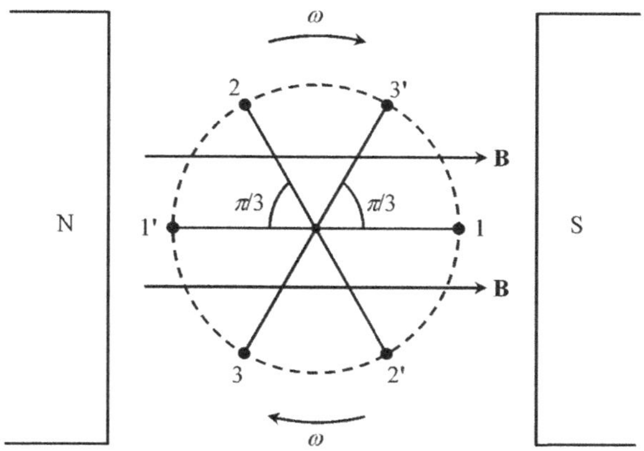

Figure 5.22 Operating principle of a three-phase AC generator. Three identical coils synchronously rotate in the space between the north and south poles of a magnet

Answers

Q_1

$$u_1(t) = U_M \cos(\omega t + \theta_0), \text{ with } U_M = AB\omega N$$
$$u_2(t) = U_M \cos(\omega t + \theta_0 - 2\pi/3)$$
$$u_3(t) = U_M \cos(\omega t + \theta_0 - 4\pi/3)$$

Q_2

$$u_1(t) + u_2(t) + u_3(t) = 0$$

Problem 5.13.7

Figure 5.23 illustrates the functioning principle of the moving-coil microphone. The coil is attached to a diaphragm (not shown) on which sound waves impinge. When the diaphragm vibrates, the coil is set in motion. The coil moves in a region where a radial stationary magnetic field, produced by a magnet, exists. The terminals of the coil are left open ($i = 0$). Due to magnetic induction, a voltage signal at the moving-coil terminals appears.

To simplify matters, assume that the coil is a single-turn coil and that the magnetic induction field in the (very small) air gap between the north and south poles of the magnet is approximately uniform, $B = 0.5\,\text{T}$.

Also assume that the coil movement is described by $x(t) = l/2 + X\sin(\omega t)$, with $X < l/2$.

Q_1 Find the magnetic flux $\phi(x)$ across the single-turn coil positioned at x.

Q_2 By applying the generalized induction law, determine the voltage $u(t)$ between the terminals a and b of the moving coil.

Q_3 Consider that the coil contains N turns tightly packed.
Take $N = 5$, $X = 10\,\mu\text{m}$, $R = 5\,\text{mm}$, $\omega = 2\pi f$ and $f = 3\,\text{kHz}$.
Determine $u(t)$ numerically.

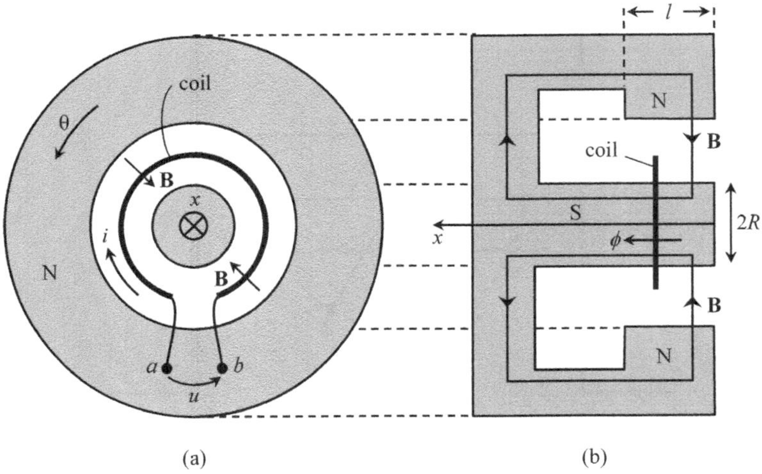

Figure 5.23 Moving-coil microphone. (a) Front view. (b) Transverse cross-section

Answers

Q_1 From div $\mathbf{B} = 0$, and noting that the radial **B** field in the air gap is redirected and transformed into an x-oriented field along the south pole piece, you find $\phi(x) = 2\pi RBx$.

Q_2

$$\begin{cases} \oint_{S(t)} \mathbf{E} \cdot d\mathbf{s} = -u \\ -\dfrac{d}{dt} \int_{S_{s(t)}} \mathbf{B} \cdot \mathbf{n}_S \, dS = -\dfrac{d\phi}{dt} = -\dfrac{d\phi}{dx}\dfrac{dx}{dt} = -2\pi RB\, v = -2\pi RBX\omega \cos(\omega t) \end{cases}$$

where $v = dx/dt$ is the coil's instantaneous velocity along x,

$$u(t) = 2\pi RBX\omega \cos(\omega t)$$

Q_3

$$u(t) = U \cos(\omega t), \text{ with } U = (2\pi)^2 RNBXf = 14.8\,\text{mV}$$

Problem 5.13.8

Figure 5.23 illustrating the functioning principle of the moving-coil microphone can also be used to explain the functioning principle of the loudspeaker.

Assume that a sinusoidal current $i(t) = I \cos(\omega t)$ is made to flow in the coil, with $I = 100\,\text{mA}$.

Q_1 Determine the magnetic Lorentz force $\mathbf{F} = F\vec{e}_x$ that puts the loudspeaker membrane into motion (producing sound waves).

Q_2 Evaluate the force numerically, considering the data specified in Problem 5.13.7.

Answers

Q_1

$$\mathbf{F} = \oint_{S} i \, d\mathbf{s} \times \mathbf{B} = i \int_{0}^{2\pi N} \left(-R \, d\theta \, \vec{e}_\theta\right) \times \left(-B \, \vec{e}_r\right) = \underbrace{2\pi NBRI \cos(\omega t)}_{F(t)} \vec{e}_x$$

Q_2

$$F(t) = F_{\max} \cos(\omega t), \quad F_{\max} = 2\pi NBRI = 7.85 \, \mathrm{mN}$$

6

Electric Induction Phenomena

6.1 Fundamental Equations

The fundamental laws governing electric induction problems are those in (PIII.4)

$$\begin{cases} \text{curl } \mathbf{H} = \mathbf{J} + \dfrac{\partial \mathbf{D}}{\partial t} \\ \text{div } \mathbf{D} = \rho \\ \text{curl } \mathbf{E} \approx 0 \end{cases} \tag{6.1}$$

together with the constitutive relations concerning $\mathbf{D}(\mathbf{E})$ and $\mathbf{J}(\mathbf{E})$.

While in Chapter 5 we were concerned with **E** fields originated by time-varying **B** fields, now we focus on the reverse situation, that is **B** fields originated by time-varying **E** fields.

6.2 Displacement Current, Generalized Ampère's Law

The reason why Maxwell's equations bear his name is essentially due to his speculative theoretical consideration that the equation curl $\mathbf{H} = \mathbf{J}$ (based on Ampère's previous discoveries) was an incomplete statement.

Magnetic induction phenomena (which had been known since Faraday's experiments) should have, expectedly, a natural dual counterpart consisting of electric induction phenomena. To accommodate this expectation Maxwell added to the conduction current density **J** a new contribution, the so-called displacement current density $\partial \mathbf{D}/\partial t$. With the introduction of this new term into the electromagnetic equations, the latter acquired a type of symmetry that was clearly missing. Since time-varying magnetic fields can give rise to electric fields, time-varying electric fields will likewise give rise to magnetic fields.

The speculative introduction by Maxwell of the displacement current density came to have far-reaching consequences. If the term $\partial \mathbf{D}/\partial t$ did not really exist, our world would not be like it is. It would be dark with no light at all; TV, mobile phones and the Internet would be meaningless words. (We will return to this topic in Part IV when dealing with electromagnetic waves.) The experimental proof that the displacement current was not simply an interesting speculation but real came only 35 years after Maxwell conceived it. The experimental work that revealed the existence of $\partial \mathbf{D}/\partial t$ was conducted by Hertz in 1889.

The equation

$$\text{curl } \mathbf{H} = \mathbf{J} + \frac{\partial \mathbf{D}}{\partial t} \tag{6.2}$$

shows that a magnetic field can be produced indistinctly by both current densities $\mathbf{J}$ and $\partial \mathbf{D}/\partial t$ – see Figure 6.1.

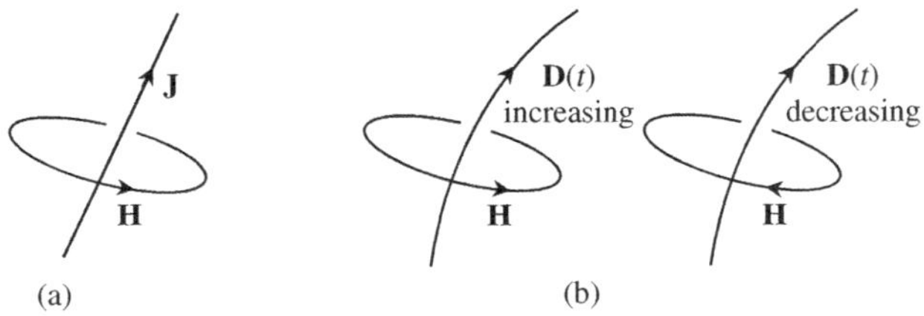

Figure 6.1 Magnetic fields can be originated either by conduction currents (a), or by displacement currents (b)

For purely stationary fields (Chapter 4), the integration of curl $\mathbf{H} = \mathbf{J}$ led in (4.3) to Ampère's law:

$$\oint_{\mathbf{s}} \mathbf{H} \cdot d\mathbf{s} = \int_{S_s} \mathbf{J} \cdot \mathbf{n}_S \, dS$$

For time-varying fields the above result can be generalized by writing

$$\oint_{\mathbf{s}} \mathbf{H} \cdot d\mathbf{s} = \int_{S_s} \left(\mathbf{J} + \frac{\partial \mathbf{D}}{\partial t} \right) \cdot \mathbf{n}_S \, dS \tag{6.3}$$

where the geometric entities $\mathbf{s}$, S_s and $\mathbf{n}_S$ should now be rather familiar to you (so often have we made use of the Stokes theorem).

6.3 Charge Continuity Equation

Let us apply the *divergence* operator to (6.2). Taking into account that div curl $\equiv 0$, and that div $\mathbf{D} = \rho$, you can easily get

$$0 = \text{div} \left(\mathbf{J} + \frac{\partial \mathbf{D}}{\partial t} \right) \;\rightarrow\; \text{div } \mathbf{J} + \frac{\partial}{\partial t} \text{div } \mathbf{D} \;\rightarrow\; \text{div } \mathbf{J} = -\frac{\partial \rho}{\partial t} \tag{6.4}$$

The result in (6.4) – the so-called charge continuity equation – shows that $\mathbf{J}$ lines being closed is not an intrinsic property of conduction currents (div $\mathbf{J} \neq 0$). Conduction currents can actually be interrupted as far as they can be continued through displacement currents.

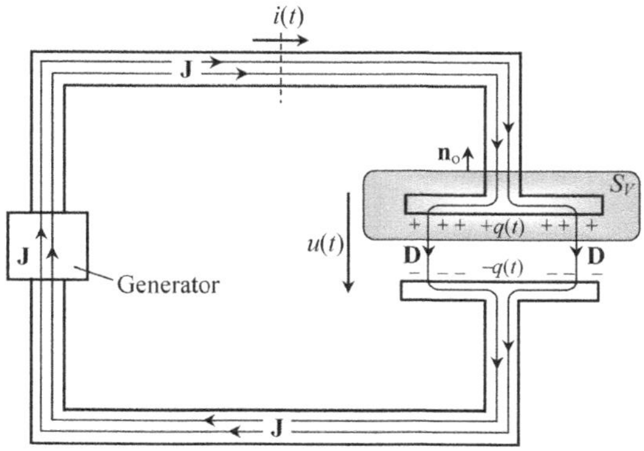

Figure 6.2 Time-varying conduction currents in the conductors are interrupted at the capacitor where they continue in the form of displacement currents

This idea is illustrated in Figure 6.2 where a closed circuit consisting of a generator, a pair of wires and a capacitor is shown.

Now, let us proceed to the volume integration of (6.4) considering the volume V encircling the upper plate of the capacitor in Figure 6.2:

$$\int_V \operatorname{div}\left(\mathbf{J}+\frac{\partial \mathbf{D}}{\partial t}\right) dV = 0 \rightarrow \int_V \operatorname{div} \mathbf{J}\, dV = -\frac{d}{dt}\int_V \rho\, dV \tag{6.5a}$$

Making use of the Gauss theorem we get

$$\int_{S_V} \mathbf{J}\cdot\mathbf{n}_o\, dS = -\frac{d}{dt}\underbrace{\int_V \rho\, dV}_{q(t)} \tag{6.5b}$$

where $q(t)$ is the electric charge accumulated on the conductor inside volume V.

According to the reference direction assigned to the conduction current $i(t)$, we finally get

$$i(t) = \frac{dq(t)}{dt} \tag{6.6}$$

Making use of the capacitance concept, $q = Cu$, we can rewrite (6.6) as

$$i(t) = C\frac{du(t)}{dt} \tag{6.7}$$

A word of caution is necessary here. If, for some reason, one of the reference directions for $i(t)$ or $u(t)$, shown in Figure 6.2, happens to be reversed, then a minus sign will need to be incorporated into (6.7). This is usually forgotten and may be the source of several mistakes.

A second word of caution: you have probably heard that capacitors can be used to block direct currents, or, in other words, that stationary currents are not allowed to flow in a circuit where a capacitor is included. Well, this is only half true!

If a continuously rising voltage is applied to a capacitor $u(t) = kt$, with k a constant, then from (6.7) you will get $i(t) = I = Ck$ = constant. Therefore you can see that while the capacitor voltage is allowed to increase steadily, you will get a direct current in the circuit. The problem is that, in reality, $u(t)$ cannot increase indefinitely (especially because of breakdown phenomena).

6.4 Revisiting the Current Intensity Concept

Contrary to what happened with stationary fields, we now have div $\mathbf{J} \neq 0$.

This new circumstance does affect the concept of current intensity in a conductor as originally introduced in Chapter 3. To illustrate the problem, consider Figure 6.3, where one conductor runs parallel to a conducting ground. The surrounding medium is a perfect insulator. A time-varying voltage is applied between the conductors.

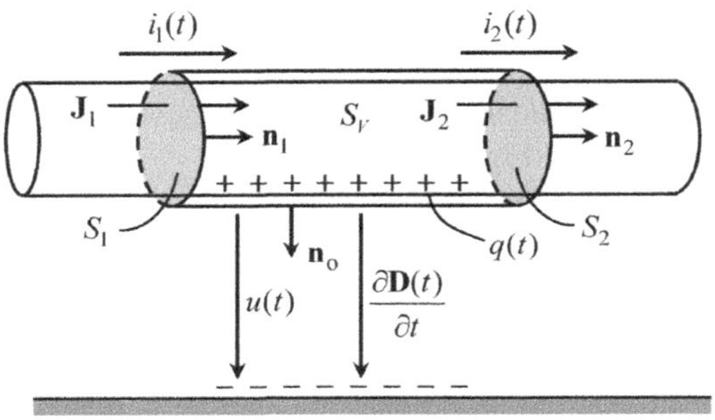

Figure 6.3 For time-varying regimes, the current intensity in a conductor varies along the length of the conductor, $i_1(t) \neq i_2(t)$. For evaluating $i(t)$, a clear specification of the conductor's cross-section being considered is required

If you apply (6.5) to the closed surface S_V intersecting the conductor at S_1 and S_2, you get

$$\begin{cases} \int_{S_V} \mathbf{J} \cdot \mathbf{n}_o \, dS = \int_{S_2} \mathbf{J}_2 \cdot \mathbf{n}_2 \, dS - \int_{S_1} \mathbf{J}_1 \cdot \mathbf{n}_1 dS = i_2(t) - i_1(t) \\ -\int_{S_V} \frac{\partial \mathbf{D}}{\partial t} \cdot \mathbf{n}_o \, dS = -\frac{dq(t)}{dt} \end{cases}$$

By equating the equations above you will see that $i_1(t) \neq i_2(t)$, that is

$$i_1(t) - i_2(t) = dq(t)/dt \tag{6.8}$$

where $q(t)$ is the conductor's time-varying charge in the region between S_1 and S_2.

The difference between the current intensities in the same conductor, but measured at two distinct cross-sections, is justified by the flow of displacement current between the conductor and the ground.

As the magnitude and time rate of the applied voltage increase, the term dq/dt increases, and as a result the discrepancy between i_1 and i_2 becomes more and more important.

Hence, you can see that for an unambiguous definition of the current intensity $i(t)$ in a conductor, you ought to specify clearly the particular cross-section S where its evaluation takes place. For slow time-varying phenomena this requirement is ordinarily dropped, since $dq/dt \approx 0$.

6.5 Application Example (Capacitor Self-Discharge)

A parallel-plate capacitor of area S and thickness δ is connected to a DC generator with voltage U_0. The capacitor's dielectric medium is characterized by its permittivity ε and conductivity σ. At $t = 0$ the capacitor is disconnected from the generator and from there on the capacitor starts its self-discharge process.

Assume that the electric field inside the dielectric medium is uniform (small δ).

Questions

Q_1 Find the initial value of the electric field intensity E_0.

Q_2 Find the evolution of $E(t)$ for $t > 0$.

Q_3 Determine the conduction current density and the displacement current density inside the capacitor for $t > 0$.

Q_4 Find the magnetic field inside the discharging capacitor.

Solutions

Q_1 For the case of uniform fields we have $U_0 = E_0\,\delta \;\rightarrow\; E_0 = U_0/\delta$.

Q_2 Consider the application of (6.5b) to the closed surface S_V containing the upper plate of the capacitor (Figure 6.4).

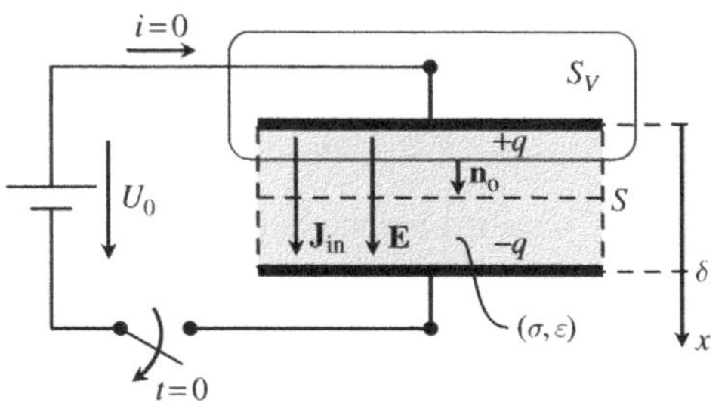

Figure 6.4 Application of the charge continuity equation to the analysis of the capacitor self-discharge process. Note that the insulation medium that fills the capacitor is imperfect ($\sigma \neq 0$)

For $t > 0$ the external conduction current density $\mathbf{J}_{\text{ex}}$ is zero; however, inside the imperfect dielectric medium an internal conduction current density $\mathbf{J}_{\text{in}}$ exists,

$$\mathbf{J}_{\text{in}}(t) = \sigma \mathbf{E}(t) = \sigma E(t)\vec{e}_x$$

then we have

$$\underbrace{\int_{S_V} \mathbf{J}_{\text{in}} \cdot \mathbf{n}_{\text{o}}\, dS}_{\sigma S E(t)} = -\frac{d}{dt} q(t)$$

For the case of uniform fields, the electric charge $q(t)$ is the product of the surface charge density $w(t)$ and the plate area $q = wS$. But, from Chapter 2, $w = \varepsilon E$. Therefore we find, for the electric field inside the capacitor,

$$\sigma E(t) + \varepsilon \frac{dE(t)}{dt} = 0$$

This is a homogeneous first-order linear equation with constant coefficients whose solution is a decaying exponential function

$$E(t) = E_0 \exp\left(-\frac{\sigma}{\varepsilon} t\right) \tag{6.9}$$

The constant $\tau = \varepsilon/\sigma$ is usually termed the relaxation time. Depending on the dielectric medium, this relaxation time can go from seconds to hours.

Q_3 Internal conduction current density:

$$\mathbf{J}_{\text{in}} = \sigma E_0 \exp\left(-\frac{\sigma}{\varepsilon} t\right) \vec{e}_x$$

Displacement current density:

$$\frac{d\mathbf{D}}{dt} = \varepsilon \frac{dE}{dt} \vec{e}_x = -\sigma E_0 \exp\left(-\frac{\sigma}{\varepsilon} t\right) \vec{e}_x$$

Q_4 As for the total current density inside the dielectric medium, we find

$$\mathbf{J}_{\text{in}} + \frac{\partial \mathbf{D}}{\partial t} = 0$$

and, consequently, from (6.2), no net magnetic field is originated, $\mathbf{H} = 0$.

6.6 Voltages and Currents in Electrically Multicoupled Systems

Consider a set of $N + 1$ wires immersed in a perfectly insulating dielectric medium, one of them being taken as the reference conductor (0) – see cross-sectional view in Figure 6.5.

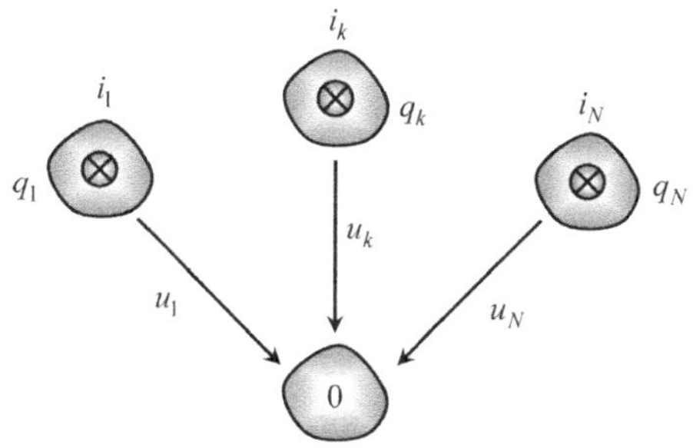

Figure 6.5 In a multi-wire system, the voltages of all the conductors contribute to the current intensity in any given conductor of the system, $i_k = \sum C_{kj}\, du_j/dt$

If time-varying voltages are applied between conductors, conduction currents will flow along them, and displacement currents will flow across the dielectric medium.

Denoting the conductor charges by $q_1(t), \ldots, q_k(t), \ldots, q_N(t)$, we have from (6.6)

$$i_k(t) = \frac{dq_k(t)}{dt} \tag{6.10a}$$

Since charges and voltages, in multiconductor systems, are related by capacitance coefficients –recall (2.43) and (2.44) from Chapter 2 – then we can rewrite (6.10a) as

$$i_k(t) = C_{k1}\frac{du_1(t)}{dt} + \cdots + C_{kk}\frac{du_k(t)}{dt} + \cdots + C_{kN}\frac{du_N(t)}{dt} \tag{6.10b}$$

In compact matrix format we get

$$[i(t)] = [C]\frac{d}{dt}[u(t)] \tag{6.11}$$

where the column matrix $[i(t)]$ gathers the conduction currents flowing along the wires, and the column matrix $[u(t)]$ gathers the wire voltages. The square symmetric matrix $[C]$ is the capacitance matrix whose properties were discussed in Chapter 2.

If the dielectric is not a perfectly insulating medium, leakage conduction currents will also flow in the dielectric. In that case we must substitute (6.12) for (6.11):

$$[i(t)] = [G][u(t)] + [C]\frac{d}{dt}[u(t)] \tag{6.12}$$

where the first term on the right-hand side of (6.12) accounts for the perturbation due to leakage currents. The square symmetric matrix $[G]$ is the system conductance matrix.

If the dielectric medium is homogeneous, characterized by parameters ε and σ, the relationship between $[G]$ and $[C]$ is trivial:

$$[G] = \frac{\sigma}{\varepsilon}[C] \tag{6.13}$$

See (3.11) in Chapter 3 and Problem 3.9.6.

6.7 Proposed Homework Problems

Problem 6.7.1

A generator voltage $u(t)$ is applied to two series-connected capacitors whose capacitances are C_1 and C_2 – see Figure 6.6. The capacitors are initially discharged. The generator voltage has the shape of a trapezoidal pulse.

Assume the ordinary approximations valid for slow time-varying regimes.

Data: $C_1 = 4\,\mu\text{F}$, $C_2 = 6\,\mu\text{F}$, $T = 1\,\text{ms}$, $U = 15\,\text{V}$.

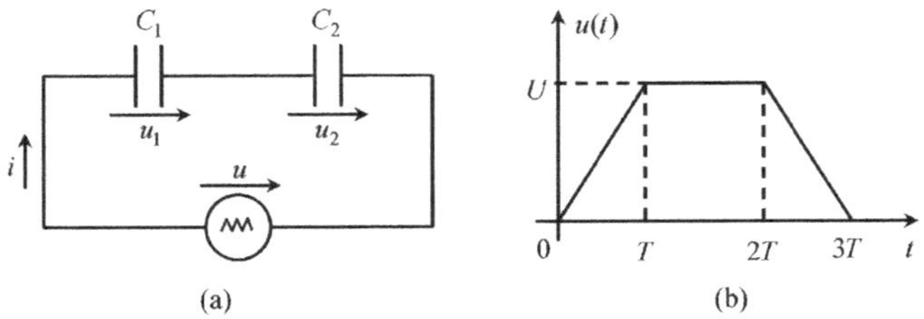

Figure 6.6 Two series-connected capacitors (a) subjected to a trapezoidal pulse voltage (b)

Q_1 Write the governing equations of the system.

Q_2 Determine $i(t)$.

Q_3 Determine $u_1(t)$ and $u_2(t)$.

Answers

Q_1 $i = dq_1/dt = C_1 du_1/dt$; $i = dq_2/dt = C_2 du_2/dt$; $u = u_1 + u_2$

$$\frac{i}{C_1} + \frac{i}{C_2} = \frac{d}{dt}\underbrace{(u_1 + u_2)}_{u} \rightarrow i(t) = \underbrace{\left(\frac{C_1 C_2}{C_1 + C_2}\right)}_{C}\frac{du(t)}{dt}$$

$$\begin{cases} C_1 u_1 = C_2 u_2 \\ u_1 + u_2 = u \end{cases} \rightarrow \begin{cases} u_1 = \dfrac{C}{C_1} u \\ u_2 = \dfrac{C}{C_2} u \end{cases}$$

Q_2

$$i(t) = \begin{cases} +CU/T = 36\,\text{mA}, \text{ for } 0 < t < T \\ 0, \text{ for } T < t < 2T \\ -CU/T = -36\,\text{mA}, \text{ for } 2T < t < 3T \end{cases}$$

Note that in the interval from T to $2T$ the electric field remains time invariant and electric induction phenomena do not occur ($\mathbf{H}=0$).

Q_3

$$\begin{cases} u_1(t) = 0.6 \times u(t) \\ u_2(t) = 0.4 \times u(t) \end{cases}$$

Problem 6.7.2

A sinusoidal voltage $u(t)$ is applied to two parallel-connected capacitors whose capacitances are C_1 and C_2 – see Figure 6.7. The capacitors are initially discharged.

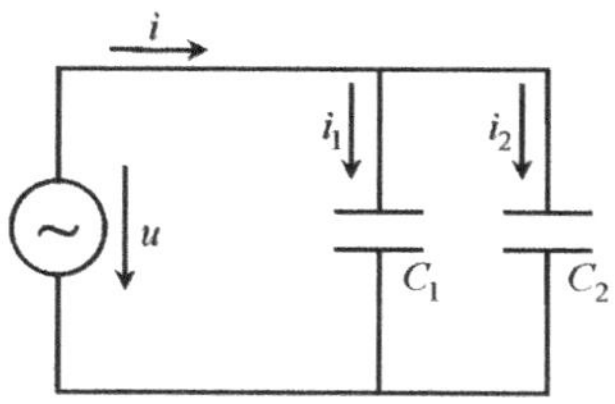

Figure 6.7 Two parallel-connected capacitors subjected to a sinusoidal voltage

The current intensity i_1 is known and is given by $i_1(t) = I_1 \cos(\omega t)$.
Assume the ordinary approximations valid for slow time-varying regimes.
Data: $C_1 = 4\,\mu\text{F}$, $C_2 = 6\,\mu\text{F}$, $I_1 = 40\,\text{mA}$, $\omega = 1\,\text{krad/s}$.

Q_1 Write the governing equations of the system.

Q_2 Determine $u(t)$.

Q_3 Determine $i_2(t)$ and $i(t)$.

Answers

Q_1 $u(t) = \dfrac{1}{C_1} \int i_1(t)dt;\ i_2(t) = C_2 \dfrac{du(t)}{dt};\ i(t) = i_1(t) + i_2(t)$

Q_2 $u(t) = U \sin(\omega t)$, with $U = I_1/(\omega C_1) = 10\,\text{V}$.

Q_3 $i_2(t) = I_2 \cos(\omega t)$, with $I_2 = \omega C_2 U = 60\,\text{mA}$.

$i(t) = I \cos(\omega t)$, with $I = I_1 + I_2 = 100\,\text{mA}$.

Problem 6.7.3

A printed circuit board consisting of three conducting lands on the surface of a dielectric board above a reference conducting ground plane is shown in Figure 6.8.

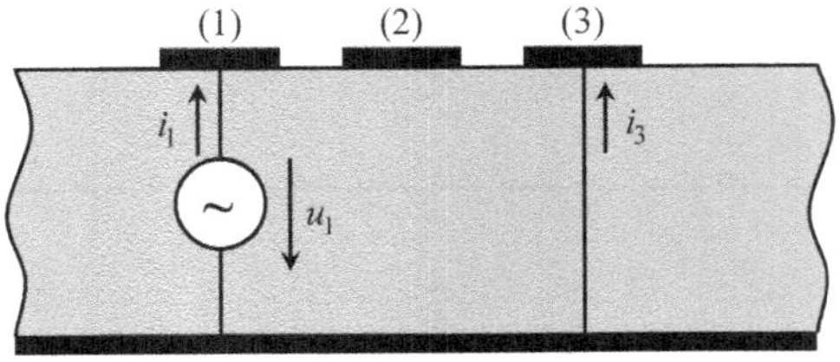

Figure 6.8 Front view of a printed circuit board with three lands on a dielectric over a ground plane

All lands are left open at the far end of the board. As for the near end, the situation is described as follows: land 1 is driven by a voltage u_1, land 2 is left open ($i_2 = 0$) and land 3 is short-circuited to ground ($u_3 = 0$). The conductor system is characterized by a symmetric capacitance matrix

$$[C] = \begin{bmatrix} C_{11} & C_{12} & C_{13} \\ C_{21} & C_{22} & C_{23} \\ C_{31} & C_{32} & C_{33} \end{bmatrix}$$

The generator voltage is a sinusoidal function, $u_1(t) = U_1 \sin(\omega t)$.

Q_1 Obtain the governing equations of the system.

Q_2 Determine $u_2(t)$. Determine $i_1(t)$ and $i_3(t)$.

Answers

Q_1 From (6.11), with $u_3 = 0$, we get

$$\begin{cases} i_1 = C_{11}\dfrac{du_1}{dt} + C_{12}\dfrac{du_2}{dt} \\ 0 = C_{21}\dfrac{du_1}{dt} + C_{22}\dfrac{du_2}{dt} \\ i_3 = C_{31}\dfrac{du_1}{dt} + C_{32}\dfrac{du_2}{dt} \end{cases}$$

Q_2

$$u_2(t) = -\frac{C_{21}}{C_{22}} u_1(t)$$

$$i_1(t) = \left(C_{11} - \frac{C_{12}^2}{C_{22}}\right)\frac{du_1(t)}{dt} = \omega U_1\left(C_{11} - \frac{C_{12}^2}{C_{22}}\right)\cos(\omega t)$$

$$i_3(t) = \left(C_{31} - \frac{C_{32}C_{21}}{C_{22}}\right)\frac{du_1(t)}{dt} = \omega U_1\left(C_{31} - \frac{C_{32}C_{21}}{C_{22}}\right)\cos(\omega t)$$

Problem 6.7.4

A section of a three-phase symmetrical shielded cable is described by its capacitance matrix

$$[C] = \begin{bmatrix} C_s & -C_m & -C_m \\ -C_m & C_s & -C_m \\ -C_m & -C_m & C_s \end{bmatrix}$$

As shown in Figure 6.9, the internal cable conductors are all short-circuited. The cable is left open at its far end. At the near end, a ramp-type voltage is applied between the reunion point of the conductors and the shield conductor.

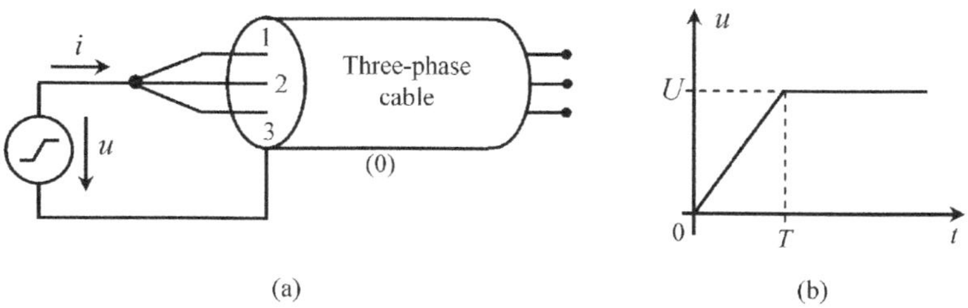

Figure 6.9 Shielded three-phase cable. (a) System connections. (b) Generator voltage

Data: $C_s = 50\,\text{nF}$, $C_m = 20\,\text{nF}$, $U = 5\,\text{kV}$, $T = 1\,\text{ms}$.

Q_1 Write the equations governing the system.

Q_2 Determine the generator current $i(t)$, showing that it has the shape of a positive pulse.

Q_3 Evaluate the energy expended by the generator to charge the cable. Compare it to the final electric energy stored in the cable.

Answers

Q_1 From (6.11) with $u_1 = u_2 = u_3 = u(t)$ you find

$$i_1(t) = i_2(t) = i_3(t) = (C_s - 2C_m)\frac{du(t)}{dt}$$

$$i(t) = i_1(t) + i_2(t) + i_3(t) = \underbrace{3(C_s - 2C_m)}_{C}\frac{du(t)}{dt}$$

where C is the equivalent capacitance observed at the generator terminals.

Q_2 Since $C = 30\,\text{nF}$ and $du/dt = 5\,\text{MV/s}$, you find

$$\begin{cases} i(t) = I = 0.15\,\text{A}, & \text{for } 0 < t < T \\ i(t) = 0, \ \text{for } t > T \end{cases}$$

Q_3

$$W_G = \int_{t=0}^{t=\infty} ui\,dt = \int_{t=0}^{t=T} ui\,dt = \frac{UIT}{2} = 375\,\text{mJ};\ W_e = \frac{1}{2}CU^2 = 375\,\text{mJ for } t > T$$

7

Lumped Parameters Circuit Analysis

7.1 Introduction

This chapter is probably the easiest one for you. Vector calculus, differential operators, the Stokes and Gauss theorems, all of which have been harsh tools in previous chapters, will be almost absent here. Nonetheless, new, but softer, difficulties may now arise with the handling of complex algebra (see Appendix C).

In this new chapter we will make extensive use of the results derived in Chapters 5 and 6 concerning slow time-varying field phenomena (quasi-stationary regimes), which is the standard framework for circuit analysis. Magnetic induction phenomena and electric induction phenomena are considered separately; while the former are taken into account when lumped inductors are analyzed, the latter are taken into account when lumped capacitors are analyzed. In the case of lumped resistors, neither induction phenomena are considered.

Again, bear in mind that this lumped parameters approach is only valid when the length of the circuit structure under analysis is much shorter than the lowest wavelength characterizing the time evolution of the field.

To make things clearer, consider the typical *RLC* series circuit in Figure 7.1.

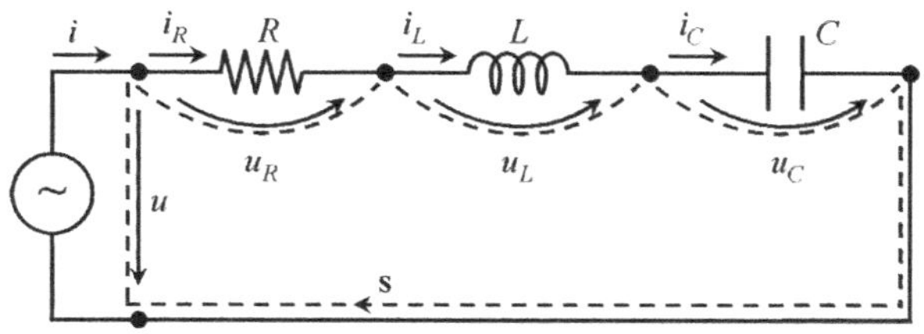

Figure 7.1 *RLC* series circuit

Electromagnetic Foundations of Electrical Engineering J. A. Brandão Faria

Assuming that the term $\partial \mathbf{B}/\partial t$ created by the time-varying currents in the connecting wires is negligibly small, the application of the induction law gives

$$\oint_{\mathbf{S}} \mathbf{E} \cdot d\mathbf{s} = -\int_{S_s} \frac{\partial \mathbf{B}}{\partial t} \cdot \mathbf{n}_s \, dS \approx 0 \rightarrow u(t) = u_R(t) + u_L(t) + u_C(t) \tag{7.1}$$

Assuming that the displacement currents $\partial \mathbf{D}/\partial t$ created by the time-varying voltages between the connecting wires is negligibly small, application of the generalized Ampère's law gives

$$\int_{S_V} \mathbf{J} \cdot \mathbf{n}_o \, dS = -\int_{S_V} \frac{\partial \mathbf{D}}{\partial t} \cdot \mathbf{n}_o \, dS \approx 0 \rightarrow i(t) = i_R(t) = i_L(t) = i_C(t) \tag{7.2}$$

Assuming that both terms $\partial \mathbf{B}/\partial t$ and $\partial \mathbf{D}/\partial t$ are negligible for the resistor analysis, we have from Ohm's law (Chapter 3)

$$u_R(t) = R \, i_R(t) \tag{7.3}$$

Neglecting $\partial \mathbf{D}/\partial t$, but taking into account the important magnetic induction phenomena in the inductor, we have, from the induction law (Chapter 5),

$$u_L = L \frac{di_L(t)}{dt} \tag{7.4}$$

Finally, neglecting $\partial \mathbf{B}/\partial t$, but taking into account the important electric induction phenomena in the capacitor, we have, from the generalized Ampère's law (Chapter 6),

$$i_C = C \frac{du_C(t)}{dt} \rightarrow u_C(t) = \frac{1}{C} \int i_C(t) \, dt \tag{7.5}$$

By using the results from (7.1) to (7.5), valid for quasi-stationary regimes, we find the time-domain equation that governs the lumped parameters circuit in Figure 7.1:

$$u(t) = Ri(t) + L \frac{di(t)}{dt} + \frac{1}{C} \int i(t) \, dt \tag{7.6}$$

7.2 Steady-State Harmonic Regimes

In this section we particularize circuit analysis for the special case of steady-state harmonic regimes, where fields, voltages, magnetic fluxes, electric charges and current intensities are described by time-varying sinusoidal functions. Moreover, it is assumed that the generator driving the circuit under analysis was turned on a long time ago (transient phenomena discarded).

In order to ensure that all the quantities referred to above have a sinusoidal description, an additional condition ought to be fulfilled: all the lumped components of the circuit must exhibit linear behavior. Note for instance that if you apply a sinusoidal voltage to a diode (nonlinear component) its current will be non-sinusoidal.

Before we proceed to the analysis of harmonic regimes a necessary comment is in order to justify the need for this type of analysis. Why are sinusoidal functions so important?

First of all, the transmission and distribution of electric power is made using AC (Alternating Current) – that is, using sinusoidal currents. In residential applications, the accessible voltage at the sockets that you use every day at home to plug in your appliances is a sinusoidal voltage (its frequency can be 50 Hz or 60 Hz depending on the country you live in).

At this stage you may be wondering about communication signals. They are certainly not sinusoidal functions!

Well, you are right. But you are missing the point.

Any regular well-behaved time-varying signal $s(t)$ can be expanded into a discrete or continuous sum of sinusoids of different frequencies, which constitute the so-called signal spectrum as follows:

$$\text{Periodic signals of period } T: \; s(t) = s_{av} + \sum_{k=1}^{\infty} S_k \cos(\omega_k t - \phi_k), \text{ with } \omega_k = 2\pi k/T$$

$$\text{Non-periodic signals}: \; s(t) = \int_0^{\infty} S(\omega) \cos(\omega t - \phi(\omega))\, d\omega$$

(See the results on Fourier series and transforms, in Appendix D.)

So, what you have to do is to analyze the time response of the circuit to each and every sinusoidal component of the signal spectrum and, at the end, superpose the results obtained. We emphasize again that this procedure with sinusoidal functions is only valid for linear circuits!

7.2.1 Characterization of Sinusoidal Quantities

In this subsection we introduce the standard terminology used to deal with sinusoidal functions.

Let $u_1(t)$ be a sinusoidal voltage given by

$$u_1(t) = U_1 \cos \underbrace{(\omega t + \alpha_1)}_{\phi_1(t)} \tag{7.7}$$

where U_1 denotes the maximum value of the voltage or amplitude, $\phi_1(t)$ denotes the time-varying phase, α_1 denotes the initial phase (for $t = 0$) and ω denotes the angular frequency $\omega = 2\pi f$, where f is the frequency in hertz.

If $\phi_1(t)$ is plotted against time you will obtain a tilted straight line (Figure 7.2). Since the cosine function repeats itself upon an angle shift of 2π, you can see that the sinusoidal time period is such that $T = 2\pi/\omega = 1/f$.

If two sinusoidal functions of the same frequency are compared, $u_1(t)$ given above and $u_2(t)$ given by

$$u_2(t) = U_2 \cos \underbrace{(\omega t + \alpha_2)}_{\phi_2(t)}$$

we will say that they are out of phase, because $\varphi = \phi_1(t) - \phi_2(t) = \alpha_1 - \alpha_2 \neq 0$.

Note that, in the specification of phase shifts, the interval $-\pi < \varphi < +\pi$ is commonly used.

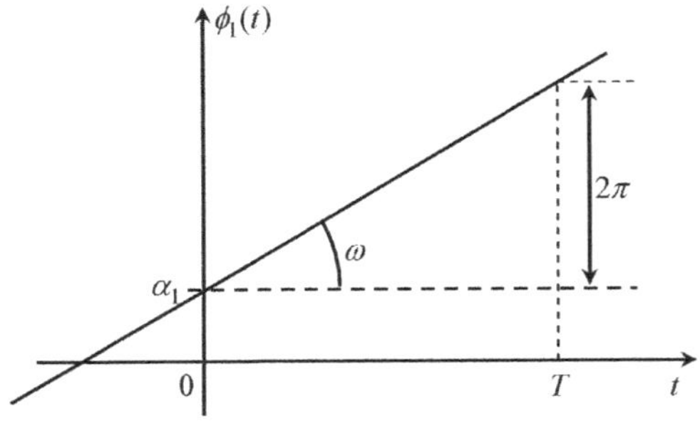

Figure 7.2 Phase against time; α_1 denotes the initial phase and ω denotes the angular frequency

Let us analyze a few particular cases:

- If $\varphi = 0$ the sinusoidal functions are said to be in phase.
- If $\varphi > 0$ we say that u_1 leads u_2; on the contrary, if $\varphi < 0$ we say that u_1 lags u_2.
- If $\varphi = \pm\pi$ the sinusoidal functions are said to be in phase opposition, which is equivalent to a $T/2$ time shift.
- If $\varphi = \pm\pi/2$ the sinusoidal functions are said to be in phase quadrature, which is equivalent to a $T/4$ time shift.

Figure 7.3 illustrates the above particular cases for you.

The specification of the average value of sinusoidal functions is absolutely useless. In fact, from (7.7) you can immediately recognize that any sinusoidal function has zero as its average value. A really important piece of information related to the amplitude of a sinusoidal function is its root-mean-square (rms) value. In general, for a time-periodic function $u(t)$, its rms value is defined as

$$U_{rms} = \sqrt{(u^2(t))_{av}} \tag{7.8}$$

The importance of this concept is linked to the evaluation of the average value of time-varying energetic quantities, like power and energy, as illustrated in the following cases.

Joule power losses in a resistor:

$$p_J(t) = Ri^2(t) \rightarrow P_J = (p_J(t))_{av} = R\left(i^2(t)\right)_{av} = RI_{rms}^2 \tag{7.9a}$$

Magnetic energy in an inductor:

$$W_m(t) = \tfrac{1}{2}Li^2(t) \rightarrow (W_m)_{av} = \tfrac{1}{2}L\left(i^2(t)\right)_{av} = \tfrac{1}{2}LI_{rms}^2 \tag{7.9b}$$

Electric energy in a capacitor:

$$W_e(t) = \tfrac{1}{2}Cu^2(t) \rightarrow (W_e)_{av} = \tfrac{1}{2}C\left(u^2(t)\right)_{av} = \tfrac{1}{2}CU_{rms}^2 \tag{7.9c}$$

In the particular situation of sinusoidal quantities, where $\left(\cos^2(\phi(t))\right)_{av} = 1/2$, the result in (7.8) simplifies to

$$U_{rms} = \sqrt{(u^2(t))_{av}} = U/\sqrt{2} \tag{7.10}$$

where U is the maximum value of $u(t)$.

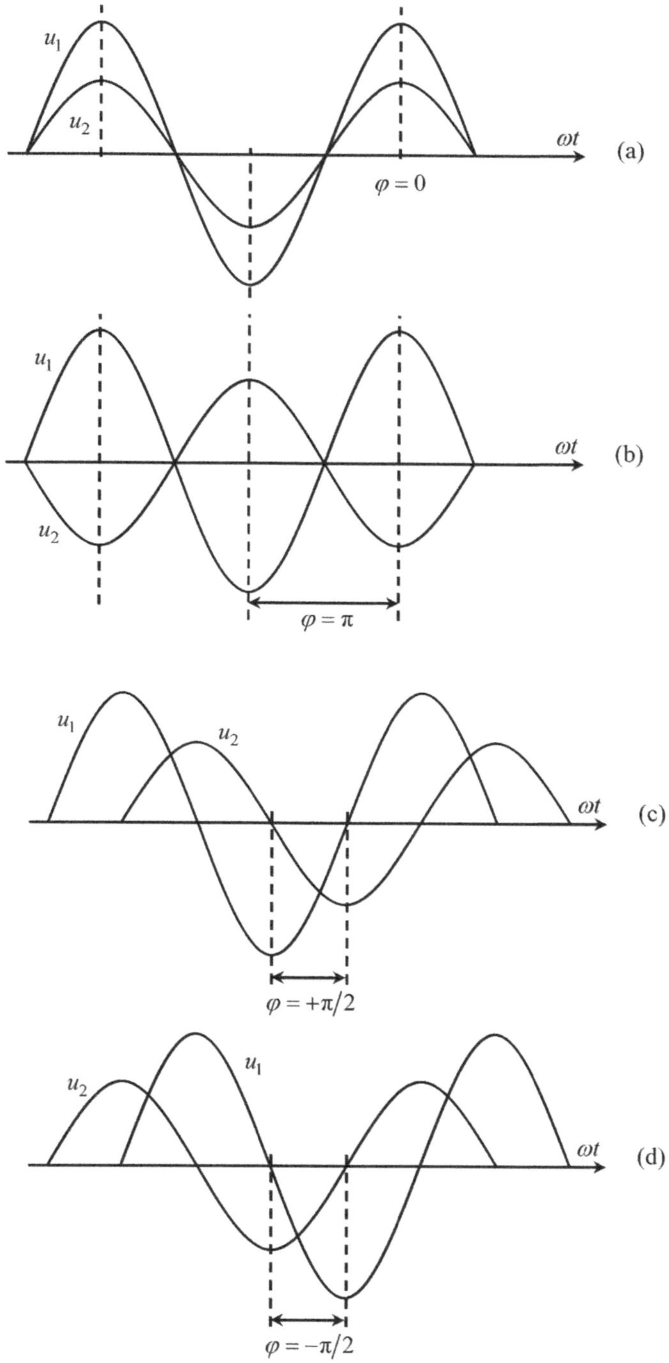

Figure 7.3 (a) Voltages u_1 and u_2 are in phase. (b) Voltages u_1 and u_2 are in phase opposition. (c) Voltages u_1 and u_2 are in phase quadrature with u_1 leading u_2. (d) Voltages u_1 and u_2 are in phase quadrature with u_1 lagging u_2

7.2.2 Complex Amplitudes or Phasors

Time-varying sinusoidal quantities of a given frequency can be represented by complex constants conveying information not only on the amplitude, but also on the initial phase of the sinusoid. These complex constants are termed complex amplitudes or phasors.

Consider, for instance, the voltage $u_1(t)$ in (7.7),

$$u_1(t) = U_1 \cos \underbrace{(\omega t + \alpha_1)}_{\phi_1(t)}$$

By using the Euler identity

$$e^{j\phi_1(t)} = \cos(\phi_1(t)) + j\sin(\phi_1(t)) \rightarrow \Re\left(e^{j\phi_1(t)}\right) = \cos(\phi_1(t))$$

we immediately recognize that $u_1(t)$ can be rewritten in the form

$$u_1(t) = \Re\left(\left(U_1 e^{j\alpha_1}\right) e^{j\omega t}\right) = \Re\left(\overline{U}_1\, e^{j\omega t}\right) \tag{7.11}$$

The time-invariant quantity $\overline{U}_1 = U_1 e^{j\alpha_1}$ is the complex amplitude of the sinusoidal voltage $u_1(t)$.

The simplicity of the preceding formulation should certainly not pose any doubt. However, some of you may be thinking what the purpose of this is. What is the usefulness of substituting complex amplitudes by the time functions they represent?

This a simple question, with a simple answer. First of all, you should note that, by this means, we are using a time-invariant quantity to represent a time-varying function. Secondly, as will show next, operations (like sum, differentiation and integration) involving sinusoidal functions can be much more easily performed in the complex domain than in the time domain.

Consider the following example.

You want to determine the sum, $u_3 = u_1 + u_2$, of two sinusoidal voltages with the same frequency, $u_1(t) = U_1\cos(\omega t + \alpha_1)$ and $u_2(t) = U_2\cos(\omega t + \alpha_2)$. You can do that by resorting to standard trigonometry but you will waste a lot a time. One thing that you should know is that the sum of two sinusoids of the same frequency will yield a resultant sinusoid with the same frequency – that is, you are expecting a result in the form $u_3(t) = U_3\cos(\omega t + \alpha_3)$. So, let us then use the phasor technique to find U_3 and α_3:

$$U_3\cos(\omega t + \alpha_3) = U_1\cos(\omega t + \alpha_1) + U_2\cos(\omega t + \alpha_2)$$
$$\Re\left(\overline{U}_3\, e^{j\omega t}\right) = \Re\left(\overline{U}_1\, e^{j\omega t}\right) + \Re\left(\overline{U}_2\, e^{j\omega t}\right) = \Re\left(\left(\overline{U}_1 + \overline{U}_2\right) e^{j\omega t}\right)$$

From the above result you immediately obtain $\overline{U}_3 = U_3\, e^{j\alpha_3} = \overline{U}_1 + \overline{U}_2$ (see Figure 7.4). Now that's simple, isn't it?

Consider another example.

Let $q(t)$ be a given sinusoidal function, $q(t) = Q\cos(\omega t + \alpha_q)$. We wish to determine its time derivative, $i(t) = dq/dt = I\cos(\omega t + \alpha_i)$. Let us then use the phasor technique to find I and α_i:

$$i(t) = \Re\left(\overline{I}\, e^{j\omega t}\right) = \frac{d}{dt}q(t) = \frac{d}{dt}\Re\left(\overline{Q}\, e^{j\omega t}\right) = \Re\left(\left(j\omega\overline{Q}\right) e^{j\omega t}\right)$$

from which you can see that $\overline{I} = j\omega\overline{Q}$ or, equivalently, $I\, e^{j\alpha_i} = \omega Q\, e^{j(\alpha_q + \pi/2)}$.

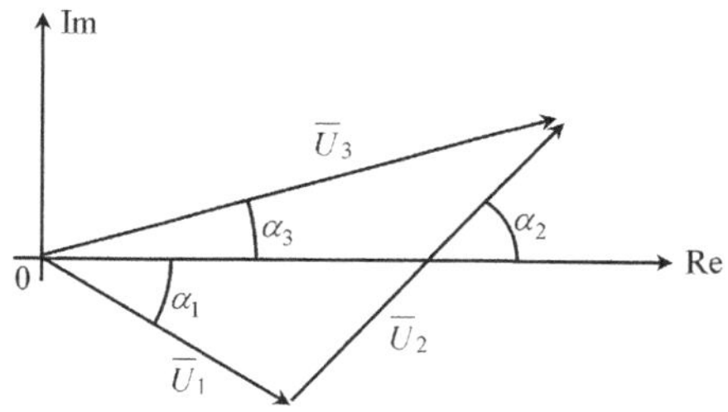

Figure 7.4 Illustration of the sum operation involving complex amplitudes

Results for time integration can be obtained from the above:

$$q(t) = \int i(t)\, dt \rightarrow \overline{Q} = \frac{1}{j\omega}\overline{I}$$

Table 7.1 summarizes the principal conclusions concerning the equivalence between time-domain operations and their corresponding phasor-domain operations.

Table 7.1 Time- and phasor-domain operations

Time domain	Phasor domain
$u_3(t) = u_1(t) + u_2(t)$	$\overline{U}_3 = \overline{U}_1 + \overline{U}_2$
$i(t) = \dfrac{d}{dt}q(t)$	$\overline{I} = j\omega\overline{Q}$
$q(t) = \int i(t)\, dt$	$\overline{Q} = \dfrac{1}{j\omega}\overline{I}$

7.2.3 Application Example (RLC Circuit)

Consider the time-domain equations for the *RLC* circuit obtained in (7.1)–(7.6). Assume that all voltages and currents are sinusoidal functions of angular frequency ω.

Questions

Q_1 Obtain the phasor-domain equations for this circuit (Figure 7.1).

Q_2 Comment on the phase relationships between voltages and currents at the resistor, inductor and capacitor terminals.

Q_3 Draw an illustrative phasor diagram showing the phasor's positions in the complex plane. Consider the following data: $i(t) = I\cos(\omega t)$, $I = 1\,\text{A}$, $\omega = 1\,\text{krad/s}$.
Take $R = 100\ \Omega$, $L = 0.2\,\text{H}$, $C = 0.1\,\mu\text{F}$.

Q_4 Several voltmeters are connected to the circuit as shown in Figure 7.5; voltmeter readings are rms values. Determine the readings of V_1, V_2, V_3, V_4 and V_5. Comment on the results.

Q_5 Write the expression for $u(t)$ and plot it against time.

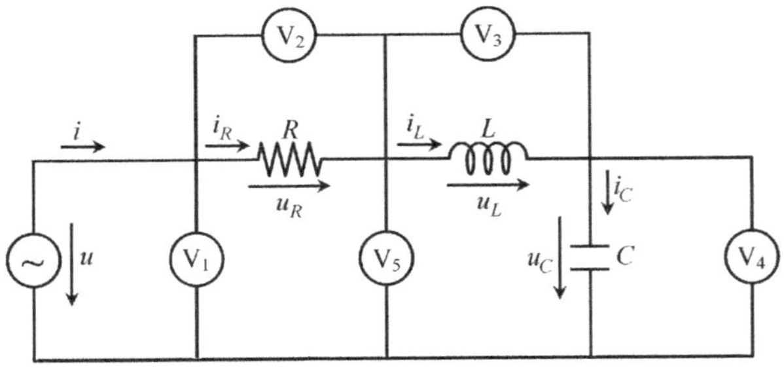

Figure 7.5 *RLC* series circuit provided with a set of voltmeters for measuring voltage rms values

Solutions

Q_1

$$u(t) = u_R(t) + u_L(t) + u_C(t) \rightarrow \overline{U} = \overline{U}_R + \overline{U}_L + \overline{U}_C \tag{7.12}$$

$$i(t) = i_R(t) = i_L(t) = i_C(t) \rightarrow \overline{I} = \overline{I}_R = \overline{I}_L = \overline{I}_C$$

$$u_R(t) = R\, i_R(t) \rightarrow \overline{U}_R = R\,\overline{I}_R \tag{7.13}$$

$$u_L = L\frac{di_L(t)}{dt} \rightarrow \overline{U}_L = j\omega L\,\overline{I}_L \tag{7.14}$$

$$u_C(t) = \frac{1}{C}\int i_C(t)\,dt \rightarrow \overline{U}_C = \frac{1}{j\omega C}\overline{I}_C \tag{7.15}$$

$$u(t) = Ri(t) + L\frac{di(t)}{dt} + \frac{1}{C}\int i(t)\,dt \rightarrow \overline{U} = \left(R + j\omega L + \frac{1}{j\omega C}\right)\overline{I} \tag{7.16}$$

Q_2 From (7.13) you can see that, in the complex plane, the phasors $\overline{U}_R$ and $\overline{I}_R$ are parallel vectors. This means that $u_R(t)$ and $i_R(t)$ are in phase.

From (7.14) you can see that the phasors $\overline{U}_L$ and $\overline{I}_L$ are orthogonal vectors. This means that $u_L(t)$ and $i_L(t)$ are in phase quadrature, with u_L leading i_L.

From (7.15) you can see that the phasors $\overline{U}_C$ and $\overline{I}_C$ are again orthogonal. This means that $u_C(t)$ and $i_C(t)$ are in phase quadrature, with u_C lagging i_C.

Q_3 $\overline{I} = 1\,\text{A}$, $\overline{U}_R = 100\,\text{V}$, $\overline{U}_L = 200\, e^{j\pi/2}\,\text{V}$, $\overline{U}_C = 100\, e^{-j\pi/2}\,\text{V}$, $\overline{U} = \sqrt{2}\;100\,e^{j\pi/4}\,\text{V}$.

See the corresponding phasor diagram in Figure 7.6.

You may be commenting to yourself that the phasor diagram is not correct, because the size of the phasor $\overline{I}$ appears to be bigger than the size of the voltage phasors. Well, you are wrong!

In the same way that you cannot compare kg to km/s, you cannot compare ampere with volt. In no case at all can you establish inequality relations between quantities of different nature. This means that, in order to draw the phasor diagram in Figure 7.6, different scales have to be adopted, one for currents and another for voltages!

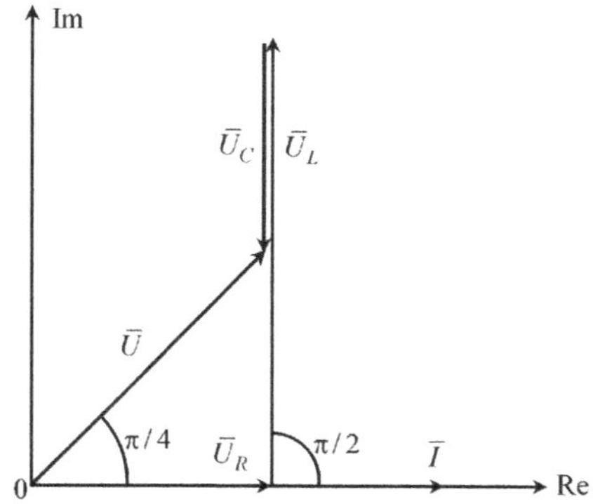

Figure 7.6 Illustrative phasor diagram for the *RLC* series circuit

Q_4

$$V_1 \equiv |\bar{U}|/\sqrt{2} = 100.0 \text{ V}$$

$$V_2 \equiv |\bar{U}_R|/\sqrt{2} = 70.7 \text{ V}, \; V_3 \equiv |\bar{U}_L|/\sqrt{2} = 141.4 \text{ V}, \; V_4 \equiv |\bar{U}_C|/\sqrt{2} = 70.7 \text{ V}$$

$$V_5 \equiv |\bar{U}_L + \bar{U}_L|/\sqrt{2} = 70.7 \text{ V}$$

It should be emphatically noted that $V_1 \neq V_2 + V_3 + V_4$!
Although the sum of the voltage phasors concerning the R, L and C circuit components is equal to the generator voltage phasor, see (7.12), the same is not true for the sum of the corresponding amplitudes or for the sum of the corresponding rms values, that is

$$\bar{U} = \bar{U}_R + \bar{U}_L + \bar{U}_C, \text{ but } \begin{cases} U \neq U_R + U_L + U_C \\ U_{rms} \neq U_{R_{rms}} + U_{L_{rms}} + U_{C_{rms}} \end{cases}$$

Summing vectors is not the same thing as summing scalars. Take care with this issue because it is a source of frequent mistakes.

Q_5 $u(t) = \Re\left(\bar{U}\, e^{j\omega t}\right) = \Re\left(\left(U\, e^{j\pi/4}\right) e^{j\omega t}\right) = U\, \cos(\omega t + \pi/4)$.
$u(t) = \sqrt{2}\; 100\, \cos(\omega t + \pi/4)$ V. See Figure 7.7.

A word of caution: you might be tempted to write $u(t)$ as $u(t) = U\, \cos(\omega t + 45°)$, but don't do it! That too is a common mistake. The problem is that the units for ωt are radians, and as you should know, radians and degrees cannot be mixed.

7.2.4 Instantaneous Power, Active Power, Power Balance Equation

As shown in Figure 7.8, a sinusoidal voltage $u(t) = U\, \cos(\omega t + \alpha_u)$ is applied across the terminals of a given linear passive circuit, that is a circuit containing resistors, inductors, capacitors, but no energy sources. Since the circuit behaves linearly, the generator current is also sinusoidal, $i(t) = I\, \cos(\omega t + \alpha_i)$.

The instantaneous power delivered by the generator, $p(t) = u(t)i(t)$, can be evaluated as

$$p(t) = u(t)i(t) = UI \cos(\omega t + \alpha_u)\cos(\omega t + \alpha_i)$$

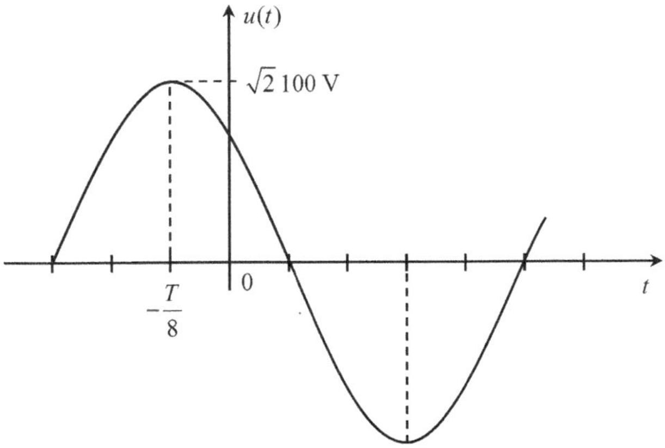

Figure 7.7 Graphical plot of generator voltage against time

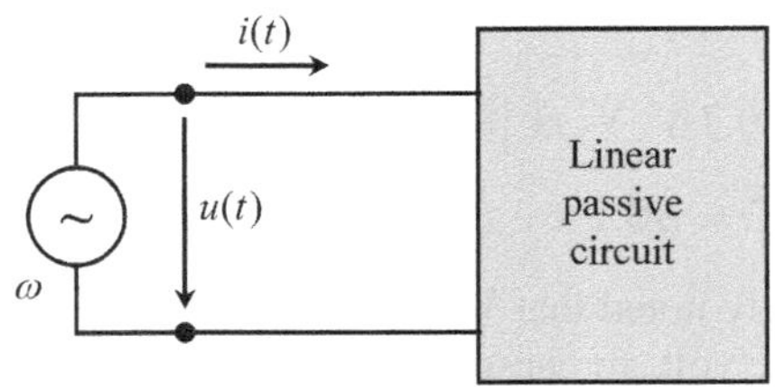

Figure 7.8 In a linear passive circuit, if $u(t)$ is sinusoidal then $i(t)$ will also be sinusoidal

From trigonometry we find

$$p(t) = \frac{UI}{2}\cos(\alpha_u - \alpha_i) + \frac{UI}{2}\cos(2\omega t + \alpha_u + \alpha_i) \tag{7.17a}$$

As an alternative, using the rms definition in (7.10), we can write

$$p(t) = U_{rms}I_{rms}\cos\underbrace{(\alpha_u - \alpha_i)}_{\varphi} + U_{rms}I_{rms}\cos(2\omega t + \alpha_u + \alpha_i) \tag{7.17b}$$

where $\varphi = \alpha_u - \alpha_i$ denotes the phase shift between the sinusoidal voltage and current.

A typical plot of the instantaneous power $p(t)$ is depicted in Figure 7.9.

From (7.17) and from Figure 7.9 we observe that the instantaneous power is not a sinusoidal function of time. Further, we see that $p(t)$ contains a time-invariant term plus a sinusoidal function of angular frequency 2ω. The constant term, representing the averaged power over time, is the so-called active power (units: W, watt)

$$\text{Active power}: \ P = (p(t))_{av} = U_{rms}I_{rms}\cos\varphi \tag{7.18}$$

From Figure 7.9 you can also see that when u and i are out of phase ($\varphi \neq 0$) the power delivered by the generator is negative during certain time intervals. The physical

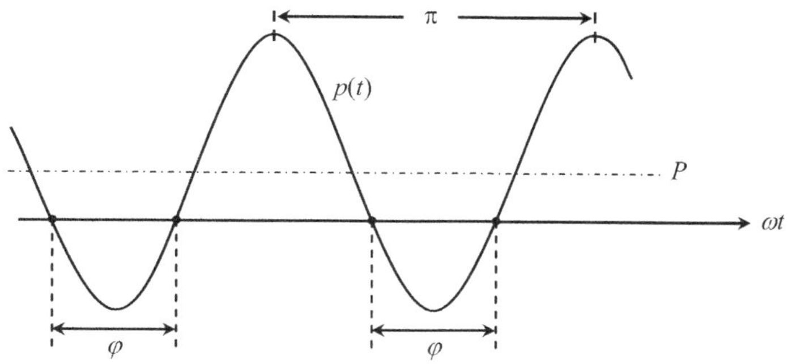

Figure 7.9 Typical plot of the instantaneous power against time. The time average of the instantaneous power, denoted by P, is the so-called active power. The angle φ indicates the difference between the phases of the voltage and current

interpretation for $p < 0$ is rather simple: during those intervals, inductors and/or capacitors discharge their stored energy towards the generator, thus reversing the normal flow of energy.

Moreover, if the circuit contains no resistors at all, the time intervals during which $p < 0$ will have the same duration as the time intervals during which $p > 0$. In this case u and i are in phase quadrature, $\varphi = \pm\pi/2$, and the active power is zero.

The preceding remarks, together with the result in (7.18), allows you to reach an important conclusion concerning the phase shift φ between $u(t)$ and $i(t)$. Since the average power delivered to a passive circuit cannot be negative, you must always have $\cos\varphi \geq 0$. This is tantamount to saying that, for passive circuits, you can only find values for φ in the range $-\pi/2 \leq \varphi \leq \pi/2$.

Next we introduce, in an intuitive way, the power balance equation in the time domain. Using the energy conservation principle we write

$$W(t) = W_J(t) + W_m(t) + W_e(t) \tag{7.19}$$

where W, W_J, W_m and W_e respectively represent the energy brought into by the generator, the energy dissipated by the Joule effect in resistors, the magnetic energy stored in inductors, and the electric energy stored in capacitors. Taking the time derivative of (7.19), we obtain the corresponding powers

$$p(t) = p_J(t) + p_m(t) + p_e(t) \leftrightarrow p(t) = p_J(t) + \frac{d}{dt}\left(W_m(t) + W_e(t)\right) \tag{7.20}$$

This last result is called the Poynting theorem. A rigorous proof of this theorem, based directly on Maxwell's equations, will be given later in Chapter 8.

The Poynting theorem can be immediately used to show that, for time-harmonic regimes, the active power is to be physically identified with Joule losses averaged over time. In fact, from (7.20), you can obtain $(p(t))_{av} = (p_J(t))_{av} + (p_m(t))_{av} + (p_e(t))_{av}$. Since the powers $p_m(t)$ and $p_e(t)$ associated to inductors and capacitors are purely sinusoidal functions, their average values are zero (do not forget that voltages and currents across inductors and capacitors are in phase quadrature), hence

$$P = (p_J(t))_{av} = P_J \tag{7.21}$$

7.2.5 Complex Power, Complex Poynting Theorem

In the framework of time-harmonic regimes, we introduced in Section 7.2.2 the phasor representation of sinusoidal voltages and sinusoidal currents. Can we do the same with the instantaneous power $p(t)$? The answer is no! Writing $p(t) = \Re\left(\overline{P}\, e^{j\omega t}\right)$ would be complete nonsense.

Phasors are only defined for sinusoidally varying quantities and, as you can see in Figure 7.9, the instantaneous power $p(t)$ is not in general a sinusoid. However, at this stage, it is customary to introduce a helpful auxiliary entity, the so-called complex power, whose definition is

$$\overline{P} = \frac{\overline{U}\,\overline{I}^*}{2} = P + jP_Q = P_S\, e^{j\varphi} \tag{7.22a}$$

which (we emphasize) it is not a phasor. In (7.22a), the asterisk on $\overline{I}$ denotes complex conjugation.

One of the main advantages of introducing this new entity is that the evaluation of its real part directly yields the active power, $P = \Re\left(\overline{P}\right)$. Another advantage is that the angle of $\overline{P}$ provides information on the existing phase shift between u and i:

$$\measuredangle \overline{P} = \varphi = \alpha_u - \alpha_i.$$

To see that this is true let us substitute the expressions for $\overline{U}$ and $\overline{I}$ into (7.22a):

$$\overline{P} = \frac{\left(\sqrt{2}U_{rms}e^{j\alpha_u}\right)\left(\sqrt{2}I_{rms}e^{-j\alpha_i}\right)}{2} = U_{rms}I_{rms}\, e^{j(\alpha_u - \alpha_i)} = P_S\, e^{j\varphi} \tag{7.22b}$$

$$\overline{P} = \underbrace{U_{rms}I_{rms}\cos\varphi}_{P} + j\,\underbrace{U_{rms}I_{rms}\sin\varphi}_{P_Q} \tag{7.22c}$$

By doing this, two new auxiliary quantities show up, the apparent power and the reactive power:

$$\text{Apparent power}: \; P_S = \left|\overline{P}\right| = U_{rms}I_{rms} \text{ (units: VA, volt ampere)}$$

$$\text{Reactive power}: \; P_Q = \Im\left(\overline{P}\right) = P_S \sin\varphi \quad \text{(units: VAr, volt ampere reactive)} \tag{7.23}$$

Several physical interpretations for the apparent power can be given. From (7.17b) you will see that P_S represents the amplitude of the sinusoidal term of frequency 2ω belonging to $p(t)$. In addition, P_S represents the averaged power over time for circuits that behave as pure resistors ($\varphi = 0$).

The relationship among the diverse powers $\overline{P}$, P, P_S and P_Q is illustrated through the triangle representation shown in Figure 7.10.

The best way to learn about the physical significance of the reactive power P_Q is through the complex Poynting theorem. This theorem is an absolutely general theorem that can be deduced directly from Maxwell's equations for time-harmonic regimes. Here, we will skip a general demonstration of the theorem, and limit ourselves to arriving to it with the help of a particular example.

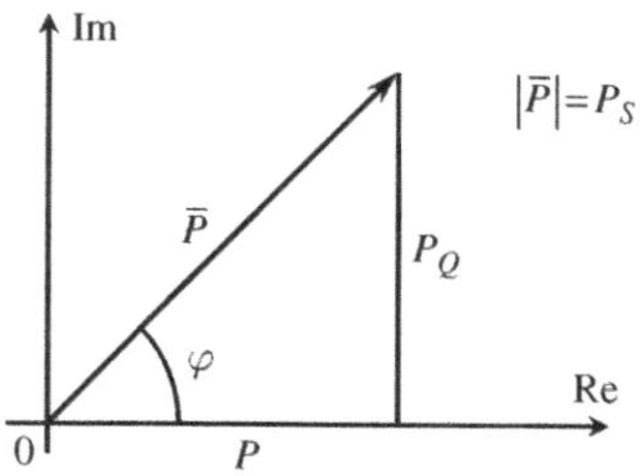

Figure 7.10 Triangle representation of powers

Take the phasor equation governing the *RLC* circuit obtained in (7.16):

$$\overline{U}=\left(R+j\omega L-j\frac{1}{\omega C}\right)\overline{I}$$

Let us multiply both sides of the above equation by $\overline{I}^*/2$. This yields on the left-hand side the complex power $\overline{P}$ introduced in (7.22). On the other hand, we have

$$\frac{\overline{I}\,\overline{I}^*}{2}=I_{rms}^2=\left(i^2(t)\right)_{av} \tag{7.24}$$

Therefore we find

$$\overline{P}=R\,I_{rms}^2+j\omega L I_{rms}^2-j\frac{1}{\omega C}I_{rms}^2 \tag{7.25}$$

Further, from (7.15), at the capacitor terminals we have $I_{rms}=\omega C U_{C_{rms}}$.

By taking this into account, the third term on the right hand side of (7.25) can be rewritten as $I_{rms}^2/(\omega C)=\omega C U_{C_{rms}}^2$. Hence, we get

$$\overline{P}=RI_{rms}^2+j2\omega\left(\tfrac{1}{2}LI_{rms}^2-\tfrac{1}{2}CU_{C_{rms}}^2\right)$$

The term $RI_{rms}^2=\left(R\,i^2(t)\right)_{av}=P_J$ is interpreted as the time-averaged power losses due to the Joule effect in the resistor.

The term $\frac{1}{2}LI_{rms}^2=\frac{1}{2}L\left(i^2(t)\right)_{av}=(W_m)_{av}$ is interpreted as the time-averaged magnetic energy stored in the inductor.

Likewise, the term $\frac{1}{2}CU_{C_{rms}}^2=\frac{1}{2}C\left(u_C^2(t)\right)_{av}=(W_e)_{av}$ is interpreted as the time-averaged electric energy stored in the capacitor.

Finally we obtain

$$\overline{P}=P_J+j2\omega\left((W_m)_{av}-(W_e)_{av}\right) \tag{7.26}$$

The general result shown in (7.26) is the complex Poynting theorem, which is of key importance in many electrical engineering applications, including rapid time-varying field phenomena.

Since the complex power, on the left-hand side of (7.26), is given by

$$\overline{P}=\frac{\overline{U}\,\overline{I}^*}{2}=P+jP_Q$$

we see that the active power P can be interpreted as the time-averaged power losses at the resistors ($P = P_J \geq 0$). This conclusion is not new, and confirms the analysis result in (7.21).

Also – and this is new – the reactive power, P_Q, can now be interpreted as a measure of the balance between the time-averaged magnetic and electric energies stored in inductors and capacitors. If magnetic energy predominates, you will have $P_Q > 0$; otherwise, if electric energy predominates, you will have $P_Q < 0$.

If, at a given frequency, the generator voltage and current are in phase, $\varphi = 0$, the reactive power is null, $P_Q = 0$, and consequently, the time-averaged values of the magnetic and electric energies stored must necessarily compensate each other, $(W_m)_{av} = (W_e)_{av}$.

At this point a few remarks are in order:

- The complex Poynting theorem in (7.26) is not to be confused with the former Poynting theorem in (7.20). These theorems are independent of each other.
- The result in (7.26) cannot be deduced from (7.20).
- The Poynting theorem applies to any kind of time regime, harmonic or non-harmonic.
- The complex Poynting theorem applies only to harmonic regimes.

7.2.6 Impedance and Admittance Operators

The concept of impedance is very simple, but very important and useful. Look again at Figure 7.8 where a linear passive circuit is driven by a generator whose sinusoidal voltage and current are given by $u(t) = U\ \cos(\omega t + \alpha_u)$ and $i(t) = I\ \cos(\omega t + \alpha_i)$. The phase shift between u and i is $\varphi = \alpha_u - \alpha_i$.

The impedance of the circuit $\overline{Z}$ (units: Ω, ohm) is the complex operator that transforms the complex amplitude of $i(t)$ into the complex amplitude of $u(t)$

$$\overline{U} = \overline{Z}\,\overline{I}$$

In other words,

$$\overline{Z} = Z\,e^{j\varphi} = \frac{\overline{U}}{\overline{I}} = \frac{U\,e^{j\alpha_u}}{I\,e^{j\alpha_i}} \rightarrow Z = |\overline{Z}| = \frac{U}{I} = \frac{U_{rms}}{I_{rms}};\ \measuredangle\overline{Z} = \varphi = \alpha_u - \alpha_i \tag{7.27}$$

The following are important notes that you should never forget:

- The impedance operator is only defined in the context of time-harmonic regimes.
- The impedance is an operator acting over complex amplitudes; it does not operate on time-varying quantities. Writing $u(t) = \overline{Z}\ i(t)$ is complete nonsense!
- The impedance operator cannot be defined for nonlinear circuits.
- For linear passive circuits the impedance angle φ is limited to the range $-\pi/2$ to $\pi/2$.
- In general, the impedance operator is frequency dependent, $\overline{Z} = \overline{Z}(\omega)$.

For exemplification purposes consider the *RLC* circuit in Figure 7.1 and recall the phasor-domain equation in (7.16):

$$\overline{U} = \left(R + j\omega L + \frac{1}{j\omega C}\right)\overline{I}$$

The term in parentheses is the impedance operator for the *RLC* circuit

$$\overline{Z} = R + j\left(\omega L - \frac{1}{\omega C}\right) = \sqrt{R^2 + \left(\omega L - \frac{1}{\omega C}\right)^2}\exp\left[j\arctan\left(\frac{\omega L}{R} - \frac{1}{\omega RC}\right)\right]$$

$$Z(\omega) = \sqrt{R^2 + \left(\omega L - \frac{1}{\omega C}\right)^2}$$

$$\varphi(\omega) = \arctan\left(\frac{\omega L}{R} - \frac{1}{\omega RC}\right) \tag{7.28}$$

The admittance operator $\overline{Y}$ (units: S, siemens), which can also be useful, is the inverse of the impedance operator

$$\overline{Y} = \overline{Z}^{-1} = \frac{\overline{I}}{\overline{U}} \tag{7.29}$$

In general, both $\overline{Z}$ and $\overline{Y}$ can be broken down into their constituent real and imaginary parts, that is $\overline{Z} = R + jX$ and $\overline{Y} = G + jS$.

R, G, X and *S* are usually named resistance, conductance, reactance and susceptance, respectively. For passive circuits, where the impedance angle φ is limited to the range $-\pi/2$ to $\pi/2$, you must always find $R \geq 0$ and $G \geq 0$, whereas both X and S can be positive, negative or zero.

7.2.7 Resonance

For time-harmonic regimes, a circuit including resistors, inductors and capacitors is said to be a resonant circuit when its impedance (or admittance) is purely real. Whenever this situation occurs, the magnitude of the driving voltage or current is at a stationary point, at a maximum or at a minimum.

Despite the presence of inductors and capacitors, the generator feeding the circuit interprets the latter as a pure resistor; the phase shift between the voltage and current at the generator terminals is zero, $\varphi = 0$.

Resonance conditions are critically dependent on the working frequency.

Just to give you a simple example, if the impedance angle of the *RLC* circuit in (7.28) is analyzed, and $\varphi = 0$ is enforced, you will at once conclude that the resonance condition for such a particular circuit is $\omega L = 1/(\omega C)$, which implies that $|\overline{Z}| = Z_{\min} = R$ and $|\overline{I}| = |\overline{I}|_{\max} = U/R$. In addition, you may note that the capacitor and inductor voltages of the *RLC* resonant circuit are in phase opposition but have identical amplitudes: $\overline{U}_L = -\overline{U}_C$. This common amplitude, depending on the circuit parameters, may become much higher than the resistor voltage amplitude; their ratio Q is ordinarily termed the quality factor of the circuit at resonance

$$Q = \frac{U_L}{U_R} = \frac{U_C}{U_R} = \frac{\omega_{\text{res}} L}{R} = \frac{1}{\omega_{\text{res}} CR} = \frac{\sqrt{L/C}}{R}$$

The fact that resonant circuits behave as pure resistors, $\varphi = 0$, signifies that the inductor and capacitor effects must cancel each other in some way. According to the complex Poynting theorem in (7.26), the reactive power is zero, and the time-averaged values of the magnetic and electric energies stored compensate for each other, $(W_m)_{av} = (W_e)_{av}$.

The concept of quality factor for an *RLC* resonant circuit was introduced above as a means to quantify circuit overvoltages. However, a more general definition of the quality factor, which applies to any resonant circuit, is usually given in the form

$$Q = \omega_{\text{res}} \frac{(W_m)_{av} + (W_e)_{av}}{(p_J)_{av}} = \omega_{\text{res}} \frac{(W_{em})_{av}}{(p_J)_{av}}$$

where W_{em} denotes the total electromagnetic energy stored.

7.2.8 Application Example (RL ∥ C Circuit)

The *RL* series circuit with a capacitor *C* in parallel finds applications in several areas. It is used in power systems to illustrate the so-called power factor compensation problem; in signal processing it is used as an example of a band reject filter; in instrumentation and measurement it also permits the illustration of the basic functioning principle of a spectrum analyzer.

Here we focus attention on the power factor compensation problem. Later, in Section 7.5, we will deal with the other applications.

Consider the circuit representation in Figure 7.11 where the *RL* branch simulates an electrical installation. The generator voltage is given by $u(t) = U\cos(\omega t)$, with $\omega = 2\pi f$, and $f = 50\,\text{Hz}$. The capacitor of capacitance *C* can be switched on or off.

Data: $U = \sqrt{2}\ 230\,\text{V},\ \ R = 5\ \Omega,\ \ L = 59.4\,\text{mH}$.

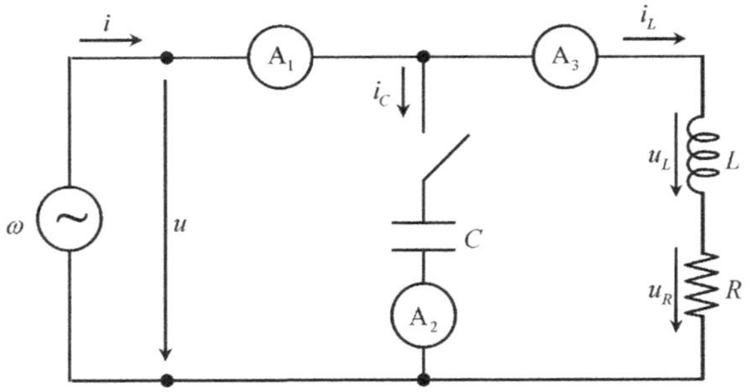

Figure 7.11 Power factor compensation problem (*RL* series circuit with a parallel-connected capacitor)

Questions

Q_1 Consider that the capacitor is switched off.

Write the time-domain and phasor-domain equations of the circuit.
Determine $i(t)$, $u_R(t)$ and $u_L(t)$. Determine the active and reactive powers.

Q_2 Consider that the capacitor is switched on.

Write the time-domain and phasor-domain equations of the circuit.

Write an expression for the admittance operator $\overline{Y} = \overline{I}/\overline{U}$ and, from it, determine C such that resonance takes place.

Determine $u_R(t)$, $u_L(t)$, $i_L(t)$, $i_C(t)$ and $i(t)$. Draw the corresponding phasor diagram.

Three ammeters A_1, A_2 and A_3 are placed in the generator, capacitor and installation branches. What are their rms readings?

Determine the active and reactive powers. Confirm the results obtained by application of the complex Poynting theorem.

Solutions

Q_1

$$i = i_L \rightarrow \overline{I} = \overline{I}_L$$

$$u = u_R + u_L \rightarrow \overline{U} = \overline{U}_R + \overline{U}_L$$

$$u_R = Ri_L \rightarrow \overline{U}_R = R\overline{I}_L$$

$$u_L = L\frac{di_L}{dt} \rightarrow \overline{U}_L = j\omega L\overline{I}_L$$

$$u = Ri_L + L\frac{di_L}{dt} \rightarrow \overline{U} = (R + j\omega L)\overline{I}_L$$

Now, evaluate the impedance $\overline{Z}_{RL}$ of the electrical installation:

$$\overline{Z}_{RL} = R + j\omega L = Z_{RL}\, e^{j\varphi} = 5.0 + j18.7 = 19.32\, e^{j75^o}\ \Omega$$

The current in the installation provided by the generator is found next:

$$\overline{I} = \overline{I}_L = \frac{\overline{U}}{\overline{Z}_{RL}} = \sqrt{2}\, 11.9\, e^{-j75^o}\ \text{A} \rightarrow i(t) = \sqrt{2}\, 11.9\, \cos(\omega t - 5\pi/12)\ \text{A}$$

The voltages across R and L are

$$\overline{U}_R = \sqrt{2}\, 59.5\, e^{-j75^o}\ \text{V} \rightarrow u_R(t) = \sqrt{2}\, 59.5\, \cos(\omega t - 5\pi/12)\ \text{V}$$

$$\overline{U}_L = \sqrt{2}\, 222.1\, e^{j15^o}\ \text{V} \rightarrow u_L(t) = \sqrt{2}\, 222.1\, \cos(\omega t + \pi/12)\ \text{V}$$

Taking into account that the phase shift φ between u and i is 75°, the active power is evaluated as

$$P = \underbrace{U_{rms} I_{rms}}_{P_S} \cos\varphi = 230 \times 11.9 \times 0.259 = 0.71\,\text{kW}$$

(Note: the power ratio $P/P_S = \cos\varphi$ is called the 'power factor'.)

The reactive power is $P_Q = U_{rms} I_{rms} \sin\varphi = 230 \times 11.9 \times 0.966 = 2.65\,\text{kVAr}$.

Q_2 Now the capacitor C is switched on and, therefore, $i \neq i_L$.

All the time-domain and phasor-domain equations obtained in Q_1 remain valid here, with the exception of the first one, $i = i_L \rightarrow \overline{I} = \overline{I}_L$, which clearly does not apply and must be replaced by

$$i = i_L + i_C \rightarrow \overline{I} = \overline{I}_L + \overline{I}_C \quad \text{and} \quad i_C = C\frac{du}{dt} \rightarrow \overline{I}_C = j\omega C\overline{U}$$

At this stage you must realize that the functioning of the RL branch is not minimally affected by the presence of the parallel-connected capacitor. The latter will simply affect the current that the generator must provide.

Because $\overline{I} = \overline{I}_L + \overline{I}_C$ you are probably guessing that the generator current is going to increase compared to the situation analyzed in Q_1. Well, your intuition is failing you! To your surprise the current is indeed going to decrease (that is exactly the goal of inserting the capacitor in this circuit. . .). The point is – we repeat – summing vectors and summing scalars are different things.

Let us evaluate the admittance of the global circuit:

$$\overline{Y} = \frac{\overline{I}}{\overline{U}} = \frac{\overline{I}_C + \overline{I}_L}{\overline{U}} = j\omega C + \frac{1}{R + j\omega L}$$
$$= j\omega C + \frac{1}{R + j\omega L} \times \frac{R - j\omega L}{R - j\omega L} = \frac{R}{Z_{RL}^2} + j\omega\left(C - \frac{L}{Z_{RL}^2}\right)$$

where $Z_{RL}^2 = R^2 + (\omega L)^2$.

Resonance occurs when the angle of $\overline{Y}$ goes through zero or, which is the same, when the imaginary part of the admittance is zero, $\Im\left(\overline{Y}\right) = 0$. Consequently, the resonance condition for this circuit is

$$C = \frac{L}{Z_{RL}^2} = \frac{L}{R^2 + (\omega L)^2} \tag{7.30}$$

Numerically, you obtain $C = 159.1\,\mu\text{F}$.

With this value for the capacitance you find $\overline{I}_C = j\omega C\overline{U} = \sqrt{2}\ 11.5\ e^{j90^\circ}$ A, from which you get $i_C(t) = \sqrt{2}\ 11.5\ \cos(\omega t + \pi/2)$ A.

The quantities $u_R(t)$, $u_L(t)$ and $i_L(t)$, remain unchanged.

Finally – and this is the most important point – the new current in the generator is

$$\overline{I} = \overline{I}_C + \overline{I}_L = \overline{Y}\,\overline{U} = \frac{R\overline{U}}{Z_{RL}^2} = \sqrt{2}\,3.08\text{ A} \rightarrow i(t) = \sqrt{2}\,3.08\cos(\omega t)\text{ A}$$

which is almost four times smaller in magnitude than the one calculated in Q_1 when the capacitor was disconnected.

Figure 7.12 shows the phasor diagram for this problem, illustrating the existing relations among the voltages and currents in the resonant circuit.

The ammeters A_1, A_2 and A_3, placed in the generator, capacitor and installation branches, read 3.08 A, 11.5 A and 11.9 A respectively.

Evaluation of the complex power $\overline{P} = \overline{U}\,\overline{I}^*/2 = P + jP_Q$ gives $P = 0.71$ kW and $P_Q = 0$.

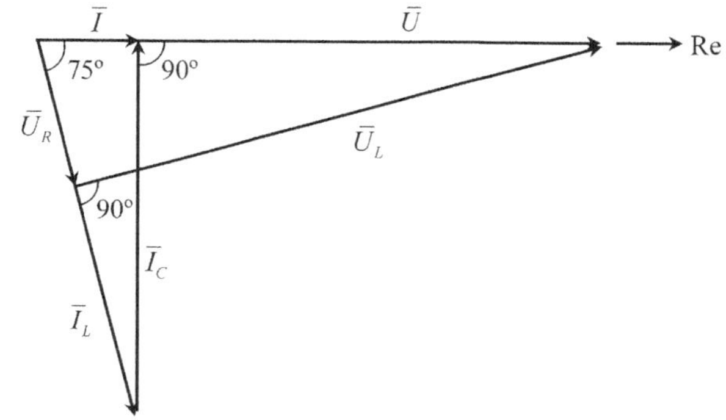

Figure 7.12 Illustrative phasor diagram for the $RL \parallel C$ circuit. When C is chosen appropriately (perfect power factor compensation), not only are the generator voltage and current in phase, but also the current magnitude is minimal

The active power P remains invariant because the Joule losses in the RL branch remain the same, $P_J = RI^2_{L_{rms}} = 0.71\,\mathrm{kW}$.

The reactive power dropped to zero because the power factor has been compensated ($\varphi = 0 \rightarrow \cos\varphi = 1,\ \sin\varphi = 0$). In other words, the time-averaged electric energy in the capacitor is equal to the time-averaged magnetic energy in the inductor:

$$(W_m)_{av} = \tfrac{1}{2}LI^2_{L\,rms} = 4.2\ \mathrm{J},\ (W_e)_{av} = \tfrac{1}{2}CU^2_{rms} = 4.2\,\mathrm{J}$$

7.3 Transformer Analysis

So far we have been dealing with circuit examples where magnetic coupling is absent. Now it is time to turn our attention to transformer circuits where magnetic coupling is a key issue. Due to space limitations, here we will only examine two-winding transformers with a linear core, such as the one shown in Figure 7.13 (remember that linearity is a prerequisite for the usage of inductance coefficients).

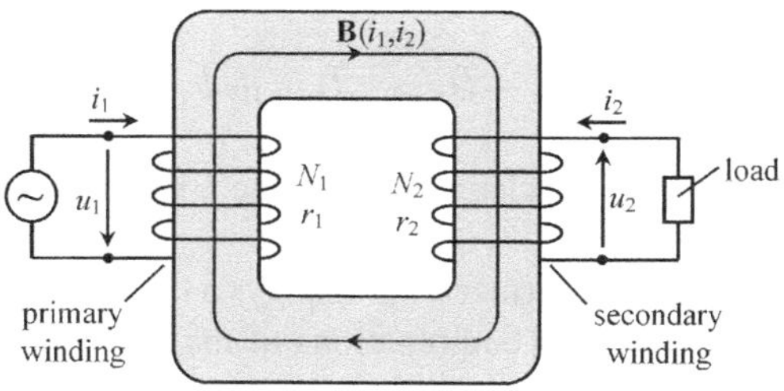

Figure 7.13 Single-core transformer representation, showing the convention for reference signs of the voltages and currents in the primary and secondary windings

Two windings are present: the primary winding is the one connected to the generator; the secondary winding is the one connected to the load (the latter is assumed to be a linear passive device). The transformer windings, with N_1 and N_2 turns, are characterized respectively by their internal resistances r_1 and r_2.

Before we proceed to the analysis of the transformer equations, a point worth emphasizing is that the writing of the equations is critically dependent on the arbitrary conventional reference directions assigned to the currents and voltages in play. Here, we adhere to the following convention. Primary and secondary currents are oriented so as to create concordant magnetic fields in the core ($L_M > 0$). Primary and secondary voltages are oriented so as to produce positive power flows from the generator to the transformer and from the transformer to the load.

Time-domain equations for the transformer circuit are obtained using the induction law from Chapter 5. Using a circulation path $\mathbf{s}_1$ oriented along i_1, going through the primary winding conductor and closing at the generator terminals, we obtain

$$\underbrace{\oint_{\mathbf{s}_1} \mathbf{E} \cdot d\mathbf{s}_1}_{r_1 i_1 - u_1} = -\frac{d\psi_1}{dt} \rightarrow u_1(t) = r_1\, i_1(t) + \frac{d\psi_1(t)}{dt} \tag{7.31}$$

where the primary linkage flux depends on both currents $\psi_1 = \psi_1(i_1, i_2)$.

Using a circulation path $\mathbf{s}_2$ oriented along i_2, going through the secondary winding conductor and closing at the load terminals, we obtain

$$\underbrace{\oint_{\mathbf{s}_1} \mathbf{E} \cdot d\mathbf{s}_2}_{r_2 i_2 + u_2} = -\frac{d\psi_2}{dt} \rightarrow -u_2(t) = r_2\, i_2(t) + \frac{d\psi_2(t)}{dt} \tag{7.32}$$

where the secondary linkage flux depends on both currents $\psi_2 = \psi_2(i_1, i_2)$.

The load equation, which at this stage cannot be written explicitly (because the load has not yet been specified), can be put in the form

$$u_2 = u_2(i_2) \tag{7.33}$$

For time-harmonic regimes, the above time-domain equations transform to the phasor domain as

$$\overline{U}_1 = r_1 \overline{I}_1 + j\omega\overline{\psi}_1; \quad -\overline{U}_2 = r_2 \overline{I}_2 + j\omega\overline{\psi}_2; \quad \overline{U}_2 = \overline{Z}_2 \overline{I}_2 \tag{7.34}$$

where $\overline{Z}_2$ is the impedance operator characterizing the load placed at secondary winding terminals.

Assuming, as before, that the transformer core displays a linear behavior, then the magnetic linkage fluxes can be written as linear combinations of the currents in play (Chapter 4):

$$\begin{cases} \psi_1(t) = L_{11} i_1(t) + L_M i_2(t) \\ \psi_2(t) = L_M i_1(t) + L_{22} i_2(t) \end{cases} \rightarrow \begin{cases} \overline{\psi}_1 = L_{11} \overline{I}_1 + L_M \overline{I}_2 \\ \overline{\psi}_2 = L_M \overline{I}_1 + L_{22} \overline{I}_2 \end{cases} \tag{7.35}$$

By substituting (7.35) into (7.34) we get the phasor-domain equations governing the transformer:

$$\overline{U}_1 = (r_1 + j\omega L_{11})\overline{I}_1 + j\omega L_M \overline{I}_2 \tag{7.36a}$$

$$-\overline{U}_2 = (r_2 + j\omega L_{22})\overline{I}_2 + j\omega L_M \overline{I}_1 \tag{7.36b}$$

$$\overline{U}_2 = \overline{Z}_2 \overline{I}_2 \tag{7.36c}$$

Note that, according to the conventions adopted, all the inductance coefficients L_{11}, L_{22} and L_M are positive quantities.

7.3.1 The Ideal Transformer

The ideal transformer does not exist. It is a fictitious device whose properties listed below cannot be fulfilled in nature. Nonetheless, some practical engineers may make use of it as a zeroth-order model to get coarse estimations of the transformer behavior.

The properties of a single-core ideal transformer are:

- Primary and secondary windings are made of perfect conductors with $\sigma \to \infty$.
- The transformer core is made of a perfect magnetic material with $\mu \to \infty$.

The first simplifying condition implies $r_1 = r_2 = 0$. The second simplifying condition ensures that the core is a perfect tube for the flux of **B** lines, that is dispersion is absent,

$$\psi_1 = N_1 \phi, \quad \psi_2 = N_2 \phi$$

and magnetic coupling is perfect ($k = 1$). Moreover, since $\mu \to \infty$, the magnetic field **H** in the transformer core is zero, $\mathbf{H} = \mathbf{B}/\mu = 0$.

By using the above properties in (7.31) and (7.32) we obtain

$$\begin{cases} u_1 = N_1 \, d\phi/dt \\ -u_2 = N_2 \, d\phi/dt \end{cases} \rightarrow \quad \frac{u_1(t)}{u_2(t)} = -\frac{N_1}{N_2} \tag{7.37}$$

On the other hand, the application of Ampère's law to a closed circulation path inside the transformer core (Chapter 4) yields

$$0 = \oint_{s} \underbrace{\mathbf{H}}_{0} \cdot d\mathbf{s} = \int_{S_s} \mathbf{J} \cdot \mathbf{n}_s \, dS = N_1 \, i_1 + N_2 \, i_2 \quad \rightarrow \quad \frac{i_1(t)}{i_2(t)} = -\frac{N_2}{N_1} \tag{7.38}$$

In short, apart from a minus sign, the ratio of the transformer voltages is equal to the corresponding winding turns ratio, whereas the ratio of the transformer currents is the reverse of the winding turns ratio.

The time-domain equations in (7.37) and (7.38) have a corresponding phasor-domain counterpart that reads as

$$\frac{\overline{U}_1}{\overline{U}_2} = -\frac{N_1}{N_2}, \quad \frac{\overline{I}_1}{\overline{I}_2} = -\frac{N_2}{N_1} \tag{7.39}$$

By using (7.39) you can readily conclude that the input impedance $\overline{Z}_1$ of the ideal transformer (measured at the primary winding terminals) is equal to the load impedance $\overline{Z}_2$ multiplied by the factor $(N_1/N_2)^2$:

$$\overline{Z}_1 = \frac{\overline{U}_1}{\overline{I}_1} = \left(\frac{N_1}{N_2}\right)^2 \frac{\overline{U}_2}{\overline{I}_2} = \left(\frac{N_1}{N_2}\right)^2 \overline{Z}_2 \tag{7.40}$$

7.3.2 Transformer Impedance, Pseudo Lenz's Law

The zeroth-order result in (7.40) is not a very reliable one as far as actual transformers are concerned, since their properties hardly match those of ideal transformers. Therefore, we need to find $\overline{Z}_1$ based on (7.36).

To start with, we determine the relationship between $\overline{I}_2$ and $\overline{I}_1$. Eliminating $\overline{U}_2$ using (7.36b) and (7.36c) yields

$$\overline{I}_2 = -\frac{j\omega L_M}{r_2 + j\omega L_{22} + \overline{Z}_2}\overline{I}_1 \tag{7.41}$$

Substituting (7.41) into (7.36a) you get

$$\overline{Z}_1 = (r_1 + j\omega L_{11}) + \left(\frac{(\omega L_M)^2}{r_2 + j\omega L_{22} + \overline{Z}_2}\right) \tag{7.42}$$

The first term on the right-hand side of this equation is to be interpreted as the primary winding self-impedance, that is the one that is observed when the secondary winding is left open ($\overline{I}_2 = 0,\ \overline{Z}_2 \to \infty$). The second term on the right-hand side, resulting from magnetic coupling, is to be interpreted as the influence of the secondary winding (load included) on the input impedance of the transformer.

It is appropriate at this stage to go back to the pseudo Lenz's law to which we referred earlier in Problem 5.13.2.

From (7.42) you can see that when the transformer load is disconnected, the inductance observed at the generator terminals coincides with $L_{11} = (1/\omega)\Im\left(\overline{Z}_1\right)_{\overline{Z}_2=\infty}$.

According to common sense (based on Lenz's law), when the load is plugged in, the transformer core is expected to demagnetize (due to the alleged counteraction of the secondary current) and, consequently, the equivalent inductance $L'_{11} = (1/\omega)\Im\left(\overline{Z}_1\right)_{\overline{Z}_2\neq\infty}$ measured at the generator terminals is expected to be smaller than L_{11} itself.

We are going to show that this is not always true.

Let us write the load impedance as $\overline{Z}_2 = R_2 + jX_2$, where $X_2 < 0$ for capacitive loads. From (7.42) we have

$$\overline{Z}_1 = (r_1 + j\omega L_{11}) + \left(\frac{(\omega L_M)^2}{(r_2 + R_2) + j(\omega L_{22} + X_2)}\right)$$

After simple algebraic manipulation we get

$$\overline{Z}_1 = (r_1 + j\omega L_{11}) + \underbrace{\left(\frac{(\omega L_M)^2}{(r_2 + R_2)^2 + (\omega L_{22} + X_2)^2}\right)}_{K \geq 0} \times [(r_2 + R_2) - j(\omega L_{22} + X_2)]$$

Then we conclude that $\Re\left(\overline{Z}_1\right) = r_1 + K(r_2 + R_2) \geq r_1$; that is, the effect of the secondary winding is always to increase the equivalent input resistance of the transformer.

As for the equivalent input inductance of the transformer, we find

$$L'_{11} = \frac{1}{\omega}\Im(\overline{Z}_1) = L_{11} - K\left(L_{22} + \frac{X_2}{\omega}\right) \tag{7.43}$$

If the transformer load $\overline{Z}_2$ includes a capacitor with reactance $X_2 = -1/(\omega C_2)$, such that $C_2 < 1/(\omega^2 L_{22})$, then the term $K(L_{22} + X_2/\omega)$ in (7.43) becomes negative, and consequently, we see that the equivalent inductance at the primary terminals increases in magnitude, $L'_{11} > L_{11}$, which is in clear contradiction with the ordinary postulate of Lenz's law.

7.3.3 Equivalent Circuits

Some computer programs devoted to circuit analysis do not handle easily problems where magnetic coupling among inductors is present, as in the case of real transformers. Fortunately, you can circumvent such a problem by making use of equivalent circuits where magnetic coupling issues are absent.

This subsection addresses the topic of transformer equivalent circuits.

Consider the 'T' circuit shown in Figure 7.14, consisting of three unknown uncoupled impedances, $\overline{Z}_\alpha$, $\overline{Z}_\beta$ and $\overline{Z}_0$, which is cascaded with an ideal transformer characterized by a transformation ratio $\upsilon = n_1/n_2$.

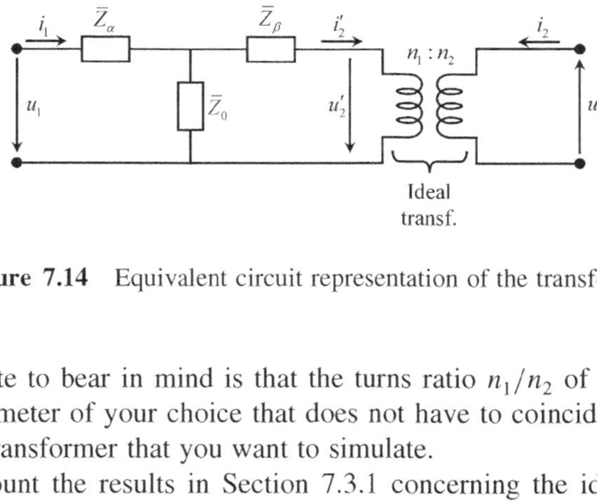

Figure 7.14 Equivalent circuit representation of the transformer

An important note to bear in mind is that the turns ratio n_1/n_2 of the ideal transformer is an arbitrary parameter of your choice that does not have to coincide with the turns ratio N_1/N_2 of the real transformer that you want to simulate.

Taking into account the results in Section 7.3.1 concerning the ideal transformer, you have here

$$\overline{U}'_2 = -\upsilon\,\overline{U}_2 \quad \text{and} \quad \overline{I}'_2 = -\frac{1}{\upsilon}\,\overline{I}_2 \tag{7.44}$$

The phasor-domain equations for the 'T' circuit are

$$\overline{U}_1 = \overline{Z}_\alpha \overline{I}_1 + \overline{Z}_0(\overline{I}_1 - \overline{I}'_2)$$
$$\overline{U}'_2 = \overline{Z}_0(\overline{I}_1 - \overline{I}'_2) - \overline{Z}_\beta \overline{I}'_2$$

By substituting (7.44) into the above results, and rearranging terms, you get

$$\overline{U}_1 = (\overline{Z}_\alpha + \overline{Z}_0)\overline{I}_1 + \frac{\overline{Z}_0}{\upsilon}\overline{I}_2 \tag{7.45a}$$

$$-\overline{U}_2 = \frac{\overline{Z}_\beta + \overline{Z}_0}{\upsilon^2}\overline{I}_2 + \frac{\overline{Z}_0}{\upsilon}\overline{I}_1 \tag{7.45b}$$

Next, all you have to do is to establish a term-by-term comparison between the preceding circuit equations and those describing the real transformer in (7.36a) and (7.36b). Then you will find

$$\overline{Z}_0 = j\omega \underbrace{\upsilon L_M}_{l_0}, \quad \overline{Z}_\alpha = r_1 + j\omega \underbrace{(L_{11} - \upsilon L_M)}_{l_1}, \quad \overline{Z}_\beta = \upsilon^2 r_2 + j\omega\upsilon^2 \underbrace{(L_{22} - L_M/\upsilon)}_{l_2} \tag{7.46}$$

The element described by $\overline{Z}_0$ is a pure inductor whose inductance is $l_0 = \upsilon L_M$. The element described by $\overline{Z}_\alpha$ is the series connection of a resistor and an inductor, whose resistance and inductance are, respectively, r_1 and l_1. Similarly, the element described by $\overline{Z}_\beta$ is the series connection of a resistor and an inductor, whose resistance and inductance are, respectively, $r_\beta = \upsilon^2 r_2$ and $l_\beta = \upsilon^2 l_2$. These results are summarized in Figure 7.15.

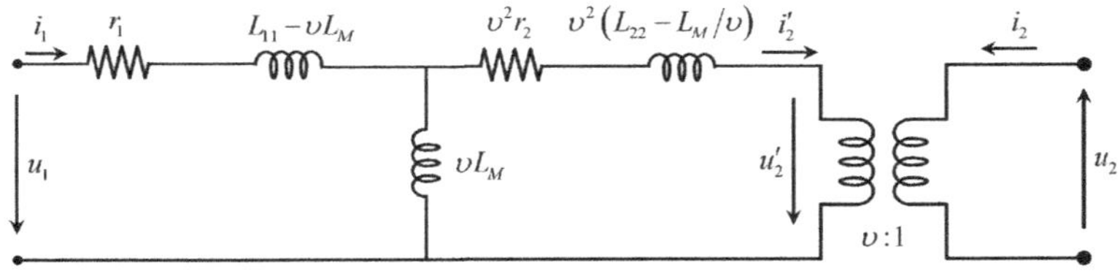

Figure 7.15 Detailed representation of the transformer equivalent circuit

At this point a few remarks are in order.

- Although the transformer's equivalent scheme in Figure 7.15 has been deduced using phasor-domain equations, it is also valid for time-domain analysis, provided that υ is chosen to be real. This is so because its internal components are described by frequency-independent parameters under the assumption of slow time-varying phenomena.
- Given the fact that υ is an arbitrary parameter, you really do not have one equivalent circuit, but an infinite number of equivalent circuits at your choice. If, for instance, you make $\upsilon = L_{11}/L_M$ then the impedance $\overline{Z}_\alpha$ will turn into a pure resistor $\overline{Z}_\alpha = r_1$; likewise, if you decide to make $\upsilon = L_M/L_{22}$ then the impedance $\overline{Z}_\beta$ will turn into a pure resistor $\overline{Z}_\beta = \upsilon^2 r_2$.
- The real transformer in Figure 7.13 and the circuit in Figure 7.15 are formally equivalent from the viewpoint of their accessible voltages (u_1, u_2) and currents (i_1, i_2). However, from an internal perspective there is no sort of correspondence between the real transformer and the equivalent circuit – it suffices to say that the internal circuit components are quite arbitrary since they depend on your own particular choice of υ.

Suppose now that your goal is to materialize an equivalent circuit so that you can run some laboratory tests on it as a way of predicting the primary and secondary voltages and currents of the real transformer. This presents you with the problem of the physical realizability of the circuit; in other words, you are going to need to buy the actual components of the circuit (resistors with positive resistances, inductors with positive inductances). This objective naturally puts some constraints on your choices regarding υ, namely:

$$l_0 > 0 \rightarrow \upsilon \in \Re,\ \ \upsilon > 0$$

$$\begin{cases} l_1 = L_{11} - \upsilon L_M \geq 0 \\ l_2 = L_{22} - L_M/\upsilon \geq 0 \end{cases} \rightarrow \quad \frac{L_M}{L_{22}} \leq \upsilon \leq \frac{L_{11}}{L_M} \tag{7.47}$$

As a parenthetical remark, you should note that the interval defined for υ is a closed interval. In fact, the condition defined in (7.47) is compatible with the inequality $L_M^2 \leq L_{11}L_{22}$, which is always true and is nothing more than a restatement that the magnetic coupling factor between any two inductors cannot exceed unity (4.42).

If you have an inquisitive mind, you must now have an additional question drumming in your head. Where are you going to buy the ideal transformer in Figure 7.15?

Nowhere is the answer. But we may add that you do not need to. In fact, all you have to do is to replace the ideal transformer and its load by the corresponding input impedance measured at the terminals where u_2' and i_2' are defined, that is, from (7.40)

$$\overline{Z}_2' = \upsilon^2 \overline{Z}_2 \tag{7.48}$$

Finally you get the physically realizable circuit shown in Figure 7.16(a).

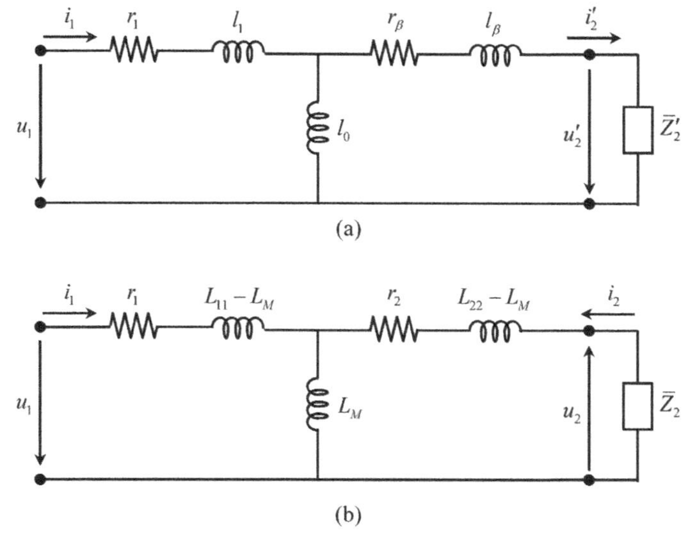

Figure 7.16 Transformer equivalent circuit with the ideal transformer removed. (a) The influence of the load impedance is taken into account through $\overline{Z}_2' = \upsilon^2\overline{Z}_2$. (b) Simplification arising from the choice $\upsilon = 1$.

Of course you can now argue that you have lost direct access to the secondary winding quantities u_2 and i_2. You are right. But the thing is that this detail is unimportant, because from (7.44) you can always retrieve such information from u_2' and i_2':

$$u_2 = -u_2'/v, \quad i_2 = -v\, i_2'$$

However, if your transformer happens to be such that $L_{11} \geq L_M$ and $L_{22} \geq L_M$, then, according to (7.47), you will be allowed to choose $v = 1$, a choice that gives you direct access to the secondary winding quantities themselves (Figure 7.16(b)).

7.3.4 Application Example (Capacitively Loaded Transformer)

Consider a given transformer whose winding resistances are negligibly small ($r_1 = r_2 = 0$). In order to evaluate the transformer induction coefficients, two laboratory experiments were conducted – see Figure 7.17.

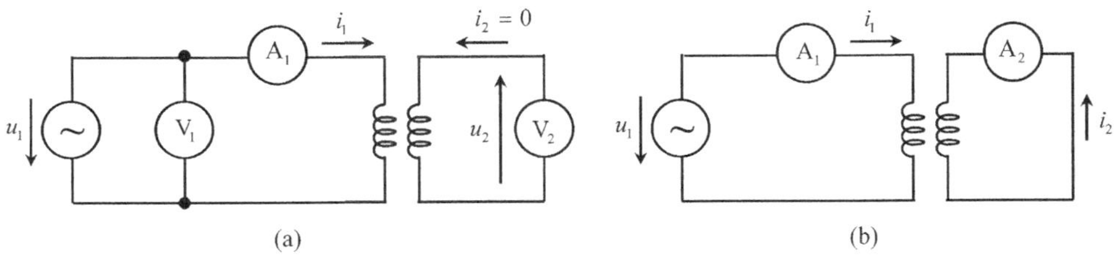

Figure 7.17 Experimental determination of the transformer inductances. (a) Secondary open. (b) Secondary short-circuited

Firstly, with the transformer left open ($i_2 = 0$), rms values for i_1, u_1 and u_2 were measured. The results obtained are, respectively, $A_1 \equiv 1\,A$, $V_1 \equiv 157.1\,V$ and $V_2 \equiv 62.8\,V$. Secondly, with the transformer short-circuited ($u_2 = 0$), rms values for i_1 and i_2 were measured. The results obtained are, respectively, $A_1 \equiv 1\,A$ and $A_2 \equiv 1.6\,A$.

At normal functioning the transformer's secondary winding is loaded with a capacitor of capacitance $C_2 = 63.66\,\mu F$. The voltage across the load is given by $u_2(t) = 200\cos(\omega t)$ V.

Assume $f = 50\,Hz$ in all your calculations.

Questions

Q_1 Find the inductance coefficients L_{11}, L_M and L_{22} of the transformer windings.

Q_2 Draw an equivalent circuit for the transformer using the choice $v = 2$ (check if this choice is a permissible one from the viewpoint of the physical realizability of the circuit).

Q_3 Determine the phasor-domain voltages and currents of the secondary and primary windings of the transformer (employ the equivalent circuit).

Q_4 Determine the input impedance of the loaded transformer and compare the equivalent self-inductance at the primary terminals to the measured one when the transformer was left open ($i_2 = 0$).

Solutions

Q_1 When the transformer is left open you obtain, from (7.36), $\overline{U}_1 = j\omega L_{11}\overline{I}_1$, and $\overline{U}_2 = -j\omega L_M \overline{I}_1$. Therefore you find:

$$L_{11} = \frac{U_{1_{rms}}}{\omega I_{1_{rms}}} = 0.5\,\text{H} \quad \text{and} \quad L_M = \frac{U_{2_{rms}}}{\omega I_{1_{rms}}} = 0.2\,\text{H}$$

When the transformer is short-circuited ($\overline{Z}_2 = 0$) you obtain, from (7.41),

$$\overline{I}_2 = -\frac{L_M}{L_{22}}\overline{I}_1$$

Therefore you find $L_{22} = L_M\, I_{1_{rms}}/I_{2_{rms}} = 0.125\,\text{H}$.

Q_2 $\dfrac{L_M}{L_{22}} \le v \le \dfrac{L_{11}}{L_M} \quad \rightarrow \quad 1.6 \le v \le 2.5$

So you see that $v = 2$ is a permissible choice.
Parameter evaluation, from (7.46):

$$\begin{cases} l_\alpha = l_1 = L_{11} - vL_M = 0.1\text{ H} \\ l_\beta = v^2 l_2 = v^2\,(L_{22} - L_M/v) = 0.1\text{ H} \\ l_0 = vL_M = 0.4\text{ H} \end{cases}$$

As for the input impedance of the ideal transformer, you have

$$\overline{Z}'_2 = \frac{1}{j\omega C'_2} = v^2 \frac{1}{j\omega C_2} \quad \rightarrow \quad C'_2 = \frac{C_2}{v^2} = 15.92\ \mu\text{F}$$

Figure 7.18 shows the particular equivalent circuit for this application example.

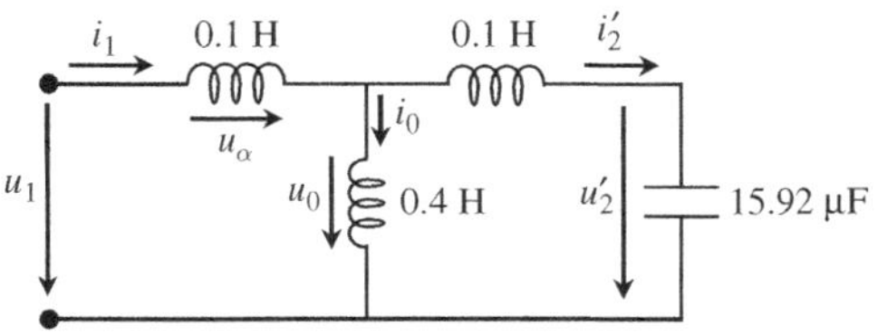

Figure 7.18 One possible equivalent circuit for the transformer examined in Section 7.3.4

Q_3

$$\overline{U}_2 = 200\text{ V} \rightarrow \overline{U}_2' = -v\overline{U}_2 = 400\, e^{j\pi}\text{ V}$$

$$\overline{I}_2 = j\omega C_2 \overline{U}_2 = 4\, e^{j\pi/2}\text{ A} \rightarrow \overline{I}_2' = -\overline{I}_2/v = 2\, e^{-j\pi/2}\text{ A}$$

$$\overline{U}_0 = j\omega l_\beta \overline{I}_2' + \overline{U}_2' = 337.2\, e^{j\pi}\text{ V}$$

$$\overline{I}_0 = \frac{\overline{U}_0}{j\omega l_0} = 2.68\, e^{j\pi/2}\text{ A}$$

$$\overline{I}_1 = \overline{I}_0 + \overline{I}_2' = 0.68\, e^{j\pi/2}\text{ A}$$

$$\overline{U}_\alpha = j\omega l_\alpha \overline{I}_1 = 21.4\, e^{j\pi}\text{ V}$$

$$\overline{U}_1 = \overline{U}_\alpha + \overline{U}_0 = 358.6\, e^{j\pi}\text{ V}$$

You may note that the primary and secondary currents are in phase, which means that they do give rise to concordant magnetizing effects.

Q_4 $\overline{Z}_1 = \overline{U}_1/\overline{I}_1 = j\omega L_{11}' = j527.4\ \Omega \rightarrow L_{11}' = 1.68\text{ H}$ (remember that $L_{11} = 0.5\text{ H}$).
When the transformer is capacitively loaded, the generator at the primary winding observes an increased inductance (contrary to expectations based on Lenz's law.)

7.4 Transient Regimes

Sections 7.2 and 7.3 have been dedicated to steady-state harmonic regimes. Now, the time has come to shift our attention to the analysis of transient phenomena, that is the phenomena subsequent to switching operations which, as time elapses, should tend to stabilize in a steady-state solution.

7.4.1 Free-Regime and Steady-State Solutions

As depicted in Figure 7.19, a generator is switched on, at $t = 0$, in a linear passive circuit containing a number of resistors, inductors and capacitors.

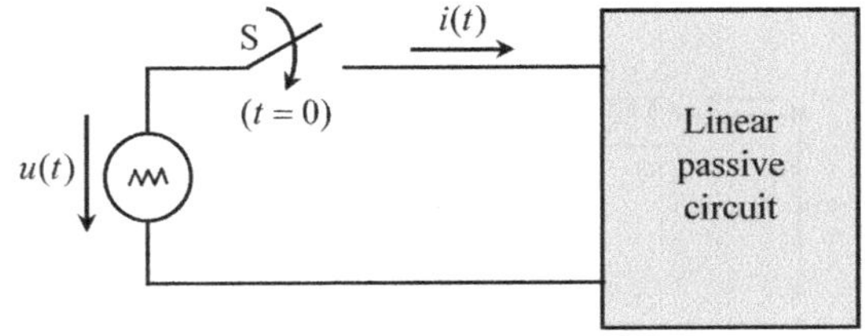

Figure 7.19 A voltage generator switched on to a linear passive circuit

By application of the time-domain fundamental laws governing both magnetic and electric induction phenomena, you will obtain a description of the generator current $i(t)$ in the typical form of a linear differential equation with constant coefficients of order n:

$$\text{For } t > 0: \ a_n \frac{d^n i}{dt^n} + \cdots + a_k \frac{d^k i}{dt^k} + \cdots + a_1 \frac{di}{dt} + a_0 i = f(u(t)) \tag{7.49}$$

where the a_k coefficients (with $k = 0$ to n) are real, and the function $f(u)$ on the right-hand side of the equation depends on the generator voltage. The order n of the equation – depending on the complexity of the circuit – has an upper limit determined by the total number of inductors and capacitors pertaining to the circuit.

The solution to (7.49) is obtained by breaking it down into two sub-solutions, the steady-state solution $i_S(t)$ and the free-regime solution $i_F(t)$.

The steady-state solution is a particular solution of the complete equation in (7.49). The steady-state solution is the one which the current $i(t)$ converges to as time goes on

$$\lim_{t \to \infty} i(t) \to i_S(t)$$

For example, when the driving voltage $u(t)$ is time harmonic, the solution for $i_S(t)$ is also time harmonic, and you can determine it by using the phasor-domain technique developed in Sections 7.2 and 7.3.

The free-regime solution is the general solution of the homogeneous equation corresponding to the one in (7.49); that is, when you make $f(u) = 0$,

$$a_n \frac{d^n i}{dt^n} + \cdots + a_k \frac{d^k i}{dt^k} + \cdots + a_1 \frac{di}{dt} + a_0 i = 0 \tag{7.50}$$

The free-regime solution derives its name from the fact that the solution of (7.50) is free from the influence of the generator.

Since the generator's influence has been removed, the solution of (7.50) must tend to zero as time elapses:

$$\lim_{t \to \infty} i_F(t) \to 0$$

Equations of the type shown in (7.50) are known to have solutions in the form of linear combinations of exponential time-decaying functions. As a matter of fact, if in (7.50) you substitute Ie^{st} for $i(t)$ you will get an algebraic polynomial equation (the so-called characteristic equation)

$$a_n s^n + \cdots + a_k s^k + \cdots + a_1 s = 0 \tag{7.51}$$

whose roots $s_1, \ldots, s_k, \ldots, s_n$ will enable you to write the free-regime solution as a sum of n independent exponentials (we are assuming that multiple roots are absent)

$$i_F(t) = \sum_{k=1}^{n} I_k e^{s_k t} \tag{7.52}$$

The roots of the characteristic equation, $s_1, \ldots, s_k, \ldots, s_n$, can be real or complex but, in either case, their real parts cannot be positive, otherwise the amplitude of the free-regime solution will increase with time, which is a physically impossible situation with generators removed.

As a parenthetical note, and for completion purposes, it should be added that in some (rare) circumstances multiple roots can occur in (7.51). When that is the case, the solution for $i_F(t)$ is more complicated. For instance, if a root s_k with multiplicity m $(m \le n)$ is found to exist then its contribution to (7.52) will take the form

$$\left(I_{k,1} + I_{k,2}t + \cdots + I_{k,m}t^{m-1}\right) e^{s_k t} \tag{7.53}$$

7.4.2 Initial Conditions

In (7.52) a total of n unknown amplitudes I_k need to be determined. For that purpose you have to take into account a set of n initial conditions that should be enforced on the so-called state variables of the problem.

Independently of the switching operations (closing or opening) that take place in a given circuit, certain quantities (state variables) can never change suddenly, namely capacitor voltages u_C and inductor currents i_L.

Time discontinuities in $u_C(t)$ would mean that the electric energy stored in the capacitor would change instantaneously; likewise, time discontinuities in $i_L(t)$ would mean that the magnetic energy stored in the inductor would change instantaneously. These energy jumps would require infinite amounts of power at those components

$$p_C = \frac{d}{dt}W_e(t) \to \infty; \quad p_L = \frac{d}{dt}W_m(t) \to \infty$$

but since this is a physical impossibility, you cannot avoid the obvious conclusion that u_C and i_L ought to remain unchanged immediately before and after the switching operation at $t = 0$,

$$u_C(0^+) = u_C(0^-); \quad i_L(0^+) = i_L(0^-) \tag{7.54}$$

The consideration of n initial conditions, as in (7.54), allows you finally to solve the original problem stated in (7.49).

7.4.3 Analysis of the Capacitor Charging Process

In order to illustrate the above theoretical considerations, we are now going to examine the very simple transient phenomena resulting from the switching on of a DC generator in an *RC* circuit as described in Figure 7.20, where the capacitor is initially discharged, $u_C(0^-) = 0$.

For $t > 0$, application of the induction law to the clockwise-oriented closed path **s** yields

$$\oint_{\mathbf{s}} \mathbf{E} \cdot d\mathbf{s} = -\frac{d\psi_{S_s}}{dt} \to -U + Ri(t) + u_C(t) \approx 0 \tag{7.55}$$

where the time derivative of the linkage magnetic flux across S_s has been neglected.

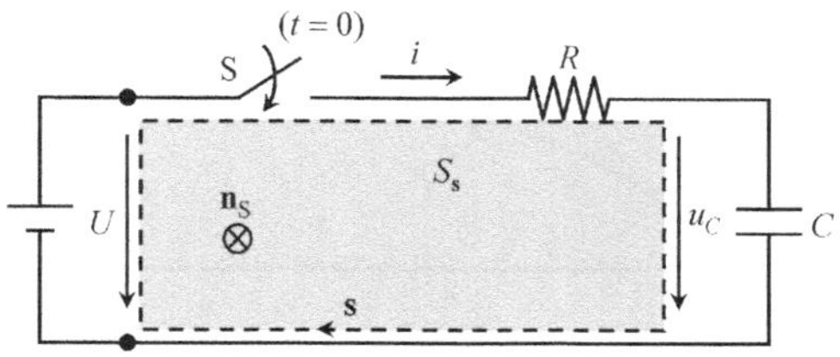

Figure 7.20 A DC voltage generator switched on in an RC series circuit. Ordinarily, the magnetic flux linkage across the shaded surface S_s is negligibly small

Since the state variable of the circuit under analysis is the capacitor voltage, we should use (6.7) to replace $i(t)$ by its expression in terms of $u_C(t)$, that is $i = C\, du_C/dt$. On doing this we find a first-order differential equation in the variable $u_C(t)$

$$RC\frac{du_C(t)}{dt} + u_C(t) = U \tag{7.56}$$

A particular solution of the preceding equation is $u_C(t) = U$, where U is a constant (the DC voltage of the generator). Therefore, the steady-state solution for the problem is

$$(u_C)_S = U \tag{7.57}$$

The free-regime solution $(u_C)_F$ is obtained upon examination of the homogeneous equation

$$RC\frac{du_C(t)}{dt} + u_C(t) = 0$$

Since the corresponding characteristic equation is $RCs + 1 = 0$, with a root

$$s = -\frac{1}{\tau} = -\frac{1}{RC} \tag{7.58}$$

where $\tau = RC$ is the so-called time constant of the circuit, we find that the free-regime solution takes the form of a time-decaying exponential function

$$(u_C(t))_F = U'\, e^{st} = U'\, e^{-t/\tau} \tag{7.59}$$

By summing the steady-state solution in (7.57) with the free-regime solution in (7.59) we get

$$u_C(t) = U + U'\, e^{-t/\tau}$$

The unknown U' is determined by enforcing the initial condition $u_C(0^+) = u_C(0^-) = 0$, that is $0 = U + U'\, e^0$, from which $U' = -U$ is obtained. Hence

$$\text{For } t > 0:\ u_C(t) = U(1 - e^{-t/\tau}) \tag{7.60}$$

If you wish to know the transient current in the circuit you merely have to find the time derivative of the above result,

$$i(t) = C\frac{du_C}{dt} = CU\left(\frac{e^{-t/\tau}}{\tau}\right) = \frac{U}{R}\, e^{-t/\tau} \tag{7.61}$$

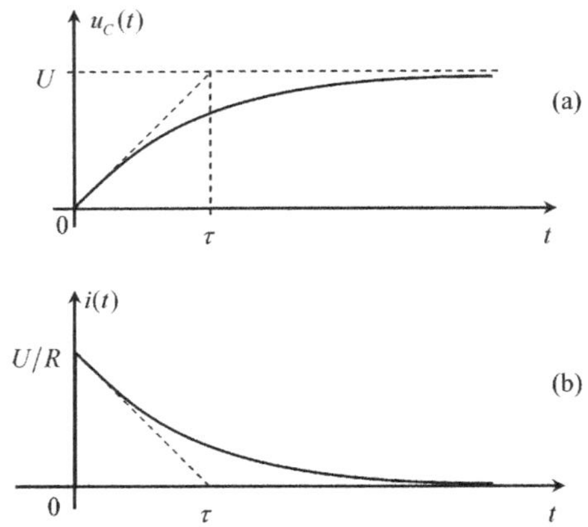

Figure 7.21 Transient response of the *RC* series circuit. The characteristic duration of transient phenomena is determined by the time constant $\tau = RC$. (a) Plot of the capacitor voltage against time. (b) Plot of the generator current against time

Graphical plots illustrating the results in (7.60) and (7.61) are shown in Figure 7.21.

An interesting point connected with the preceding results is the energy balance. As you already know, the final energy stored in the capacitor is $W_e = \frac{1}{2}CU^2$. Let us now evaluate the energy W expended by the generator as well as the energy W_J dissipated in the resistor during the charging process:

$$W = \int_0^\infty p(t)dt = U\int_0^\infty i(t)dt = CU^2; \quad W_J = R\int_0^\infty i^2(t)dt = \tfrac{1}{2}CU^2 \tag{7.62}$$

The amazing conclusion is that, irrespective of the resistance value, the DC generator always has to expend an amount of energy that is twice the value of the final energy stored in the capacitor, meaning that the yield factor of the charging process is exactly 50 %.

Suppose now that the resistor is removed from the circuit. You face a paradox, don't you? Where has the missing energy gone?

Some people will answer that it has been radiated away. But if you insist and ask for the calculation of the radiated power, you will most probably have no reply.

Well, things are simpler than they look, at first glance.

Observe again the results depicted in Figure 7.21. As you make R go to zero, the derivative *di/dt* increases greatly, the same thing happening to the derivative $d\psi_{S_s}/dt$. The real problem with the limit case $R = 0$ is that (7.56), the equation that we have used to model the problem, is no longer valid. You have to go back to (7.55) and drop the approximation $d\psi_{S_s}/dt \approx 0$.

Although the inductance L of the closed loop describing the circuit is only small, L must be taken into account when $i(t)$ undergoes rapid changes. In conclusion, for the analysis of the capacitor charging process, with $R = 0$, the governing equation to be employed is

$$\oint_s \mathbf{E}\cdot d\mathbf{s} = -\frac{d\psi_{S_s}}{dt} \rightarrow -U + u_C(t) = -L\frac{di(t)}{dt}$$

or, which is the same (using $i = C\, du_C/dt$),

$$LC\frac{d^2u_C(t)}{dt^2} + u_C(t) = U \tag{7.63}$$

As before, the steady-state solution is given by (7.57), $(u_C)_S = U$. The crucial difference resides with the free-regime solution, whose characteristic equation now reads $LCs^2 + 1 = 0$, and whose two roots are purely imaginary numbers

$$s_{1,2} = \pm j\omega_0, \quad \text{with } \omega_0 = 1/\sqrt{LC} \tag{7.64}$$

Consequently, the free-regime solution is now described by

$$(u_C(t))_F = U_1 e^{j\omega_0 t} + U_2 e^{-j\omega_0 t} \tag{7.65}$$

At this stage you should note that the left-hand side of (7.65) is a real-valued function, and, of course, the same thing should happen to the right-hand side of (7.65). For this to be possible the unknown constants U_1 and U_2 must be complex conjugates of each other

$$U_1 = \overline{U}_0 = U_0 e^{j\theta}; \; U_2 = \overline{U}_0^* = U_0 e^{-j\theta}$$

Substituting the preceding results into (7.65) we obtain

$$(u_C(t))_F = U_0\left(e^{j(\omega_0 t+\theta)} + e^{-j(\omega_0 t+\theta)}\right) = 2U_0 \cos(\omega_0 t + \theta)$$

Finally, the transient solution for u_C, that is the sum of the steady-state and free-regime solutions, is written as

$$u_C(t) = U + 2U_0 \cos(\omega_0 t + \theta) \tag{7.66}$$

The corresponding current $i = C\, du_C/dt$ is evaluated as

$$i(t) = -2CU_0\omega_0 \sin(\omega_0 t + \theta) \tag{7.67}$$

The constants U_0 and θ are to be determined using two initial conditions. One has already been stated above, $u_C(0) = 0$. The second, new, one is $i(0) = 0$, because the current in a loop containing an inductance L cannot have time discontinuities.

The latter condition, together with (7.67), leads to $\theta = 0$. The first initial condition, combined with (7.66), leads to $U_0 = -U/2$.

In conclusion, in the limit case $R = 0$, the capacitor charging process turns out to be an undamped periodic oscillation, whose periodicity is critically dependent on the self-inductance of the loop described by the circuit; the capacitor voltage and current being given by

$$u_C(t) = U\left(1 - \cos(\omega_0 t)\right); \; i(t) = CU\omega_0 \sin(\omega_0 t) \tag{7.68}$$

See the corresponding graphical plots in Figure 7.22.

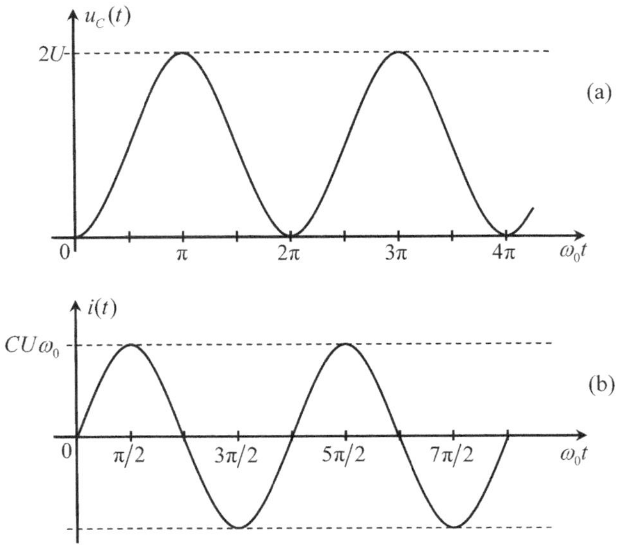

Figure 7.22 Transient response of the *RC* series circuit, with $R = 0$, but taking into account the small inductance L of the circuit's closed loop. The oscillation characterizing the transient phenomena is determined by the angular frequency $\omega_0 = 1/\sqrt{LC}$. (a) Capacitor voltage. (b) Generator current

At any given instant t' the energy W brought into play by the generator is the sum of the magnetic energy stored in the magnetic field of the loop plus the electric energy stored in the capacitor:

$$\begin{cases} W(t') = \int_0^{t'} p(t)\,dt = CU^2\,(1-\cos(\omega_0 t')) \\ W_m(t') = \dfrac{CU^2}{2}\sin^2(\omega_0 t') \\ W_e(t') = \dfrac{CU^2}{2}(1-\cos(\omega_0 t'))^2 \end{cases} \rightarrow W(t') = W_m(t') + W_e(t')$$

where, from (7.64), $\omega_0 = 1/\sqrt{LC}$.

7.4.4 Connecting an Inductive Load to an AC Generator

Next, let us examine the transient regime occurring when an AC generator is switched in an *RL* circuit – see Figure 7.23. As you will see, studying this case brings out novel aspects that we have not yet treated, providing you with further insights into transient phenomena.

As you should know already, application of the induction law to the circuit in Figure 7.23 yields the following governing equation:

$$\text{For } t > 0:\; L\frac{di(t)}{dt} + Ri(t) = u(t), \;\text{ with }\; u(t) = U\cos(\omega t + \alpha_u) \tag{7.69}$$

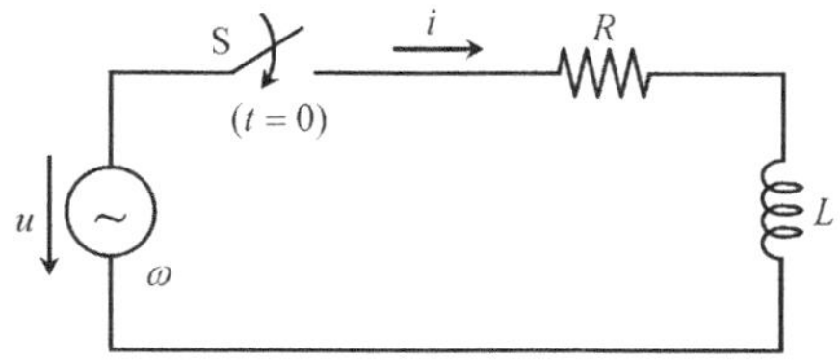

Figure 7.23 Connecting an *RL* series circuit to an AC generator

The steady-state solution is determined by resorting to the phasor-domain technique described earlier in Section 7.2. The impedance of the circuit is

$$\overline{Z} = R + j\omega L = Z\, e^{j\varphi};\;\; Z = \sqrt{R^2 + (\omega L)^2};\;\; \varphi = \arctan(\omega L/R)$$

The phasor associated with the sinusoidal current is obtained from

$$\overline{I} = \frac{\overline{U}}{\overline{Z}} = I\, e^{j\alpha_i} \rightarrow \begin{cases} I = U/Z \\ \alpha_i = \alpha_u - \varphi \end{cases} \tag{7.70}$$

From (7.70), the steady-state solution for the circuit current is written in the time domain as

$$(i(t))_S = I\cos(\omega t + \alpha_i) \tag{7.71}$$

The analysis of the free-regime starts with the homogeneous equation associated with (7.69)

$$L\frac{di(t)}{dt} + Ri(t) = 0$$

The corresponding characteristic equation is $Ls + R = 0$, and its negative root is

$$s = -\frac{1}{\tau} = -\frac{R}{L} \tag{7.72}$$

where the time constant of the circuit is $\tau = L/R$. The free-regime solution is a time-decaying exponential function

$$(i(t))_F = I'\, e^{st} = I'\, e^{-t/\tau} \tag{7.73}$$

The transient current is obtained by summing the sub-solutions in (7.71) and (7.73):

$$\text{For } t > 0:\;\; i(t) = I\cos(\omega t + \alpha_i) + I'\, e^{-t/\tau} \tag{7.74}$$

The unknown amplitude I' of the free-regime solution is found by consideration of the initial condition pertaining to this problem, which is $i(0^+) = i(0^-) = 0$. Consequently, from (7.74), we have: $0 = I\cos(\alpha_i) + I'$. Hence

$$I' = -I\cos(\alpha_i) = -I\cos(\alpha_u - \varphi) \tag{7.75}$$

Combining (7.74) and (7.75), we get the final solution:

$$\text{For } t > 0:\;\; i(t) = I\left[\cos(\omega t + \alpha_i) - \cos(\alpha_i)\, e^{-t/\tau}\right]$$

Noting, from (7.75), that the initial amplitude I' of the free-regime depends on the initial phase α_u of the generator voltage, several interesting points arise and deserve to be pointed out.

If $\alpha_u = \varphi \pm \pi/2$ then the free regime will be absent. You switch on the circuit and a purely sinusoidal current establishes itself immediately, $i(t) = I\cos(\omega t \pm \pi/2)$. Clearly, this is the most desirable situation.

On the contrary, if $\alpha_u = \varphi$ or $\alpha_u = \varphi + \pi$ then the free regime will have a maximum initial amplitude. For $\alpha_u = \varphi$ you get

$$i(t) = I\left(\cos(\omega t) - e^{-t/\tau}\right) \tag{7.76a}$$

For $\alpha_u = \varphi + \pi$ you get

$$i(t) = I\left(e^{-t/\tau} - \cos(\omega t)\right) \tag{7.76b}$$

The cases described by (7.76) are the least desirable situations, because after the closing of the switch, the circuit current displays a distorted asymmetrical shape. Furthermore, if the time constant τ happens to be much longer than the sinusoid period T, then several repetitive overcurrent peaks will occur for quite a long time. To better understand what we are talking about, look for example at (7.76b) and assume that the exponential decay is extremely slow – that is, consider the approximation $\exp(-t/\tau) \approx 1$. At instants of time such that $t = t_m = mT/2$, with m an odd number, you will get $i(t_m) \approx 2I$, which signifies that the transient current may reach an intensity that is twice the predicted one for the steady-state regime (this can not only blow protective fuses, but even endanger the equipment itself).

7.4.5 Disconnecting an Inductive Load

Have you ever tried to unplug a running inductive appliance (like a washing machine or an electric fan) from its wall socket? What did you notice? Most probably you saw an arc discharge (sparks) occurring. Do you know why that happened? If not, you will learn about it now.

Take the same circuit we analyzed earlier (see Figure 7.23), and consider that enough time has elapsed after the switching-on operation, so that the final steady state has been reached

$$i(t) = (i(t))_S = I\cos(\omega t + \alpha_i) \tag{7.77}$$

At $t = t_0$ you decide to open the switch, disconnecting the load. If you are very lucky, that is if by chance $\omega t_0 + \alpha_i = \pm\pi/2$, then you will be interrupting a null current, $i(t_0^+) = i(t_0^-) = 0$. The initial condition for inductor currents is not violated and nothing unusual happens.

However, most likely the opening of the switch will occur when $i(t_0^-) \neq 0$. In this case, although the switch is open, the circuit current refuses to go to zero immediately (inductor currents can never have time discontinuities).

By analyzing the circuit in Figure 7.24, you find for the voltage $u_S(t)$ at the switch terminals

$$\text{For } t > t_0: \; u_S = u - Ri - L\frac{di}{dt} \tag{7.78}$$

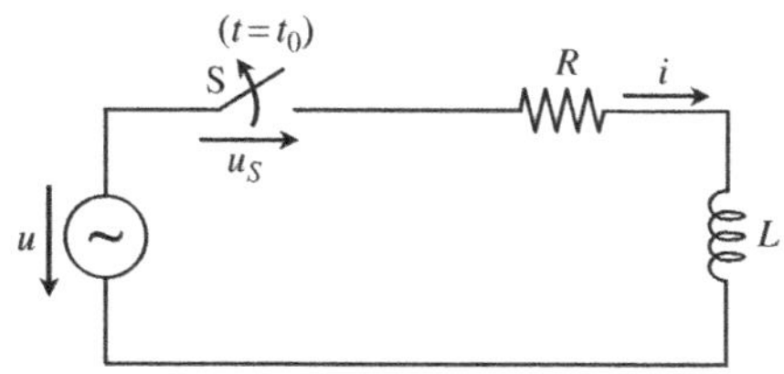

Figure 7.24 Disconnecting an inductive load

Subsequent to the switching operation, and since you are trying to interrupt $i(t)$ abruptly, the derivative *di/dt* increases dramatically. This has several consequences related to overvoltage problems. The first is that the inductor itself can be damaged (the winding insulation may fail). Secondly, a sudden overvoltage appears in u_S leading to a very intense electric field $\mathbf{E}(t)$ between the switch contacts, which in turn gives rise to breakdown phenomena in the air, originating an arc discharge channel through which conduction currents are allowed to flow until $i(t)$ reaches zero and the phenomenon ceases naturally.

A technical solution to avoid arc discharges consists of inserting the switch inside a vacuum enclosure (by definition a vacuum cannot be ionized, and therefore dielectric breakdown cannot occur). The question, for the most curious among you, is how does the electric current manage to flow in a vacuum?

Do you have any idea? Maybe you can find the answer in Chapter 6. . .

If it helps, you may imagine that a small capacitance exists between the terminals of the open switch. In fact, the current flow is ensured in the form of a displacement current, $\varepsilon_0\, \partial\mathbf{E}/\partial t$, across the open switch.

We have learnt that inductor currents cannot be interrupted without harmful consequences. Having said that, let us analyze a simple protection scheme that can be used to prevent the problems mentioned above.

As shown in Figure 7.25, a protective resistor R_0 is placed in parallel with the *RL* circuit. When the switch is open, the current $i(t)$ suffers no interruption, since it can still flow in the closed loop formed by R_0, R and L.

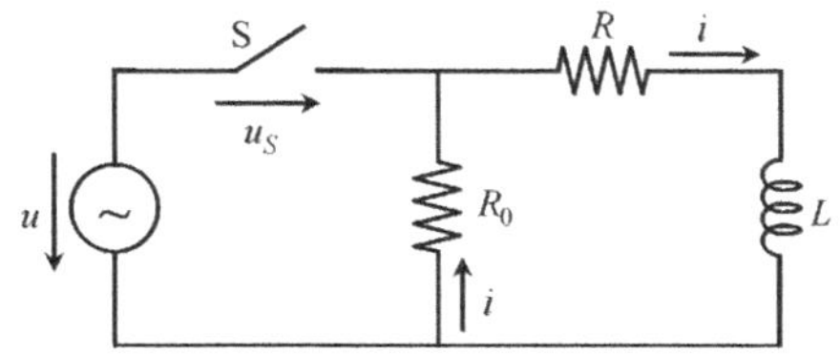

Figure 7.25 Disconnecting an inductive load protected by a parallel-connected resistor R_0

The governing equations of the new circuit for $t > t_0$ are

$$L\frac{di(t)}{dt} + (R + R_0)\, i(t) = 0 \tag{7.79}$$

$$u_S(t) = u(t) + R_0 i(t) \tag{7.80}$$

Since (7.79) is itself a homogeneous differential equation, the steady-state solution for $i(t)$ is zero. In other words, the transient regime is completely described by the free-regime solution, that is

$$\text{For } t > t_0 : \; i(t) = I_0 \, e^{-(t-t_0)/\tau_0}, \quad \text{with } \tau_0 = L/(R+R_0) \tag{7.81}$$

The unknown constant I_0 is determined from (7.81) and (7.77), by enforcing the initial condition at $t = t_0 : \; I_0 = i(t_0^+) = i(t_0^-) = I\cos(\omega t_0 + \alpha_i)$. Finally we find

$$i(t) = I\cos(\omega t_0 + \alpha_i) \, e^{-(t-t_0)/\tau_0}$$

$$u_S(t) = U\cos(\omega t + \alpha_u) + R_0 I\cos(\omega t_0 + \alpha_i) \, e^{-(t-t_0)/\tau_0} \tag{7.82}$$

As $t \to \infty$, the load current goes exponentially to zero, whereas the switch voltage tends to follow the generator's sinusoidal voltage.

If the protective resistor R_0 in Figure 7.25 is removed, $R_0 \to \infty$, you will see from (7.82) that u_S increases to infinity, leading to breakdown phenomena at the switch terminals, as we discussed at the beginning of this subsection.

7.4.6 Application Example (Switching Off a Transformer Protected by a Capacitor)

A transformer, characterized by its intrinsic parameters r_1, r_2, L_{11}, L_{22} and L_M, has its secondary winding open ($i_2 = 0$). A protective capacitor of capacitance C is connected in parallel with the primary winding – see Figure 7.26.

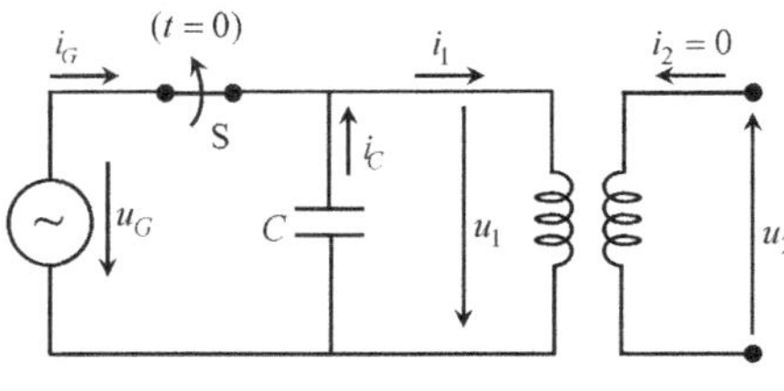

Figure 7.26 Switching off a transformer protected by a parallel-connected capacitor

The generator's applied voltage is $u_G(t) = U_G\cos(\omega_G t + \pi/2)$.

Data: $r_1 = r_2 = 10\ \Omega$, $L_{11} = L_{22} = 17.32\,\text{mH}$, $L_M = 10\,\text{mH}$, $C = 46.41\,\mu\text{F}$, $U_G = 10\,\text{V}$, $\omega_G = 1\,\text{krad/s}$.

Questions

Q_1 Assume that the switch S is closed for a long time. Determine the phasors associated with the voltages and currents marked in Figure 7.26.

Q_2 Analyze the transient regime resulting from opening the switch at $t = 0$; in particular, determine the time evolution of $i_1(t)$, $u_1(t)$ and $u_2(t)$.

Solutions

Q$_1$ $\overline{U}_G = \overline{U}_1 = 10\, e^{j90^\circ}$ V.

$$\overline{I}_1 = \frac{\overline{U}_1}{r_1 + j\omega_G L_{11}} = 500\, e^{j30^\circ} \text{ mA}; \ \overline{I}_C = -j\omega_G C\overline{U}_1 = 464.1 \text{ mA};$$

$$\overline{I}_G = \overline{I}_1 - \overline{I}_C = 252\, e^{j97^\circ} \text{ mA}; \ \overline{U}_2 = -j\omega_G L_M \overline{I}_1 = 5\, e^{-j60^\circ} \text{ V}.$$

Q$_2$ By opening the switch you make $i_G = 0$.

Because of the initial conditions that we will need to employ later, it is recommended that you first find the initial values of the state variables of the problem, that is the voltage u_1 across the capacitor and the current i_1 in the primary winding:

$$u_1(0) = U_{10} = \Re\left(\overline{U}_1\right) = 0; \quad i_1(0) = I_{10} = \Re\left(\overline{I}_1\right) = 500 \cos(30^o) = 433 \text{ mA} \quad (7.83)$$

The capacitor current i_C (which is not a state variable) coincides with i_1, for $t > 0$.

By application of the induction law to the primary winding you find

$$u_1(t) = r_1 i_1(t) + L_{11}\frac{di_1(t)}{dt} \quad (7.84)$$

Further, for $t > 0$, at the capacitor terminals you have

$$i_1(t) = -C\frac{du_1(t)}{dt} \quad (7.85)$$

By combining the two preceding equations you immediately obtain a second-order homogeneous differential equation in the state variable $u_1(t)$:

$$\frac{d^2}{dt^2}u_1(t) + \frac{r_1}{L_{11}}\frac{d}{dt}u_1(t) + \frac{1}{L_{11}C}u_1(t) = 0$$

This result can be rewritten in the canonical form

$$\frac{d^2}{dt^2}u_1(t) + 2\beta\frac{d}{dt}u_1(t) + \omega_0^2 u_1(t) = 0 \quad (7.86)$$

where

$$\beta = \frac{r_1}{2L_{11}}, \quad \omega_0 = \frac{1}{\sqrt{L_{11}C}} \quad (7.87)$$

The constant β is called the damping factor (units: Np/s), and ω_0 is called the undamped angular frequency (units: rad/s). The reason for this terminology will become clear very soon.

The characteristic equation corresponding to (7.86), as well as its roots s_1 and s_2, is

$$s^2 + 2\beta s + \omega_0^2 = 0, \text{ with roots}: s_{1,2} = -\beta \pm \sqrt{\beta^2 - \omega_0^2} \quad (7.88)$$

Three distinct situations can now happen:

1. Two negative real roots are found when $\beta > \omega_0$. The solution, which is said to be overdamped, takes the form: $u_1(t) = U'\ e^{-t/\tau_1} + U''\ e^{-t/\tau_2}$, where $\tau_1 = -1/s_1$ and $\tau_2 = -1/s_2$.
2. One double real root is found when $\beta = \omega_0$. The solution, which is said to be critically damped, takes the form $u_1(t) = (U' + tU'')\, e^{-\beta t}$.
3. Two complex conjugate roots are found when $\beta < \omega_0$. The solution, which is said to be underdamped, takes the oscillatory form

$$u_1(t) = U\ e^{-\beta t}\cos(\omega t + \theta) \tag{7.89}$$

In our problem we have, from (7.87), $\beta = 288.7\,\mathrm{Np/s}$ and $\omega_0 = 1115.4\,\mathrm{rad/s}$. Two complex conjugate characteristic roots with negative real part are encountered:

$$s = -\beta + j\omega = \omega_0\ e^{j\delta} \text{ and } s^* = -\beta - j\omega = \ \omega_0\ e^{-j\delta}$$

where $\omega = \sqrt{\omega_0^2 - \beta^2}$, the so-called damped angular frequency, is $\omega = 1077.4\,\mathrm{rad/s}$ and $\delta = 105° = 7\pi/12$.

The transient regime for $u_1(t)$, which is a purely free regime, can then be written as $u_1(t) = U'\ e^{st} + U''\ e^{s^*t}$, where U' and U'' must necessarily be complex conjugate numbers (otherwise $u_1(t)$ would not be a real-valued function).

By making $U' = (U'')^* = \frac{1}{2}\overline{U}$, with $\overline{U} = Ue^{j\theta}$, we obtain

$$u_1(t) = \left(\frac{\overline{U}}{2}\ e^{st}\right) + \left(\frac{\overline{U}}{2}\ e^{st}\right)^* = \Re\left(\overline{U}\ e^{st}\right)$$

Substituting $(-\beta + j\omega)$ for s, and substituting $Ue^{j\theta}$ for $\overline{U}$, we find

$$\Re\left(\overline{U}\ e^{st}\right) = U\ e^{-\beta t}\cos(\omega t + \theta)$$

which confirms the expression in (7.89). In conclusion, we can write the transient oscillatory solution for $u_1(t)$ as

$$u_1(t) = \begin{cases} \Re\left(\overline{U}\ e^{st}\right) \\ U\ e^{-\beta t}\cos(\omega t + \theta) \end{cases} \tag{7.90}$$

The result on the top, with complex quantities, is more compact and more useful for calculations (since the variable t appears a single time). The result on the bottom, involving real functions, is, however, more appealing from a physical point of view, as it permits a straightforward interpretation of the phenomenon – an oscillation of angular frequency ω whose amplitude decreases with time according to the damping factor β.

Going back to the problem under analysis, the initial condition $u_1(0) = 0$ applied to (7.90) leads to $\theta = \pi/2$. For the determination of the unknown U we have to resort to the second initial condition in (7.83), $i_1(0) = I_{10} = 433\,\text{mA}$, which leads us to an examination of $i_1(t)$ in (7.85):

$$i_1(t) = -C\frac{du_1(t)}{dt} = -C\,\Re\left(\frac{d}{dt}Ue^{j\pi/2}e^{st}\right) = -CU\,\Re\,(jse^{st}) \tag{7.91a}$$

Taking into account that $s = -\beta + j\omega = \omega_0\, e^{j\delta}$, we may rewrite (7.91a) as

$$i_1(t) = \omega_0 CU\,\Re\left(e^{-\beta t}\, e^{j(\omega t+\delta-\pi/2)}\right) = \omega_0 CU\, e^{-\beta t}\cos(\omega t + \delta - \pi/2) \tag{7.91b}$$

For $t = 0$, we get

$$I_{10} = -CU\,\Re\,(js) = \omega_0 CU \sin\delta$$

from which the constant U is evaluated,

$$U = \sqrt{\frac{L_{11}}{C}}\frac{I_{10}}{\sin\delta} = 8.66\ \text{V}$$

The voltage across the secondary winding is finally obtained from $u_2 = -L_M\, di_1/dt$. By using (7.91a) we obtain, for $t > 0$,

$$u_2(t) = L_M CU\,\Re\left(js\frac{d}{dt}e^{st}\right) = L_M CU\,\Re\left(js^2e^{st}\right) = \omega_0^2 L_M CU\,\Re\left(e^{-\beta t}e^{j(\omega t+2\delta+\pi/2)}\right)$$

$$u_2(t) = \frac{L_M}{L_{11}}U\, e^{-\beta t}\cos\left(\omega t + 2\delta + \tfrac{\pi}{2}\right)$$

Let us summarize the results obtained for the transient regime subsequent to the opening of the switch:

$$\text{For } t > 0: \begin{cases} u_1(t) = U_1\, e^{-\beta t}\cos(\omega t + \pi/2), & \text{with } U_1 = 8.66\ \text{V} \\ i_1(t) = I_1\, e^{-\beta t}\cos(\omega t + \pi/12), & \text{with } I_1 = 448.3\ \text{mA} \\ u_2(t) = U_2\, e^{-\beta t}\cos(\omega t - \pi/3), & \text{with } U_2 = 5.00\ \text{V} \end{cases}$$

where $\beta = 288.7\,\text{Np/s}$ and $\omega = 1077.4\,\text{rad/s}$.

The nature of the oscillatory regime is explained by the flowing back and forth of the energy between the capacitor and the transformer. At $t = 0$ the system's energy resides entirely in the transformer core (magnetic energy). After a few instants have elapsed, at $\omega t = 5\pi/12$, the energy stored in the transformer vanishes; most of the initial energy is now found to be stored in the capacitor (electric energy). During the transfer process some energy is transformed into heat due to Joule losses in the resistor (damping). The interchange of energy between the two reactive components of the circuit repeats itself periodically, lasting until no more electromagnetic energy exists. This idea is schematically represented in Figure 7.27.

A final remark: although the voltage u_2 at the transformer's secondary winding is not a state variable, you can check that $u_2(0^+) = u_2(0^-) = 2.5\,\text{V}$. Can you figure out why this must indeed happen in our circuit (Figure 7.26)?

(Hint: Analyze the continuity of the function di_1/dt.)

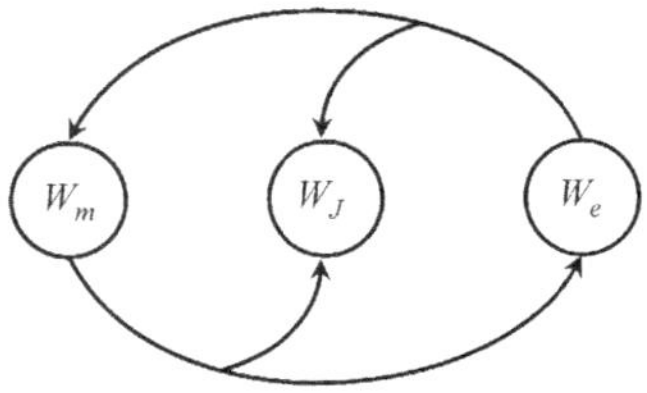

Figure 7.27 The oscillatory nature of the transient regime can be interpreted as the result of the periodic interchange of electromagnetic energy between the transformer and the capacitor, the resistor being responsible for the damping effect (Joule losses)

7.5 Proposed Homework Problems

Problem 7.5.1

Figure 7.28(a) shows the basic idea of a radio receiver tuning circuit, which essentially consists of two coupled inductors and a variable capacitor. The primary inductor current $i_a(t)$ is fed from a receiving antenna. The mutual inductance between inductors is L_M. The secondary inductor is characterized by its self-inductance L and by its internal resistance R.

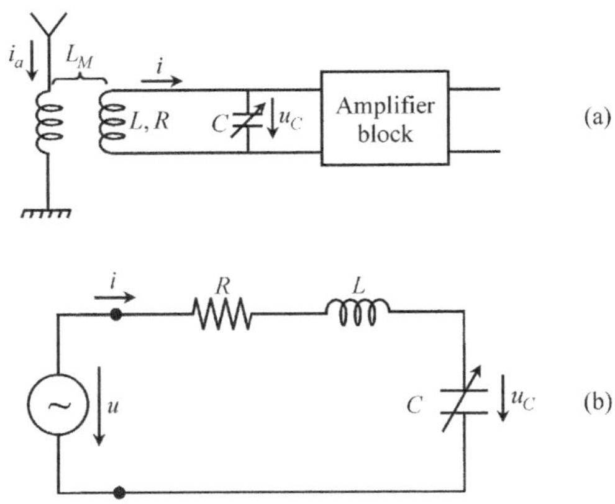

Figure 7.28 Radio receiver tuning circuit. (a) Schematic diagram. (b) Equivalent *RLC* series circuit, where the applied voltage is given by $u(t) = -L_M \, di_a/dt$

The antenna current includes contributions from distinct broadcasting stations, $i_a = \sum_k i_k$, but we are interested in tuning only to the station operating at $f_1 = 1\,\text{MHz}$ (medium-frequency band). The 1 MHz component of $i_a(t)$ is given by $i_1(t) = I_1 \cos(\omega_1 t)$, with $I_1 = 2.25\,\mu\text{A}$.

Assume that the amplification block is ideal ($\overline{Z}_{\text{amp}} \to \infty$).

Data: $L_M = 40\,\mu\text{H}$, $L = 84.4\,\mu\text{H}$, $R = 10\ \Omega$.

Q_1 Show that the *RLC* circuit in Figure 7.28(b) is equivalent to the tuning circuit in Figure 7.28(a). In particular, determine the equivalent generator voltage $u(t)$.

Q_2 For $f = f_1$, determine the tuning capacitance of the circuit which maximizes the amplitude of the voltage u_C.

Q_3 For $f = f_1$, determine the resonance capacitance of the circuit which maximizes the amplitude of the current i.

Q_4 Show that the situations in Q_2 and Q_3 converge to each other when $R \ll \omega L$.

Q_5 For the situation in Q_3, considering $i_a(t) = i_1(t) = I_1 \cos(\omega_1 t)$, determine the phasors, $\overline{U}$, $\overline{I}$ and $\overline{U}_C$.

Answers

Q_1

$$u(t) = -L_M \frac{di_a(t)}{dt}$$

Q_2

$$C_{tuning} = \frac{L}{R^2 + (\omega_1 L)^2} = 300.01\,\text{pF}$$

Q_3

$$C_{res} = \frac{1}{\omega_1^2 L} = 300.12\ \text{pF}$$

Q_4

$$C_{res} = \lim_{R \to 0} C_{tuning}$$

Q_5

$$\overline{U} = 0.566\, e^{-j\pi/2}\ \text{mV};\ \overline{I} = 56.6\, e^{-j\pi/2}\ \mu\text{A};\ \overline{U}_C = 30\, e^{j\pi}\ \text{mV}.$$

Problem 7.5.2

As shown in Figure 7.29, an $L \parallel C$ circuit is inserted between an AC voltage generator and a resistive load R. Data: $L = 10\,\text{mH}$, $C = 100\,\text{pF}$.

Q_1 Determine analytically the transfer function $\overline{T}(\omega) = \overline{U}_2/\overline{U}_1$.

Q_2 Find the frequency f_0 for which $i = 0$, and therefore $T(\omega) = 0$.

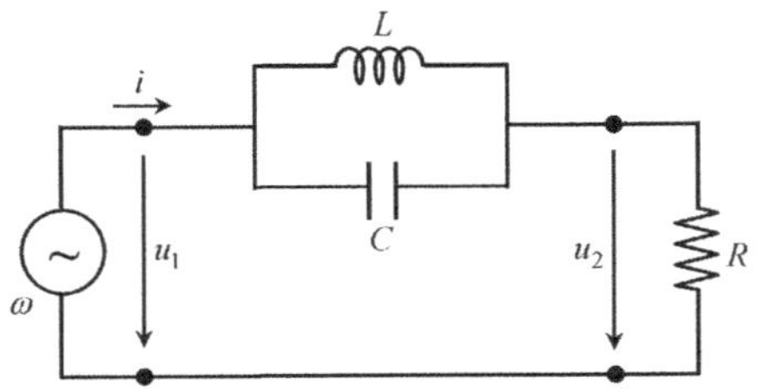

Figure 7.29 Notch filter made of an $L \parallel C$ circuit

Q_3 Draw a sketch of $T(\omega)$, for $\omega \in [0; \infty[$, showing that the circuit under analysis behaves as a notch filter (a notch filter passes all frequencies except those in a stopband centered at ω_0).

Answers

Q_1

$$\overline{T}(\omega) = \frac{1}{1 + j\dfrac{\omega L/R}{1 - \omega^2 LC}}$$

Q_2

$$\omega_0 = 1/\sqrt{LC} = 1\,\text{Mrad/s}; \;\; f_0 = \omega_0/(2\pi) = 159.15\,\text{kHz}. \tag{7.92}$$

Q_3

$$T(0) = T(\infty) = 1; \;\; T(\omega_0) = 0. \quad \text{See sketch in Figure 7.30.} \tag{7.93}$$

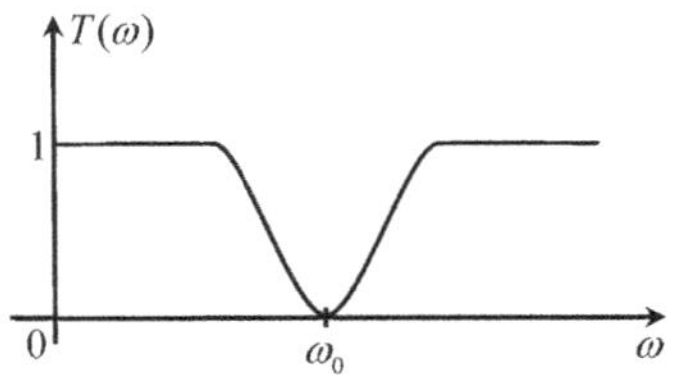

Figure 7.30 Magnitude plot of the filter transfer function, showing the center frequency $\omega_0 = 1/\sqrt{LC}$

Problem 7.5.3

Consider the circuit shown in Figure 7.31, which illustrates the basic principle of a spectrum analyzer. The input voltage signal is defined as

$$u(t) = \sum_{k=1}^{5} u_k(t) = \sum_{k=1}^{5} U_k \cos(k\omega t + \alpha_k)$$

where $\omega = 10\,\text{krad/s}$.

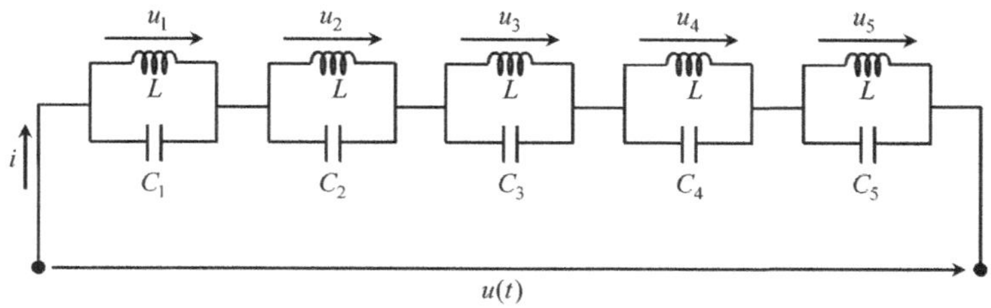

Figure 7.31 Basic principle of a spectrum analyzer

Q_1 Taking into account the results from Problem 7.5.2, show that the voltage components u_1, u_2, u_3, u_4 and u_5 appear separately across each $L \parallel C$ block, provided that each block respectively resonates at $\omega_1 = \omega$, $\omega_2 = 2\omega$, $\omega_3 = 3\omega$, $\omega_4 = 4\omega$ and $\omega_5 = 5\omega$.

Q_2 Assuming that the uncoupled inductors are equal, with $L = 10\,\text{mH}$, determine the required values for C_1, C_2, C_3, C_4 and C_5.

Answers

Q_1 By using the superposition principle (valid for linear circuits) each frequency component can be analyzed separately. For ω_k the resonant $L \parallel C_k$ block behaves as an open circuit ($i = 0$), but the other blocks, having a finite impedance, will show zero voltage across them. Consequently, the voltage across the resonant $L \parallel C_k$ block is $u_k(\omega_k)$.

Q_2 $C_k = 1/(\omega_k^2 L)$; $C_1 = 1000\,\text{nF}$; $C_2 = 250\,\text{nF}$; $C_3 = 111.1\,\text{nF}$; $C_4 = 62.5\,\text{nF}$; $C_5 = 40\,\text{nF}$.

Problem 7.5.4

Consider the circuit depicted in Figure 7.32, where the ideal ammeters A_R, A_C, A_L and A_0 read rms values. The applied voltage is $u(t) = \sqrt{2}\, U_{rms} \cos(\omega t)$.

Data: $U_{rms} = 5\,\text{V}$, $\omega = 10\,\text{krad/s}$, $R = 50\ \Omega$ and $L = 5\,\text{mH}$.

Q_1 Write the time-domain and phasor-domain equations that govern the steady-state harmonic regime of the circuit.

Q_2 Determine the capacitor's capacitance $C = C_{\text{res}}$ that brings the circuit to a resonant situation.

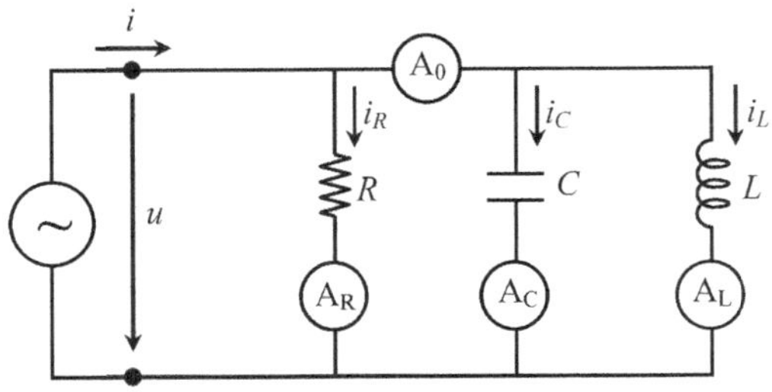

Figure 7.32 An *RLC* parallel circuit with ideal ammeters placed for measuring rms current intensities

Q_3 Assume $C = C_{res}$.

Determine numerically the phasor representation of u, i_R, i_C, i_L and i.
Indicate the readings of all the ammeters included in the circuit.
Determine the active and reactive powers at the generator terminals. Check the results using the complex Poynting theorem.

Answers

Q_1 $i_R = u/R \rightarrow \overline{I}_R = \overline{U}/R$; $i_C = C\, du/dt \rightarrow \overline{I}_C = j\omega C\overline{U}$; $u = L\, di_L/dt \rightarrow \overline{I}_L = \overline{U}/(j\omega L)$.

$$i = i_R + i_C + i_L \rightarrow \overline{I} = \overline{I}_R + \overline{I}_C + \overline{I}_L = \overline{U}\left(\frac{1}{R} + j\omega C + \frac{1}{j\omega L}\right)$$

Q_2 $C = C_{res} = 2\,\mu\text{F}$.

Q_3 $\overline{U} = \sqrt{2}\,5\,\text{V}$; $\overline{I}_R = \sqrt{2}\,0.1\,\text{A}$; $\overline{I}_L = \sqrt{2}\,0.1\,e^{-j\pi/2}\,\text{A}$; $\overline{I}_C = \sqrt{2}\,0.1\,e^{+j\pi/2}\,\text{A}$; $\overline{I} = \sqrt{2}\,0.1\,\text{A}$

$A_R = A_C = A_L = 100\,\text{mA}$; $A_0 \equiv 0$ (note that $i_C(t) + i_L(t) = 0$).

$P = 5\,\text{W}$; $P_Q = 0$; $P_J = RI^2_{R_{rms}} = 5\,\text{W}$; $(W_m)_{av} = (W_e)_{av} = 25\,\mu\text{J}$.

Problem 7.5.5

A residential four-wire installation is fed by a three-phase voltage generator (see Problem 5.13.6). As shown in Figure 7.33, both the generator and the load are star connected. The generator phase-to-neutral voltages are defined as $u_k(t) = \sqrt{2}\ U_{rms}\ \cos(\omega t - \alpha_k)$, where $U_{rms} = 230\,\text{V}$, and $\alpha_k = (k-1)2\pi/3$, for $k = 1, 2, 3$. The frequency is $f = 50\,\text{Hz}$.

Q_1 Determine the phase-to-phase generator voltages.

Q_2 Assume an unbalanced load whose impedances are $\overline{Z}_1 = 20\ \Omega$, $\overline{Z}_2 = 30\ e^{j\pi/3}\ \Omega$ and $\overline{Z}_3 = 40\ e^{j\pi/6}\ \Omega$. Determine the phasors characterizing the installation currents, including the neutral, that is $\overline{I}_1, \overline{I}_2, \overline{I}_3$ and $\overline{I}_N$.

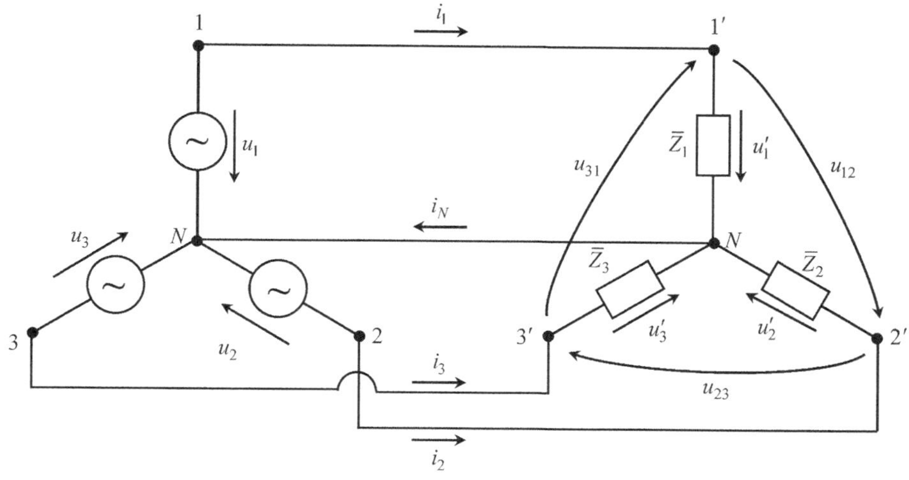

Figure 7.33 Four-wire, three-phase installation. The generator and the unbalanced load are star connected

Q_3 Keep considering the unbalanced load described in Q_2 but suppose that, by accident, the neutral wire is interrupted, $\overline{I}_N = 0$. Find the new voltages $\overline{U}_1'$, $\overline{U}_2'$ and $\overline{U}_3'$ applied to the load impedances. Show that the load voltage u_2' exceeds its nominal value (230 V) by about 30 %.

Q_4 Assume that the protective circuit breaker associated with load $\overline{Z}_2$ opens, a situation that from the generator's point of view is equivalent to making $\overline{Z}_2 = \infty$. Recompute the new voltage phasors $\overline{U}_1'$, $\overline{U}_2'$ and $\overline{U}_3'$. Check that a subsequent new overvoltage appears across $\overline{Z}_3$.

Answers

Q_1 $\overline{U}_{12} = \overline{U}_1 - \overline{U}_2 = \sqrt{2}\left(\sqrt{3}\ 230\right)\ e^{j\pi/6}\ \text{V} \rightarrow u_{12}(t) = \sqrt{2}\ 398.4\ \cos(\omega t + \pi/6)\ \text{V}.$

$\overline{U}_{23} = \overline{U}_2 - \overline{U}_3 = \sqrt{2}\left(\sqrt{3}\ 230\right)\ e^{-j\pi/2}\ \text{V} \rightarrow u_{23}(t) = \sqrt{2}\ 398.4\ \cos(\omega t - \pi/2)\ \text{V}.$

$\overline{U}_{31} = \overline{U}_3 - \overline{U}_1 = \sqrt{2}\left(\sqrt{3}\ 230\right)\ e^{j5\pi/6}\ \text{V} \rightarrow u_{31}(t) = \sqrt{2}\ 398.4\ \cos(\omega t + 5\pi/6)\ \text{V}.$

Q_2 With the neutral wire present we have $\overline{U}_1' = \overline{U}_1$, $\overline{U}_2' = \overline{U}_2$, $\overline{U}_3' = \overline{U}_3$.

$$\overline{I}_k = \frac{\overline{U}_k'}{\overline{Z}_k} = \begin{cases} \overline{I}_1 = \sqrt{2}\ 11.50\ \text{A} \\ \overline{I}_2 = \sqrt{2}\ 7.67\, e^{j180^o}\,\text{A}, \\ \overline{I}_3 = \sqrt{2}\ 5.75\, e^{j90^o}\,\text{A} \end{cases} \qquad \overline{I}_N = \overline{I}_1 + \overline{I}_2 + \overline{I}_3 = \sqrt{2}\ 6.91\, e^{j56.3^o}\,\text{A}$$

Q_3

$$\begin{bmatrix} \overline{Z}_1 & -\overline{Z}_2 & 0 \\ 0 & \overline{Z}_2 & -\overline{Z}_3 \\ 1 & 1 & 1 \end{bmatrix} \begin{bmatrix} \overline{I}_1 \\ \overline{I}_2 \\ \overline{I}_3 \end{bmatrix} = \begin{bmatrix} \overline{U}_{12} \\ \overline{U}_{23} \\ 0 \end{bmatrix} \rightarrow \begin{bmatrix} \overline{I}_1 \\ \overline{I}_2 \\ \overline{I}_3 \end{bmatrix} = \begin{bmatrix} \overline{Z}_1 & -\overline{Z}_2 & 0 \\ 0 & \overline{Z}_2 & -\overline{Z}_3 \\ 1 & 1 & 1 \end{bmatrix}^{-1} \begin{bmatrix} \overline{U}_{12} \\ \overline{U}_{23} \\ 0 \end{bmatrix}$$

$$\overline{U}_1' = \overline{Z}_1 \overline{I}_1 = \sqrt{2}\ 230.3\, e^{-j17.7^o}\ \text{V};\ \overline{U}_2' = \overline{Z}_2 \overline{I}_2 = \sqrt{2}\ 297.1\, e^{-j115.0^o}\ \text{V};$$
$$\overline{U}_3' = \overline{Z}_3 \overline{I}_3 = \sqrt{2}\ 180.1\, e^{j134.2^o}\ \text{V}$$

Q_4

$$\overline{U}_1' = \sqrt{2}\ 136.9\, e^{-j50.1^\circ}\ \text{V};\ \overline{U}_2' = \sqrt{2}\ 398.4\, e^{-j130.2^\circ}\ \text{V};\ \overline{U}_3' = \sqrt{2}\ 273.9\, e^{j159.9^\circ}\ \text{V}.$$

Problem 7.5.6

A three-phase generator feeds a perfectly balanced load $\overline{Z}_1 = \overline{Z}_2 = \overline{Z}_3 = \overline{Z} = Z\ e^{j\varphi}$. The load is star connected – see Figure 7.34(a).

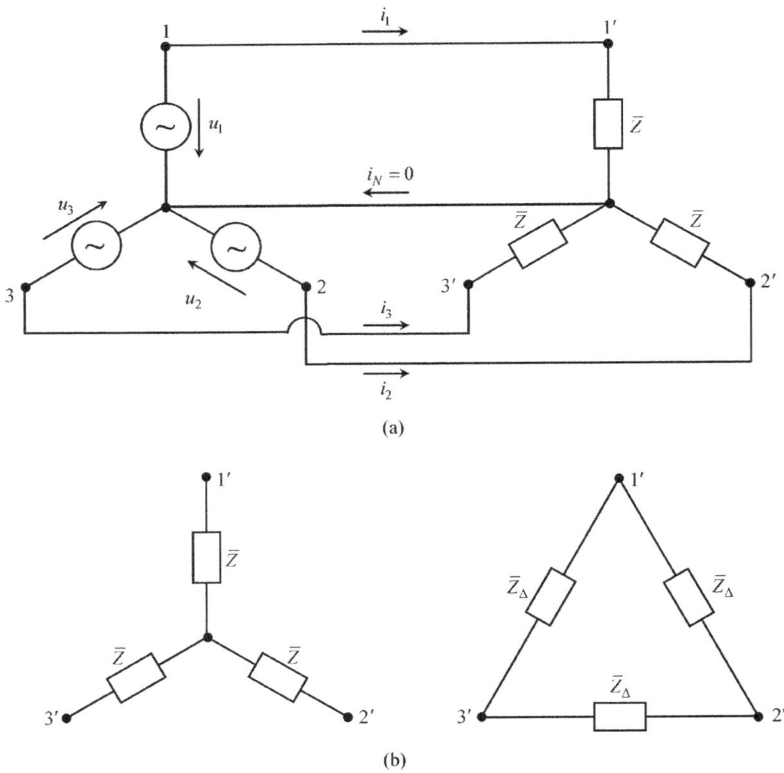

Figure 7.34 (a) Four-wire, three-phase installation with a star-connected balanced load (the neutral current is zero). (b) Equivalence between a star-connected balanced load and a delta-connected balanced load, $\overline{Z}_\Delta = 3\overline{Z}$

Q_1 Check that the neutral wire is not needed, by showing that $\overline{I}_N = 0$.

Q_2 Show that the instantaneous three-phase power $p(t) = p_1(t) + p_2(t) + p_3(t)$ is time invariant.

Q_3 The star load connection can be replaced by an equivalent delta connection as shown in Figure 7.34(b). Show that the relationship between $\overline{Z}_\Delta$ and $\overline{Z}$ is given by $\overline{Z}_\Delta = 3\overline{Z}$.

Answers

Q_1

$$\overline{I}_N = \frac{\overline{U}_1 + \overline{U}_2 + \overline{U}_3}{\overline{Z}} = 0$$

Q_2

$$p(t) = \sum_{k=1}^{3} u_k(t) i_k(t) = 2\frac{U_{rms}^2}{Z} \sum_{k=1}^{3} \cos(\omega t - \alpha_k)\cos(\omega t - \alpha_k - \varphi)$$

$$p(t) = \frac{U_{rms}^2}{Z}\left(\sum_{k=1}^{3} \cos\varphi + \underbrace{\sum_{k=1}^{3} \cos(2\omega t - 2\alpha_k - \varphi)}_{0}\right) = 3\frac{U_{rms}^2}{Z}\cos\varphi$$

Q_3 At node 1′:

$$\begin{cases} \text{Star connection}: \overline{I}_1 = \dfrac{\overline{U}_1}{\overline{Z}} \\ \\ \text{Delta connection}: \overline{I}_1 = \dfrac{\overline{U}_{12} - \overline{U}_{31}}{\overline{Z}_\Delta} \end{cases} \rightarrow \quad \overline{Z}_\Delta = \overline{Z} \times \frac{\overline{U}_{12} - \overline{U}_{31}}{\overline{U}_1} = 3\overline{Z}$$

At the remaining nodes the same result is obtained, $\overline{Z}_\Delta = 3\overline{Z}$.

Problem 7.5.7

The transformer shown in Figure 7.35 is subjected to a steady-state harmonic regime with $f = 50\,\text{Hz}$.

In order to determine its constitutive parameters two experiments were conducted. Firstly, in Figure 7.36(a), the secondary winding was left open ($i_2 = 0$). Secondly, in Figure 7.36(b), the primary and secondary windings were connected in series ($i_1 = i_2$).

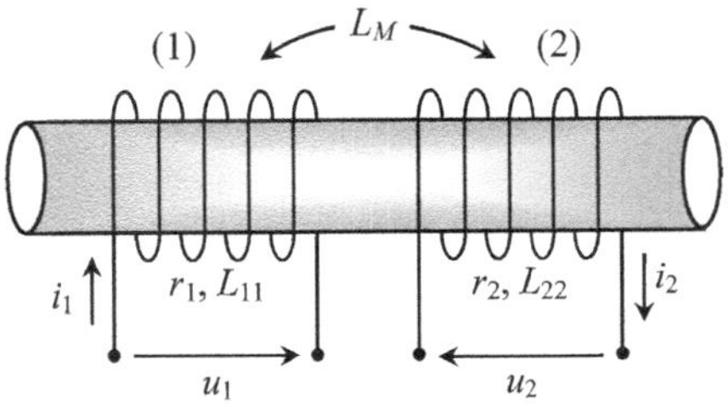

Figure 7.35 Two-windings transformer

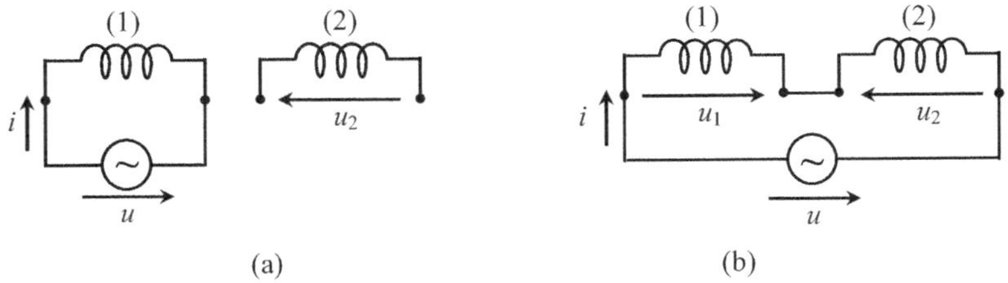

Figure 7.36 Two experiments for the characterization of the transformer parameters. (a) Secondary open. (b) Primary and secondary windings connected in series

In both experiments, rms readings of the generator voltage, generator current and secondary voltage were recorded; the generator active power P was also measured in both situations:

first experiment: $U_{rms} = 230\,\text{V},\ I_{rms} = 6.976\,\text{A},\ U_{2rms} = 109.6\,\text{V},\ P = 486.7\,\text{W};$

second experiment: $U_{rms} = 230\,\text{V},\ I_{rms} = 2.838\,\text{A},\ P = 161.1\,\text{W}.$

Q_1 Write the phasor-domain equations concerning the first experiment. Determine the parameters r_1, L_{11} and L_M.

Q_2 Write the phasor-domain equations concerning the second experiment. Determine the parameters r_2 and L_{22}.

Q_3 For the second experiment, obtain $U_{2_{rms}}$.

Answers

Q_1

$$\overline{U} = \overline{U}_1;\ \overline{I} = \overline{I}_1;\ \overline{I}_2 = 0 \rightarrow \overline{U} = (r_1 + j\omega L_{11})\overline{I};\ -\overline{U}_2 = j\omega L_M \overline{I}.$$

$$P = r_1 I_{rms}^2 \rightarrow r_1 = \frac{P}{I_{rms}^2} = 10\ \Omega;\ \frac{U_{rms}}{I_{rms}} = \sqrt{r_1^2 + (\omega L_{11})^2} \rightarrow L_{11} = 100\,\text{mH}$$

$$U_{2_{rms}} = \omega L_M I_{rms} \rightarrow L_M = U_{2_{rms}}/(\omega I_{rms}) = 50\,\text{mH}.$$

Q_2 $$\overline{U}=\overline{U}_1-\overline{U}_2;\ \overline{I}=\overline{I}_1=\overline{I}_2\ \rightarrow\ \overline{U}=[(r_1+r_2)+j\omega(L_{11}+2L_M+L_{22})]\overline{I}.$$

$$P=(r_1+r_2)I_{rms}^2\ \rightarrow\ r_2=\frac{P}{I_{rms}^2}-r_1=10\ \Omega$$

$$\frac{U_{rms}}{I_{rms}}=\sqrt{(r_1+r_2)^2+\omega^2(L_{11}+2L_M+L_{22})^2}\ \rightarrow\ L_{22}=50\,\text{mH}$$

Q_3 $$-\overline{U}_2=(r_2+j\omega L_{22})\overline{I}+j\omega L_M\overline{I}\ \rightarrow\ U_{2_{rms}}=|r_2+j\omega(L_M+L_{22})|\,I_{rms}=93.6\,\text{V}.$$

Problem 7.5.8

The transformer shown in Figure 7.37(a) is driven by a 50 Hz voltage u_1. The load connected to the secondary winding is a capacitor of capacitance $C_2=55.36\,\mu\text{F}$, across which a voltage u_2, given by $u_2(t)=\sqrt{2}\ 115\ \cos(\omega t)$ V, is found to exist.

Figure 7.37(b) shows an equivalent circuit of the real transformer, where the following parameters are defined $R=5\ \Omega,\ L=732\,\text{mH}$ and $C_2'=13.84\,\mu\text{F}$.

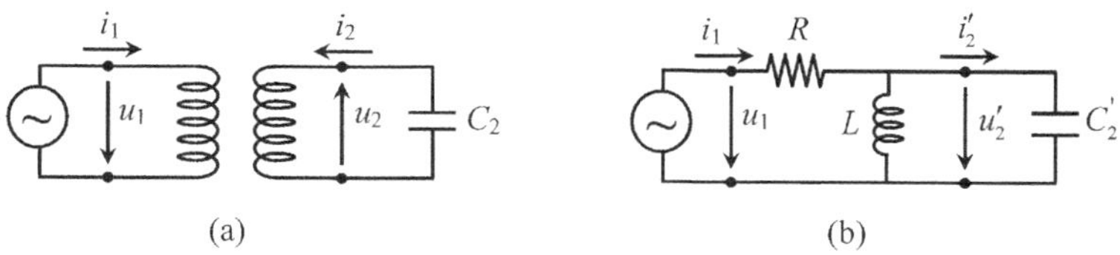

Figure 7.37 A capacitively loaded transformer. (a) Schematic diagram. (b) Equivalent circuit

Q_1 Obtain the value of the turns ratio v of the ideal transformer associated with the equivalent circuit. Next, determine the constitutive parameters of the real transformer, $r_1,\ r_2,\ L_{11},\ L_M$ and L_{22}.

Q_2 Determine the primary and secondary voltages and currents in the phasor domain. In addition, determine the auxiliary quantities $\overline{U}_2'$ and $\overline{I}_2'$.

Q_3 Compute the active and reactive powers brought into play by the generator. Verify the results obtained using the complex Poynting theorem.

Answers

Q_1 $C_2/C_2'=v^2\rightarrow v=2.$

$$r_1=5\ \Omega,\ r_2=0;\ L_{11}=732\,\text{mH};\ L_M=366\,\text{mH};\ L_{22}=183\,\text{mH}.$$

Q_2 $\overline{U}_2=\sqrt{2}\ 115\,\text{V};\ \overline{U}_2'=\sqrt{2}\ 230\,e^{j\pi}\ \text{V};\ \overline{I}_2=\sqrt{2}\ 2\,e^{j\pi/2}\ \text{A};\ \overline{I}_2'=\sqrt{2}\ e^{-j\pi/2}\ \text{A}.$

$$\overline{I}_1=0;\ \overline{U}_1=\sqrt{2}\ 230\ e^{j\pi}\ \text{V}.$$

Q_3 $P = P_Q = 0$.

$$P = P_J = 0;\ P_Q = 2\omega[(W_m)_{av} - (W_e)_{av}];\ (W_m)_{av} = (W_e)_{av} = 366\,\text{mJ}.$$

Problem 7.5.9

The circuit in Figure 7.38 refers to the transient phenomena of discharging a capacitor over an RL circuit. The switch S is closed at $t = 0$ when $u(0) = U_0 = 86.6\,\text{V}$.

Data: $R = 10\ \Omega,\ L = 1\,\text{mH},\ C = 10\,\mu\text{F}$.

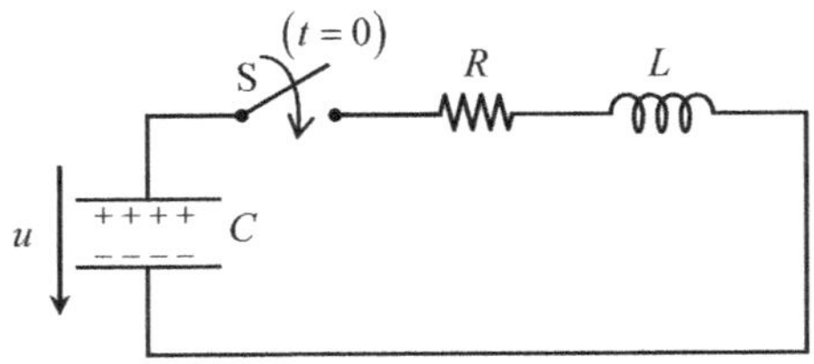

Figure 7.38 A charged capacitor is discharged over an RL circuit

Q_1 Determine the differential equation that governs $u(t)$ for $t > 0$.

Q_2 Making use of the initial conditions pertaining to this problem, find $u(0^+)$ and

$$\left(\frac{du}{dt}\right)_{t=0^+}$$

Q_3 Show that the transient regime solution for $u(t)$ is a damped periodic oscillation, that is $u(t) = Ue^{-\beta t}\cos(\omega t + \theta)$. Obtain $\beta,\ \omega,\ \theta$ and U.

Answers

Q_1

$$\frac{d^2u}{dt^2} + 2\beta\frac{du}{dt} + \omega_0^2 u = 0;\ \beta = \frac{R}{2L};\ \omega_0 = \frac{1}{\sqrt{LC}}$$

Q_2

$$u(0^+) = U_0 = 86.6\,\text{V};\ i(0^-) = i(0^+) = 0 \to \left(\frac{du}{dt}\right)_{t=0^+} = 0$$

Q_3 $\beta = 5000\,\text{Np/s};\ \omega_0 = 10\,\text{krad/s}.\ \beta < \omega_0 \to$ damped periodic oscillation.
$\omega = 8.66\,\text{krad/s};\ \theta = -\pi/6;\ U = 100\,\text{V}.$

Problem 7.5.10

Consider the circuit represented in Figure 7.39, whose parameters R, L and C are known. The DC generator voltage is $U_G = 12\,\text{V}$. The circuit, which is operating for a long time, is interrupted at $t = 0$.

Data: $R = 120\ \Omega$, $L = 1\,\text{H}$, $C = 10\,\text{nF}$.

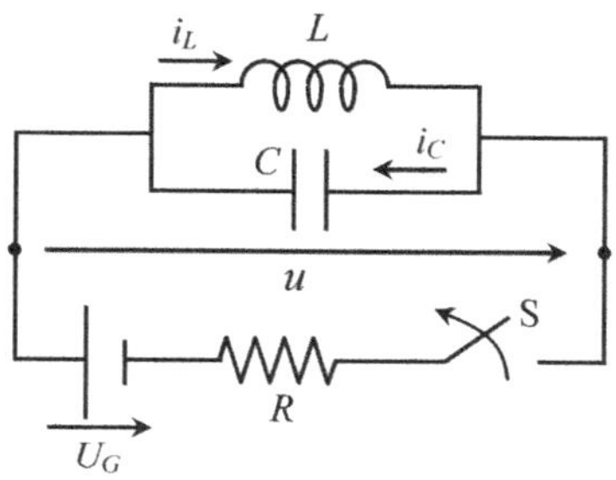

Figure 7.39 The opening of the switch S gives rise to a high-voltage harmonic regime

Q_1 Determine the differential equation that governs $u(t)$ for $t > 0$.

Q_2 Making use of the initial conditions pertaining to this problem, find $u(0^+)$ and

$$\left(\frac{du}{dt}\right)_{t=0^+}$$

Q_3 Show that the transient regime solution for $u(t)$ is a purely periodic oscillation, that is $u(t) = U\ \cos(\omega_0 t + \theta)$. Obtain ω_0, θ and U.

Q_4 Check the result obtained for U using energy balance considerations.

Answers

Q_1

$$\frac{d^2u}{dt^2} + \frac{1}{LC}u = 0; \quad (\beta = 0)$$

Q_2

$$u(0^+) = u(0^-) = 0;\ i_C(0^+) = i_L(0^+) = i_L(0^-) = 0.1\text{A} \rightarrow \left(\frac{du}{dt}\right)_{t=0^+} = -10^7\,\text{V/s}$$

Q_3

$$u(t) = U\ \cos(\omega_0 t + \theta);\ \omega_0 = 1/\sqrt{LC} = 10\,\text{krad/s};\ \theta = \pi/2;\ U = 1\,\text{kV}.$$

Q_4

$$(W_e)_{max} = \tfrac{1}{2}CU^2 = (W_m)_{max} = 5\,\text{mJ} \rightarrow U = \sqrt{\frac{2\,(W_m)_{max}}{C}} = 1\,\text{kV}$$

Problem 7.5.11

Take the transformer in Figure 7.40(a) whose constitutive parameters are known:

$$r_1 = r_2 = 25\ \Omega,\ L_{11} = L_{22} = 300\,\text{mH and } L_M = 50\,\text{mH}$$

Now consider the transformer connections shown in Figure 7.40(b), where $U = 100\,\text{V}$ and $C = 50\,\mu\text{F}$. The capacitor is initially discharged. The switch S closes at $t = 0$.

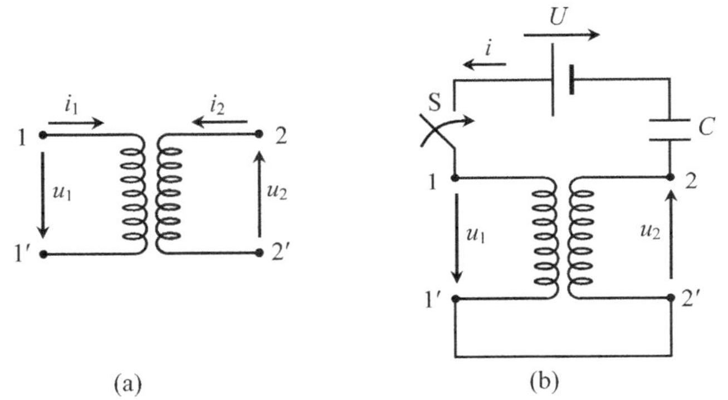

Figure 7.40 (a) Transformer characterized by r_1, r_2, L_{11}, L_{22} and L_M. (b) Switching on a DC voltage generator over a circuit containing a capacitor and a transformer whose windings are connected in series

Q_1 Bearing in mind the time-domain equations of the transformer, and taking into account that $i_1 = -i_2 = i$, determine the differential equation that governs $i(t)$ for $t > 0$.

Q_2 Define the initial conditions of the problem.

Q_3 Show that the transient regime solution for $i(t)$ is a damped periodic oscillation, that is $i(t) = I\ e^{-\beta t}\ \cos(\omega t - \theta)$. Obtain β, ω, θ and I.

Answers

Q_1

$$U = Ri + L\frac{di}{dt} + \frac{1}{C}\int i dt \rightarrow L\frac{d^2 i}{dt^2} + R\frac{di}{dt} + \frac{i}{C} = 0$$

where $R = r_1 + r_2 = 50\ \Omega$ and $L = L_{11} + L_{22} - 2L_M = 0.5\,\text{H}$.

Q_2

$$i(0) = 0; \;\; u_C(0) = 0 \rightarrow \left(\frac{di}{dt}\right)_{t=0^+} = \frac{U}{L} = 200\,\text{A/s}$$

Q_3

$$\beta = \frac{R}{2L} = 50\ \text{Np/s}; \;\; \omega_0 = \frac{1}{\sqrt{LC}} = 200\,\text{rad/s}$$

$\beta < \omega_0 \rightarrow$ damped periodic oscillation.
$\omega = 193.7\,\text{rad/s}; \;\; \theta = \pi/2; \;\; I = 1.033\,\text{A}.$

Part IV

Rapid Time-Varying Fields

Introduction

For rapid time-varying electromagnetic phenomena you need to consider the original Maxwell's equations exactly as they are:

$$\begin{cases} \text{curl } \mathbf{E} = -\dfrac{\partial \mathbf{B}}{\partial t} \\ \text{div } \mathbf{B} = 0 \\ \text{curl } \mathbf{H} = \mathbf{J} + \dfrac{\partial \mathbf{D}}{\partial t} \\ \text{div } \mathbf{D} = \rho \end{cases} \tag{PIV.1}$$

The strategy used in Part III to deal with slow time-varying fields, consisting of treating separately the capacitive effects associated with $\partial\mathbf{D}/\partial t$ from the inductive effects associated with $\partial\mathbf{B}/\partial t$, no longer applies, in general.

Because of the interdependence of the electric and magnetic fields, they, together, should be considered as a single coherent entity – the electromagnetic field.

As already referred to in the introductory section of Part III, rapid time-varying field phenomena are those for which the length of the structures under analysis is comparable or larger than the lowest wavelength characterizing the electromagnetic field dynamics.

Part IV is subdivided into two chapters.

Chapter 8 is fundamentally concerned with electromagnetic waves, energy flow, field polarization, and skin effect phenomena. Chapter 9 is entirely devoted to transmission-line guided-wave analysis in both the time and frequency domain.

8

Electromagnetic Field Phenomena

8.1 Electromagnetic Waves

You most certainly have heard about waves in previous physics courses. The simplest example you may be acquainted with is the vibrating string. Assume you have a long stretched string aligned with the x axis. If you grab the free end of the string and shake it up down, you will notice that the initial movement you transferred to the string $\psi(0, t)$ will travel at constant speed v down the string, but conserving its shape – see Figure 8.1.

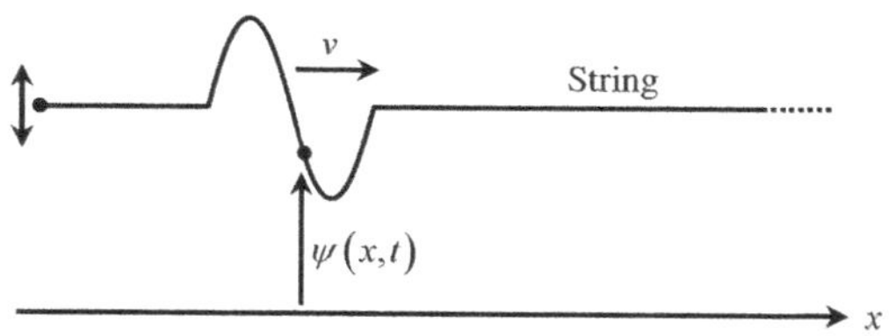

Figure 8.1 Propagation along a vibrating string

The mathematical equation describing the propagation of the oscillation (wave) along x is the well-known one-dimensional wave equation

$$\frac{\partial^2 \psi(x,t)}{\partial x^2} = \frac{1}{v^2}\frac{\partial^2 \psi(x,t)}{\partial t^2}$$

For a three-dimensional problem, where $\psi = \psi(x, y, z, t)$, the above result transforms into

$$\underbrace{\frac{\partial^2 \psi}{\partial x^2} + \frac{\partial^2 \psi}{\partial y^2} + \frac{\partial^2 \psi}{\partial z^2}}_{\text{lap}\,\psi} = \frac{1}{v^2}\frac{\partial^2 \psi}{\partial t^2} \tag{8.1}$$

which is the scalar three-dimensional (3D) wave equation.

Electromagnetic Foundations of Electrical Engineering J. A. Brandão Faria

After this brief introduction, let us return to Maxwell's equations in (PIV.1).

In free space ($\varepsilon = \varepsilon_0$, $\mu = \mu_0$) where currents and charges are absent ($\mathbf{J} = 0$, $\rho = 0$) Maxwell's equations assume the simplified version

$$\operatorname{curl} \mathbf{E} = -\frac{\partial \mathbf{B}}{\partial t} \tag{8.2}$$

$$\operatorname{curl} \mathbf{B} = \mu_0 \varepsilon_0 \frac{\partial \mathbf{E}}{\partial t} \tag{8.3}$$

$$\operatorname{div} \mathbf{B} = 0 \tag{8.4}$$

$$\operatorname{div} \mathbf{E} = 0 \tag{8.5}$$

By applying the curl operator on the left to (8.2), and taking into account the result in (8.3), we get

$$\operatorname{curl} \operatorname{curl} \mathbf{E} = -\frac{\partial}{\partial t} \operatorname{curl} \mathbf{B} = -\mu_0 \varepsilon_0 \frac{\partial^2 \mathbf{E}}{\partial t^2}$$

From vector calculus we know that $\operatorname{curl}\operatorname{curl} \equiv \operatorname{grad}\operatorname{div} - \operatorname{lap}$, but since $\operatorname{div} \mathbf{E} = 0$, we find

$$\operatorname{lap} \mathbf{E} = \mu_0 \varepsilon_0 \frac{\partial^2 \mathbf{E}}{\partial t^2}$$

Likewise, had we applied the curl operator to (8.3), and taken (8.2) into account, we would get

$$\operatorname{lap} \mathbf{B} = \mu_0 \varepsilon_0 \frac{\partial^2 \mathbf{B}}{\partial t^2}$$

In short, in free space, the electromagnetic field (**E**, **B**) satisfies the equation

$$\operatorname{lap} \begin{Bmatrix} \mathbf{E} \\ \mathbf{B} \end{Bmatrix} = \frac{1}{v^2} \frac{\partial^2}{\partial t^2} \begin{Bmatrix} \mathbf{E} \\ \mathbf{B} \end{Bmatrix}, \quad v = \frac{1}{\sqrt{\mu_0 \varepsilon_0}} \tag{8.6}$$

Comparing this result to (8.1), you can readily see that (8.6) is nothing but the 3D wave equation of the electromagnetic field. In other words, we came to the conclusion that Maxwell's equations are compatible with wave solutions whose velocity in free space is equal to the speed of light in a vacuum, $v = 3 \times 10^8\,\mathrm{m/s}$. This fact ultimately pointed out that light itself is nothing but electromagnetic radiation; the science of optics is not an independent area of knowledge, but rather the contrary – it belongs with electromagnetism.

The first experimental proof that electromagnetic waves are real dates back to 1889 (Hertz). A remark is in order here: if the displacement current density introduced by Maxwell did not exist, (8.6) would not hold – electromagnetic waves would not exist.

For homogeneous media with $\mu \neq \mu_0$, $\varepsilon \neq \varepsilon_0$, the wave velocity in (8.6) is $v = 1/\sqrt{\mu\varepsilon}$.

Before finishing this section it may help you to have a pictorial representation of the wave phenomena. From (8.2) you can see that a time-varying **B** field may give rise to a time-varying **E** field; likewise, from (8.3), you can also see that a time-varying **E** field may give rise to a time-varying **B** field. The consequence of this is that the electromagnetic field is a self-sustainable entity

$$\cdots \rightarrow \mathbf{B}(t) \rightarrow \mathbf{E}(t) \rightarrow \mathbf{B}(t) \rightarrow \mathbf{E}(t) \rightarrow \mathbf{B}(t) \rightarrow \mathbf{E}(t) \rightarrow \cdots$$

Once it has been started at a certain region in space (for example, at an antenna) it will propagate and spread through space indefinitely and for ever, the wave process progressing at a constant speed. The sketch in Figure 8.2 gives you a glimpse of the ideas we have been talking about.

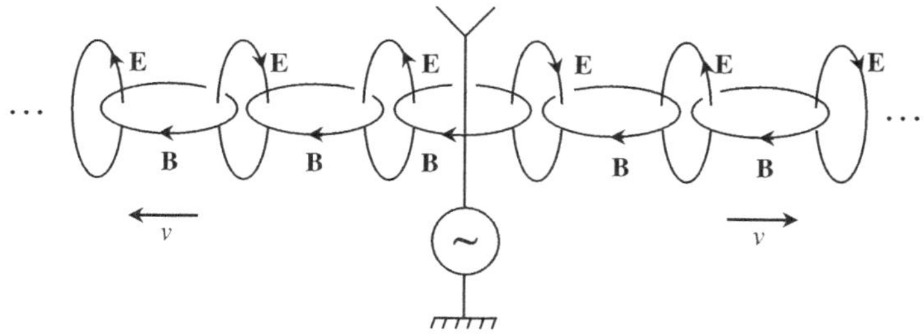

Figure 8.2 A pictorial sketch of the self-sustainable electromagnetic field radiated by an antenna, and propagating through space with velocity $v = 1/\sqrt{\mu_0 \varepsilon_0}$

8.2 Poynting Theorem, Poynting Vector, Power Flow

Based on intuitive energy balance considerations, the Poynting theorem has already been introduced in the framework of quasi-stationary regimes (Chapter 7). Here, we are going to show you a proof of such a theorem based directly on Maxwell's equations. In addition, a new interpretation for the concept of instantaneous power $p(t)$ is provided.

Take $\operatorname{curl}\mathbf{H} = \mathbf{J} + \partial\mathbf{D}/\partial t$ and multiply (inner product) both sides of the equation by $\mathbf{E}$. Then you get

$$\mathbf{E}\cdot\operatorname{curl}\mathbf{H} = \mathbf{E}\cdot\mathbf{J} + \mathbf{E}\cdot\frac{\partial\mathbf{D}}{\partial t} = \mathbf{E}\cdot\mathbf{J} + \frac{\partial}{\partial t}\left(\frac{\mathbf{E}\cdot\mathbf{D}}{2}\right) \tag{8.7a}$$

From (2.26) in Chapter 2, you will recognize that $(\mathbf{E}\cdot\mathbf{D})/2$ represents the per-unit-volume electric energy stored in the electric field, $\hat{w}_e$. Also, from (3.14) in Chapter 3, you will recognize that $(\mathbf{E}\cdot\mathbf{J})$ represents the per-unit-volume power losses associated with the Joule effect, $\hat{p}_J$. Hence, we can rewrite (8.7a) as

$$\mathbf{E}\cdot\operatorname{curl}\mathbf{H} = \hat{p}_J + \frac{\partial\hat{w}_e}{\partial t} \tag{8.7b}$$

Similarly, take $\operatorname{curl}\mathbf{E} = -\partial\mathbf{B}/\partial t$ and multiply (inner product) both sides of the equation by $\mathbf{H}$. Then you get

$$\mathbf{H}\cdot\operatorname{curl}\mathbf{E} = -\mathbf{H}\cdot\frac{\partial\mathbf{B}}{\partial t} = -\frac{\partial}{\partial t}\left(\frac{\mathbf{H}\cdot\mathbf{B}}{2}\right) \tag{8.8a}$$

From (4.36) in Chapter 4, you will recognize that $(\mathbf{H}\cdot\mathbf{B})/2$ represents the per-unit-volume magnetic energy stored in the magnetic field, $\hat{w}_m$. Hence, we can rewrite (8.8a) as

$$\mathbf{H}\cdot\operatorname{curl}\mathbf{E} = -\frac{\partial\hat{w}_m}{\partial t} \tag{8.8b}$$

By subtracting the equations in (8.7b) and (8.8b), we obtain

$$\mathbf{E} \cdot \operatorname{curl} \mathbf{H} - \mathbf{H} \cdot \operatorname{curl} \mathbf{E} = \hat{p}_J + \frac{\partial \hat{w}_m}{\partial t} + \frac{\partial \hat{w}_e}{\partial t} \tag{8.9}$$

But, from vector calculus we have $\mathbf{E} \cdot \operatorname{curl} \mathbf{H} - \mathbf{H} \cdot \operatorname{curl} \mathbf{E} = -\operatorname{div}(\mathbf{E} \times \mathbf{H})$. The vector entity

$$\mathbf{S} = \mathbf{E} \times \mathbf{H} \tag{8.10}$$

named the Poynting vector (units: $\mathrm{W/m^2}$, watt per square meter), is to be physically interpreted as the per-unit-area power flow carried by the electromagnetic field.

Therefore, the Poynting theorem, in local form, can be established as

$$-\operatorname{div} \mathbf{S} = \hat{p}_J + \frac{\partial \hat{w}_m}{\partial t} + \frac{\partial \hat{w}_e}{\partial t} \tag{8.11}$$

If this result is integrated over a volume V, bounded by a closed surface S_V, we obtain, using the already familiar Gauss theorem,

$$\underbrace{\int_{S_V} \mathbf{S} \cdot \mathbf{n}_\mathrm{i} \, dS}_{p(t)} = p_J(t) + \frac{d}{dt}\left(W_m(t) + W_e(t)\right) \tag{8.12}$$

where $\mathbf{n}_\mathrm{i}$ is the inward unit normal to S_V.

Let us emphasize the following aspects:

- The result in (8.12) is a restatement of the result in (7.20).
- The result in (8.12) is not valid when hysteresis phenomena are present.
- The instantaneous power $p(t)$ in play in a given volume of space is to be interpreted as the inward flux of the Poynting vector across the volume border.
- The Poynting vector is the carrier of electromagnetic energy; it is ultimately responsible for the flow of power in any electromagnetic system.
- Electromagnetic energy is not carried by wires. Wires are simply used to guide electromagnetic waves, but (apart from wire losses) the energy flow is essentially external to the wires.
- At any chosen point of an electromagnetic wave, the power flow density is totally determined, in both in magnitude and direction, by $\mathbf{S}$ – see Figure 8.3.

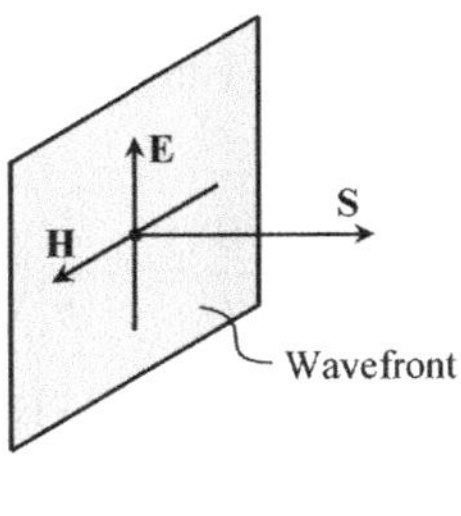

Figure 8.3 Wavefront. The Poynting vector $\mathbf{S} = \mathbf{E} \times \mathbf{H}$ determines the flow of electromagnetic energy

Well before the introduction of the Poynting theorem we utilized the instantaneous power definition $p = ui$. Now, we know that such a definition is not general; it pertains to slow time-varying field phenomena. Here, we are going to show that for stationary and quasi-stationary phenomena we have

$$\int_{S_V} \mathbf{S} \cdot \mathbf{n}_i \, dS = p(t) = ui \tag{8.13}$$

For that purpose, consider the general situation depicted in Figure 8.4(a) where a generator is connected to a load using a pair of wires. Define S_V as the global surface enclosing the load, and define S_T as the transversal plane surface intersecting the two wires (which are assumed to be perfect conductors). On the transverse plane, the field lines of $\mathbf{E}$ are circumferential arcs starting and ending at the wires, and the lines of $\mathbf{H}$ are non-coaxial circumferences around the wires – see Figure 8.4(b). The two families of curves intersect perpendicularly, originating Poynting vectors oriented along z, $\mathbf{S} = \mathbf{E} \times \mathbf{H} = S\vec{e}_z$.

For stationary or quasi-stationary regimes, at any point belonging to S_V, we have $\mathbf{E} = -\text{grad}\, V$ and $\text{curl}\, \mathbf{H} = \mathbf{J}$.

Hence, the Poynting vector is given by $\mathbf{S} = \mathbf{E} \times \mathbf{H} = -\text{grad}\, V \times \mathbf{H}$.

Taking into account the vector identity

$$\text{curl}\,(V\mathbf{H}) = V\,\text{curl}\,\mathbf{H} + \text{grad}\, V \times \mathbf{H}$$

we get $\mathbf{S} = V\mathbf{J} - \text{curl}(V\mathbf{H})$. The inward flux of $\mathbf{S}$ across S_V gives the instantaneous power that flows from the generator to the load

$$p = \int_{S_V} \mathbf{S} \cdot \mathbf{n}_i \, dS = \int_{S_V} V\mathbf{J} \cdot \mathbf{n}_i \, dS + \int_{S_V} \text{curl}\,\underbrace{(V\mathbf{H})}_{\mathbf{K}} \cdot \mathbf{n}_o \, dS$$

where $\mathbf{n}_o$ is the outward unit normal.

The last term on the right is zero. In fact, from the Gauss theorem, we have

$$\int_{S_V} \text{curl}\,\mathbf{K} \cdot \mathbf{n}_o \, dS = \int_{V} \text{div}\,\text{curl}\,\mathbf{K}\, dV = 0$$

since the operator div curl $\equiv 0$. Therefore, we have

$$p = \int_{S_V} V\mathbf{J} \cdot \mathbf{n}_i \, dS \tag{8.14}$$

The flux of $V\mathbf{J}$ across S_V in (8.14) is zero everywhere except at the regions S_1 and S_2 where the current-carrying wires intersect the transverse plane. Hence, we finally find

$$p = \int_{S_V} V\mathbf{J} \cdot \mathbf{n}_i \, dS = \int_{S_T} V\mathbf{J} \cdot \mathbf{n}_i \, dS = V_1 \underbrace{\int_{S_1} \mathbf{J} \cdot \mathbf{n}_i \, dS}_{+i} + V_2 \underbrace{\int_{S_2} \mathbf{J} \cdot \mathbf{n}_i \, dS}_{-i} = \underbrace{(V_1 - V_2)}_{u}\, i = ui$$

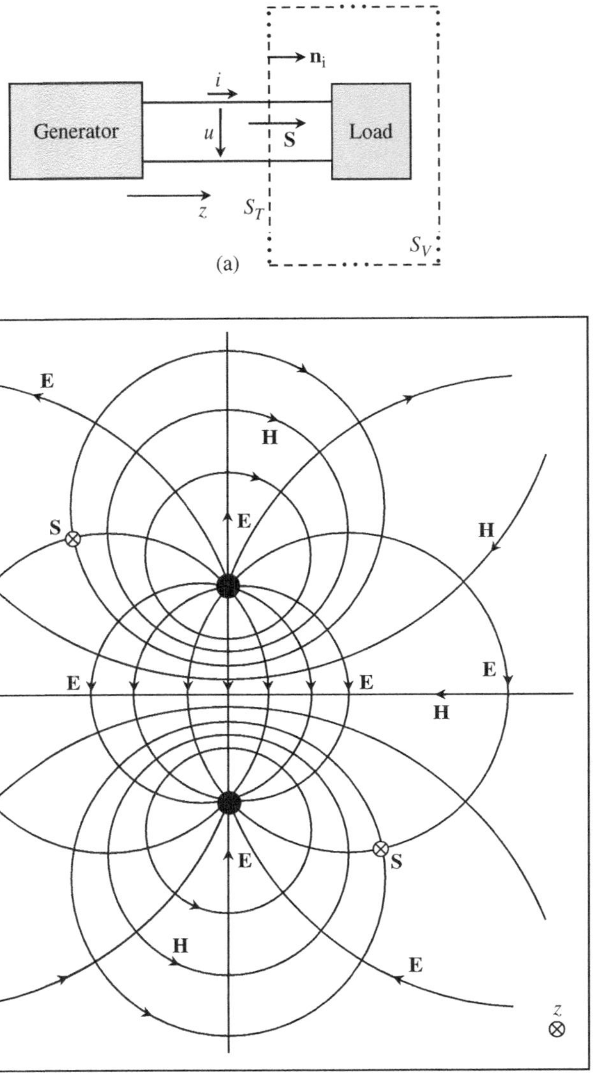

Figure 8.4 (a) The instantaneous power transmitted to the load $p = ui$ can be interpreted as the Poynting vector flow across the transverse plane S_T. (b) Electromagnetic field lines in the transverse plane

Note that, for low- or high-frequency regimes, in a transmission system made of two parallel perfect conductors, both electric and magnetic field lines belong to transverse planes. On these planes, because $\mathbf{B} \perp \mathbf{n}_S$ and $\mathbf{D} \perp \mathbf{n}_S$, you have

$$\oint_{S} \mathbf{E} \cdot ds = -\int_{S_s} \frac{\partial \mathbf{B}}{\partial t} \cdot \mathbf{n}_S \, dS = 0 \rightarrow \operatorname{curl} \mathbf{E} = 0 \rightarrow \mathbf{E} = -\operatorname{grad} V$$

$$\oint_{\mathbf{S}} \mathbf{H} \cdot d\mathbf{s} = \int_{S_S} \left(\mathbf{J} + \frac{\partial \mathbf{D}}{\partial t} \right) \cdot \mathbf{n}_S \, dS = \int_{S_S} \mathbf{J} \cdot \mathbf{n}_S \, dS \rightarrow \text{curl}\, \mathbf{H} = \mathbf{J}$$

That is, you encounter the same conditions we talked about when stationary regimes were dealt with, and, consequently, you can still use

$$p = \int_{S_T} \mathbf{S} \cdot \mathbf{n} \, dS = ui$$

where S_T is the transverse plane surface where u and i are defined.

8.3 Time-Harmonic Fields, Field Polarization, RMS Field Values

In Chapter 7 we defined complex amplitudes (phasors) of sinusoidal time-varying scalar quantities. Now we do the same with vector fields. To start with, let us consider the simplest case of a unidirectional electric field vector, for example $\mathbf{E}(x, y, z, t) = E_x \cos(\omega t + \alpha_x)\, \vec{e}_x$, where $E_x = E_x(x, y, z)$ and $\alpha_x = \alpha_x(x, y, z)$. At a given point P in space, such a field is said to be linearly polarized along x because the tip of vector $\mathbf{E}$ defines a straight line aligned with the x axis. The above field can be written in complex form as

$$\mathbf{E}(x, y, z, t) = \Re\left(\overline{\mathbf{E}}(x, y, z)\, e^{j\omega t} \right), \text{ with } \overline{\mathbf{E}} = E_x\, e^{j\alpha_x}\, \vec{e}_x \tag{8.15}$$

where $\overline{\mathbf{E}}$ is the time-independent phasor associated with the harmonic field $\mathbf{E}(x, y, x, t)$.

Let us complicate matters a little by assuming that $\mathbf{E}$ has two components of the same frequency:

$$\mathbf{E}(x, y, z, t) = E_x \cos(\omega t + \alpha_x)\, \vec{e}_x + E_y \cos(\omega t + \alpha_y)\, \vec{e}_y \tag{8.16a}$$

At a given point P in space, if $\alpha_x \neq \alpha_y$ the tip of vector $\mathbf{E}$ will move with time and describe an ellipse in the xy plane. In this case the field is said to be elliptically polarized.

In the particular case $E_x = E_y$ and $\alpha_x - \alpha_y = \pm\pi/2$, the ellipse degenerates into a circumference and the field is said to be circularly polarized.

The complex representation of $\mathbf{E}(x, y, z, t)$ in (8.16a) is given by

$$\mathbf{E}(x, y, z, t) = \Re\left(\overline{\mathbf{E}}(x, y, z)\, e^{j\omega t} \right), \text{ with } \overline{\mathbf{E}} = E_x\, e^{j\alpha_x}\, \vec{e}_x + E_y\, e^{j\alpha_y}\, \vec{e}_y \tag{8.16b}$$

where, again, $\overline{\mathbf{E}}$ denotes the complex amplitude (phasor) of the harmonic field $\mathbf{E}(x, y, z, t)$.

How do you generalize the foregoing results when the harmonic field $\mathbf{E}$ is described by three components with $\alpha_x \neq \alpha_y \neq \alpha_z$?

$$\mathbf{E}(x, y, z, t) = E_x \cos(\omega t + \alpha_x)\, \vec{e}_x + E_y \cos(\omega t + \alpha_y)\, \vec{e}_y + E_z \cos(\omega t + \alpha_z)\, \vec{e}_z \tag{8.17a}$$

How does the tip of vector $\mathbf{E}$ move in space, at a given point P, as time elapses?

Most likely, your intuition will tell you that the tip of $\mathbf{E}$ will describe a 3D ellipsoid. Well, that is not true. Intuition has misled you!

To help you find the correct answer let us rewrite (8.17a) using the phasor-domain representation

$$\mathbf{E}(x, y, z, t) = \Re\left(\overline{\mathbf{E}}(x, y, z)\, e^{j\omega t}\right), \text{ with } \overline{\mathbf{E}} = \underbrace{E_x\, e^{j\alpha_x}}_{\overline{E}_x}\, \vec{e}_x + \underbrace{E_y\, e^{j\alpha_y}}_{\overline{E}_y}\, \vec{e}_y + \underbrace{E_z\, e^{j\alpha_z}}_{\overline{E}_z}\, \vec{e}_z$$

Breaking down $\overline{\mathbf{E}}$ into its real and imaginary parts we obtain

$$\overline{\mathbf{E}} = \left(\sum_{k=x,y,z} E_k \cos\alpha_k\, \vec{e}_k\right) + j\left(\sum_{k=x,y,z} E_k \sin\alpha_k\, \vec{e}_k\right) = E_1\, \vec{e}_1 + jE_2\, \vec{e}_2$$

where the field magnitudes E_1 and E_2 are time independent.

From $\mathbf{E} = \Re\left(\overline{\mathbf{E}}\, e^{j\omega t}\right)$ we find

$$\mathbf{E} = E_1 \cos(\omega t)\, \vec{e}_1 - E_2 \sin(\omega t)\, \vec{e}_2 \tag{8.17b}$$

Again, as shown in Figure 8.5, we find an elliptically polarized field in the plane defined by the unit vectors $\vec{e}_1$ and $\vec{e}_2$ (which, in general, are not mutually orthogonal).

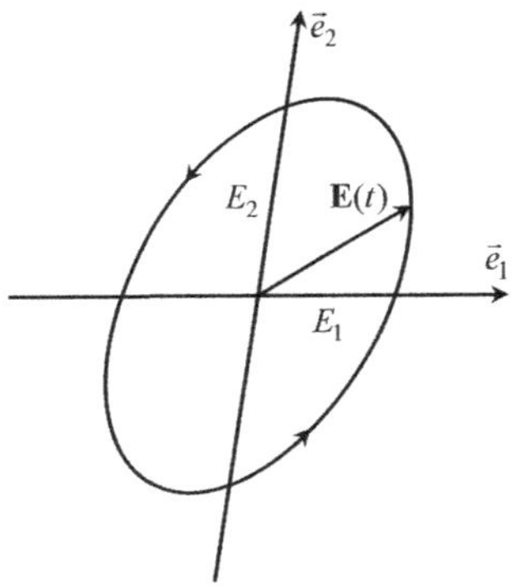

Figure 8.5 The elliptically polarized electric field vector can be constructed from two linearly polarized fields directed along $\vec{e}_1$ and $\vec{e}_2$

To conclude this section we must address the important topic of the evaluation of rms values for time-harmonic vector fields.

For sinusoidal time-varying scalar quantities we showed, in Chapter 7, that

$$U_{rms} = \sqrt{(u^2(t))_{av}} = \frac{\overline{U}\,\overline{U}^*}{2} \tag{8.18}$$

yielding $U_{rms} = U/\sqrt{2}$.

For vector fields, we have a definition similar to the one given in (8.18), that is

$$E_{rms} = \sqrt{(E^2(t))_{av}} = \sqrt{(\mathbf{E}(t) \cdot \mathbf{E}(t))_{av}} = \sqrt{\frac{E_x^2 + E_y^2 + E_z^2}{2}}$$

In addition, noting that

$$E_x^2+E_y^2+E_z^2=\overline{E}_x\overline{E}_x^*+\overline{E}_y\overline{E}_y^*+\overline{E}_z\overline{E}_z^*=\overline{\mathbf{E}}\cdot\overline{\mathbf{E}}^*$$

we can also write, using complex amplitudes,

$$E_{rms}=\sqrt{\frac{\overline{\mathbf{E}}\cdot\overline{\mathbf{E}}^*}{2}} \tag{8.19}$$

A mistake that you should avoid is to write $E_{rms}=E_{max}/\sqrt{2}$. For time-harmonic vector fields, the rms value is not obtained from the maximum value E_{max} by dividing it by $\sqrt{2}$ (as happens with scalar quantities) – that is only true for the especial case of linearly polarized fields.

Remember, for example, that in the case of circular polarization, the field intensity remains constant with time and, consequently, you get $E_{rms}=E_{max}$.

In short, depending on the field's polarization state, you can have $E_{max}/\sqrt{2}\leq E_{rms}\leq E_{max}$.

8.4 Phasor-Domain Maxwell's Equations, Material Media Constitutive Relations

The correspondence between operative rules in the time domain and in the phasor domain summarized in Table 7.1 also applies to sinusoidal (ω) time-varying vector fields. Therefore, as far as Maxwell's equations are concerned, we can write

Time domain		Phasor domain	
$\operatorname{curl}\mathbf{E}=-\frac{\partial\mathbf{B}}{\partial t}$	$\leftrightarrow$	$\operatorname{curl}\overline{\mathbf{E}}=-j\omega\overline{\mathbf{B}}$	(8.20)
$\operatorname{curl}\mathbf{H}=\mathbf{J}+\frac{\partial\mathbf{D}}{\partial t}$	$\leftrightarrow$	$\operatorname{curl}\overline{\mathbf{H}}=\overline{\mathbf{J}}+j\omega\overline{\mathbf{D}}$	(8.21)
$\operatorname{div}\mathbf{B}=0$	$\leftrightarrow$	$\operatorname{div}\overline{\mathbf{B}}=0$	(8.22)
$\operatorname{div}\mathbf{D}=\rho$	$\leftrightarrow$	$\operatorname{div}\overline{\mathbf{D}}=\overline{\rho}$	(8.23)

In addition, the material media constitutive relations, analyzed in Chapters 2, 3 and 4, translate in the phasor domain as

$$\mathbf{D}=\varepsilon\mathbf{E} \leftrightarrow \overline{\mathbf{D}}=\varepsilon\overline{\mathbf{E}} \tag{8.24}$$

$$\mathbf{J}=\sigma\mathbf{E} \leftrightarrow \overline{\mathbf{J}}=\sigma\overline{\mathbf{E}} \tag{8.25}$$

$$\mathbf{B}=\mu\mathbf{H} \leftrightarrow \overline{\mathbf{B}}=\mu\overline{\mathbf{H}} \tag{8.26}$$

Rigorously speaking, the time-domain equations in (8.24) and (8.26) are valid for stationary and quasi-stationary regimes. The corresponding phasor-domain equations on the right can be further generalized for rapid time-varying fields, considering the possibility that losses may eventually be present in the linear medium (note that linearity is a prerequisite).

In order to take into account dielectric polarization losses, you can introduce a complex permittivity, $\overline{\varepsilon} = \varepsilon' - j\varepsilon''$. Likewise, magnetic losses can be accounted for by making use of a complex permeability, $\overline{\mu} = \mu' - j\mu''$. Hence, (8.24) and (8.26) can be generalized through

$$\overline{\mathbf{D}} = \overline{\varepsilon}\,\overline{\mathbf{E}} \quad \text{and} \quad \overline{\mathbf{B}} = \overline{\mu}\,\overline{\mathbf{H}} \tag{8.27}$$

The time-domain interpretation of the equations in (8.27) is that, for sinusoidal regimes, the existence of material losses makes **D** lag **E**, and **B** lag **H**. The parametric curves $D = D(E)$ and $B = B(H)$ are ellipses where the operating point moves counterclockwise as time goes on – see Figure 8.6. The areas circumscribed by the ellipses denote the energy loss per period, per unit volume of the material.

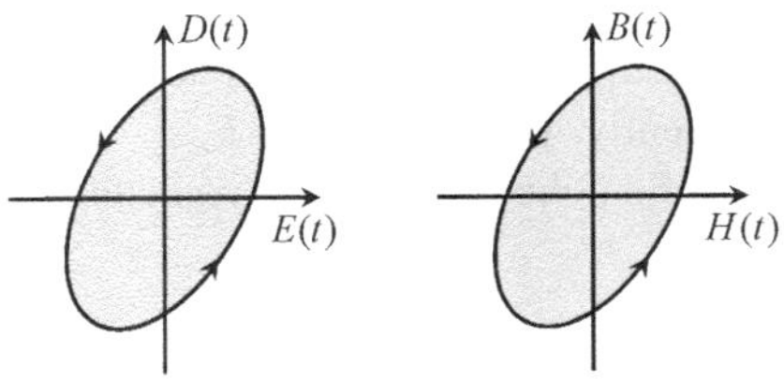

Figure 8.6 For sinusoidal regimes the parametric curves $D(E)$ and $B(H)$ change from straight lines to ellipses, when dielectric and magnetic losses are present, respectively

8.5 Application Example (Uniform Plane Waves)

A uniform plane wave is one where fields **E** and **H** belong to planes perpendicular to the propagation direction, the fields in those planes being invariant from point to point. A monochromatic wave is one where fields are sinusoidal with time with a given frequency ω.

Uniform plane waves are a physical abstraction, but their analysis is very important because any type of wave can be synthesized via a summation of uniform plane waves (Fourier space transforms).

Consider the propagation of a wave produced by a light source. Assume that such a wave can be approximately described by a uniform monochromatic plane wave, propagating along the positive z axis in free space, where the electric field is given by

$$\mathbf{E}(z, t) = E_0 \cos(\omega t - \theta(z))\,\vec{e}_x$$

where $E_0 = 20\,\text{V/m}$, $\omega = 2\pi f$, $f = 474\,\text{THz}$.

Questions

Q_1 Write the complex amplitude corresponding to $\mathbf{E}(z, t)$.

Q_2 Write the phasor-domain equation corresponding to the wave equation in (8.6), and determine the function $\theta(z)$. Assume $\theta(0) = 0$.

Q_3 Determine the operating wavelength λ of the light source.

Q_4 Find the complex amplitude, as well as the time-domain representation, of the magnetic field **H**.

Q_5 Determine the Poynting vector and evaluate its time-averaged value.

Solutions

Q_1 $\overline{\mathbf{E}} = \overline{E}(z)\,\vec{e}_x$, where $\overline{E}(z) = E_0\, e^{-j\theta(z)}$.

Q_2

$$\frac{\partial^2 \mathbf{E}}{\partial z^2} - \frac{1}{v^2}\frac{\partial^2 \mathbf{E}}{\partial t^2} = 0 \rightarrow \frac{d^2\overline{E}(z)}{dz^2} + \left(\frac{\omega}{v}\right)^2 \overline{E}(z) = 0$$

where the wave velocity is $v = c = 1/\sqrt{\mu_0 \varepsilon_0} = 3 \times 10^8\,\mathrm{m/s}$.
Noting that

$$\frac{d^2\overline{E}(z)}{dz^2} = -E_0\, e^{-j\theta(z)}\left[\left(\frac{d\theta}{dz}\right)^2 + j\frac{d^2\theta}{dz^2}\right]$$

we obtain

$$\left(\frac{d\theta}{dz}\right)^2 + j\frac{d^2\theta}{dz^2} = \left(\frac{\omega}{v}\right)^2 + j0 \rightarrow \left\{\begin{array}{l} \dfrac{d^2\theta}{dz^2} = 0 \\ \left(\dfrac{d\theta}{dz}\right)^2 = \left(\dfrac{\omega}{v}\right)^2 \end{array}\right\} \rightarrow \theta(z) = \pm\beta z$$

where β , the so-called phase constant, is given by

$$\beta = \frac{\omega}{v} \tag{8.28}$$

The plus or minus sign in $\theta(z) = \pm\beta z$ means that two solutions are available. The plus sign, which we will adopt here, corresponds to a wave propagating along the positive z axis (check the answer to question Q_5 showing that such an option indeed leads to an energy flow along the positive z axis). The choice $\theta(z) = -\beta z$ would describe a propagating wave along the negative z axis.

Q_3 Taking into account the result $\theta(z) = \beta z$, the expression for the electric field wave is

$$\mathbf{E}(z, t) = E_0 \cos(\omega t - \beta z)\,\vec{e}_x$$

If at a given moment $t = t_0$ you take a snapshot of the electric field wave, you will find that $\mathbf{E}(z,\, t_0)$ is periodic along z. The space period, or wavelength λ, is such that $\beta\lambda = 2\pi$. Hence you have

$$\lambda = \frac{2\pi}{\beta} = \frac{2\pi v}{\omega} = \frac{v}{f} \tag{8.29}$$

Numerically you find $\lambda = 0.633\,\mu\mathrm{m}$ (red light).

Q_4 From

$$\begin{cases} \operatorname{curl} \overline{\mathbf{E}} = -j\omega\mu_0 \overline{\mathbf{H}} \\ \operatorname{curl} \overline{\mathbf{E}} = \dfrac{d\overline{E}(z)}{dz} \vec{e}_y \end{cases}$$

we see that **H** is oriented along the y axis, $\overline{\mathbf{H}} = \overline{H}(z)\,\vec{e}_y$,

$$\overline{H}(z) = \frac{j}{\omega\mu_0}\frac{d\overline{E}(z)}{dz} = \frac{\overline{E}(z)}{R_w}$$

where R_w, the so-called characteristic wave resistance of free space, is

$$R_w = \frac{\omega\mu_0}{\beta} = \sqrt{\frac{\mu_0}{\varepsilon_0}} = 120\pi = 377\,\Omega \tag{8.30}$$

$\mathbf{H}(z, t) = \Re\left(\overline{\mathbf{H}}\, e^{j\omega t}\right) = H_0 \cos(\omega t - \beta z)\,\vec{e}_y$, with $H_0 = E_0/R_w = 53\,\text{mA/m}$.

Q_5 $\mathbf{S} = \mathbf{E} \times \mathbf{H} = S\,\vec{e}_z$.

$$S(z, t) = \frac{E_0^2}{R_w}\cos^2(\omega t - \beta z),\; (S)_{av} = \frac{E_0^2}{2R_w} = 530\,\text{mW/m}^2$$

8.6 Complex Poynting Vector

The complex Poynting theorem, valid for time-harmonic regimes, has already been dealt with in (7.26) in Chapter 7,

$$\overline{P} = P_J + j2\omega\left((W_m)_{av} - (W_e)_{av}\right) \tag{8.31}$$

where the complex power $\overline{P}$ is obtained as $\overline{P} = \overline{U}\,\overline{I}^*/2$ for quasi-stationary regimes.

As mentioned earlier, the validity of the complex Poynting theorem is not limited to quasi-stationary regimes. It also holds in the analysis of rapid time-varying fields.

In fact, the theorem can be logically deduced from the phasor-domain Maxwell's equations in (8.20) to (8.23), following a procedure very similar to the one that led us to the Poynting theorem in Section 8.2. There is no point in again repeating the derivation steps. However, a salient aspect that must be emphasized is the general interpretation that should be given to the complex power $\overline{P}$.

When deducing the complex Poynting theorem from the phasor-domain Maxwell's equations, the complex power appears as

$$\overline{P} = \int_{S_V} \left(\frac{\overline{\mathbf{E}} \times \overline{\mathbf{H}}^*}{2}\right) \cdot \mathbf{n}_i\, dS \tag{8.32}$$

where S_V is a closed surface bounding a volume V where electromagnetic phenomena occur (energy dissipation, electric energy storage, magnetic energy storage). The complex vector under the integration symbol is the complex Poynting vector $\overline{\mathbf{S}}$ (units: W/m^2)

$$\overline{\mathbf{S}} = \frac{\overline{\mathbf{E}} \times \overline{\mathbf{H}}^*}{2} \tag{8.33}$$

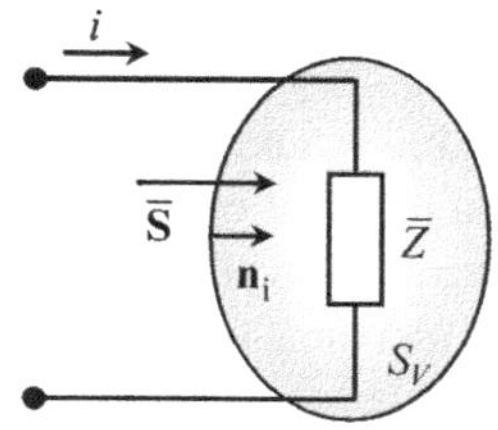

Figure 8.7 The inward flux of the complex Poynting vector across a closed surface containing a linear device allows for the evaluation of the device's impedance

The inward flux of $\overline{\mathbf{S}}$ across a closed surface containing a linear device permits the evaluation of its impedance (Figure 8.7),

$$\overline{P} = \int_{S_V} \left(\frac{\overline{\mathbf{E}} \times \overline{\mathbf{H}}^*}{2} \right) \cdot \mathbf{n}_i \, dS = \overline{Z} I_{rms}^2 \rightarrow \overline{Z} = \frac{\overline{P}}{I_{rms}^2} \tag{8.34}$$

In Part III we showed that the instantaneous power $p(t)$ and the complex power $\overline{P}$ were related through $(p(t))_{av} = \Re\left(\overline{P}\right)$. Likewise, for the Poynting vector we also have

$$(\mathbf{S}(t))_{av} = \Re\left(\overline{\mathbf{S}}\right)$$

Electromagnetic power losses are included in (8.31) in the form of conductor losses associated with the Joule effect P_J

$$P_J = \int_V \sigma E_{rms}^2 \, dV$$

However, the complex Poynting theorem can be easily extended to account for two additional frequency-dependent loss mechanisms that may be present, namely the dielectric polarization losses and the magnetization losses. For linear media, where the constitutive relations (8.27) apply, you have

$$P_{Polarization} = \int_V \omega \varepsilon'' E_{rms}^2 \, dV$$

$$P_{Magnetization} = \int_V \omega \mu'' H_{rms}^2 \, dV$$

Consequently the most general form for the complex Poynting theorem will read as

$$\overline{P} = \underbrace{\left(P_{Joule} + P_{Polarization} + P_{Magnetization}\right)}_{P_{loss}} + j2\omega\left((W_m)_{av} - (W_e)_{av}\right)$$

8.7 Application Example (Skin Effect)

Consider a cylindrical conductor of radius r_0 and indefinite length l. In Chapter 3 dedicated to stationary currents we concluded, as a result of curl $\mathbf{E} = 0$, that the $\mathbf{J}$-field lines had constant intensity at any point inside the current-carrying conductor, that is $J = I/(\pi r_0^2)$, where I denotes the current intensity. The per-unit-length DC resistance of the conductor was then evaluated as $R_{DC} = 1/(\sigma \pi r_0^2)$, where σ is the conductivity. For the case of rapid time-varying currents, the current density is no longer uniformly distributed inside the conductor – most of it tends to flow near to the conductor's surface, the so-called skin-effect phenomenon.

Consider a time-varying sinusoidal current $i(t) = I\cos(\omega t + \alpha_i)$, and assume, as shown in Figure 8.8, that conduction currents are longitudinally oriented. According to the geometry of the problem, a particular field solution to Maxwell's equations must be invariant under rotation operations around z and translation operations along z; in other words, all vector fields are independent of ϕ and z. All vector fields will, however, depend on the radial coordinate. In particular, we will have $\mathbf{J} = J(r)\,\vec{e}_z$ for the current density.

Take the permittivity and permeability of the conductor as ε_0 and μ_0, respectively.

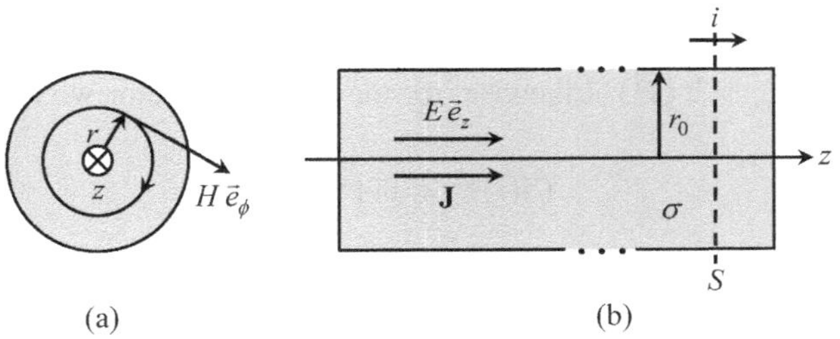

Figure 8.8 Skin-effect phenomena in a cylindrical conductor of radius r_0. (a) Conductor transverse cross-section showing the internal magnetic field $H_\phi(r)$. (b) Conductor longitudinal cross-section showing the internal longitudinal electric field $E_z(r)$

Questions

Q_1 Show that the displacement current density inside the conductor is negligibly small for the case of good conductors.

Q_2 By using a cylindrical coordinate reference frame, write the phasor-domain Maxwell's *curl* equations for the problem.

Q_3 Obtain a solution for $\overline{\mathbf{H}}$ and $\overline{\mathbf{E}}$.

Q_4 Determine the solution for $\overline{\mathbf{J}}$ and particularize it for very-low frequencies.

Q_5 Find the complex Poynting vector $\overline{\mathbf{S}}(r_0)$ at any point belonging to the conductor surface.

Q_6 Determine the per-unit-length frequency-dependent impedance $\overline{Z}(\omega)$ of the conductor.

Q_7 Particularize the results concerning $\overline{Z}(\omega)$ for low- and high-frequency regimes.

Solutions

Q_1 Conduction current density: $\overline{\mathbf{J}} = \sigma\,\overline{\mathbf{E}}$. Displacement current density: $j\omega\overline{\mathbf{D}} = j\omega\varepsilon_0\overline{\mathbf{E}}$. Their absolute ratio is

$$\eta = \left|\frac{j\omega\overline{\mathbf{D}}}{\sigma\overline{\mathbf{E}}}\right| = \frac{\omega\varepsilon_0}{\sigma} \ll 1$$

For ordinary good conductors (where $\sigma \approx 10^7\,\mathrm{S/m}$) the ratio η approaches unity for frequencies above $10^{17}\,\mathrm{Hz}$ (X-ray band). This means that for typical applications up to the gigahertz range, the displacement current can be neglected inside good conductors.

Q_2

$$\mathrm{curl}\,\overline{\mathbf{E}} = \mathrm{curl}\left(\overline{E}_z(r)\,\vec{e}_z\right) = -j\omega\mu_0\overline{\mathbf{H}} \rightarrow \frac{d}{dr}\left(\overline{E}_z(r)\right)\vec{e}_\phi = j\omega\mu_0\overline{\mathbf{H}}$$

From this, you can see that the magnetic field is purely azimuthal, $\overline{\mathbf{H}} = \overline{H}_\phi(r)\,\vec{e}_\phi$, its field lines being coaxial circumferences.

$$\mathrm{curl}\,\overline{\mathbf{H}} = \mathrm{curl}\left(\overline{H}_\phi(r)\,\vec{e}_\phi\right) = \sigma\overline{\mathbf{E}} \rightarrow \frac{1}{r}\frac{d\left(r\overline{H}_\phi(r)\right)}{dr}\vec{e}_z = \sigma\overline{E}_z(r)\,\vec{e}_z$$

To abbreviate the notation let us put $\overline{E} = \overline{E}_z(r)$ and $\overline{H} = \overline{H}_\phi(r)$.
In conclusion, the governing equations for the problem are

$$\frac{d\overline{E}}{dr} = j\omega\mu_0\overline{H} \tag{8.35}$$

$$\frac{d\overline{H}}{dr} + \frac{\overline{H}}{r} = \sigma\overline{E} \tag{8.36}$$

Q_3 Take the derivative of (8.36) with respect to r:

$$\frac{d^2\overline{H}}{dr^2} + \frac{1}{r}\frac{d\overline{H}}{dr} - \frac{\overline{H}}{r^2} = \sigma\frac{d\overline{E}}{dr}$$

Making use of (8.35), you find

$$\frac{d^2\overline{H}}{dr^2} + \frac{1}{r}\frac{d\overline{H}}{dr} + \left(\overline{p}^2 - \frac{1}{r}\right)\overline{H} = 0, \text{ with } \overline{p} = \sqrt{-j\omega\mu_0\sigma} = \frac{1-j}{\delta} \tag{8.37}$$

where δ is the so-called penetration depth

$$\delta = \sqrt{\frac{2}{\omega\mu_0\sigma}} \tag{8.38}$$

The differential equation in (8.37) is a particular case of the more general Bessel equation[1]

$$\frac{d^2\overline{H}}{dr^2} + \frac{1}{r}\frac{d\overline{H}}{dr} + \left(\overline{p}^2 - \frac{n^2}{r^2}\right)\overline{H} = 0, \text{ with } n \geq 0$$

[1] For details on Bessel functions refer to G. N. Watson, *A Treatise on the Theory of Bessel Functions*, Cambridge University Press, 1995.

whose solution is of the type $\overline{H} = H_1 J_n(\overline{p}r) + H_2 Y_n(\overline{p}r)$, and where J_n is the Bessel function of the first kind of order n, and Y_n is the Neumann function of order n. H_1 and H_2 are constants to be determined using the boundary conditions of the problem.

Since in our case we have $n = 1$, we write for the magnetic field solution

$$\overline{H}(r) = H_1 J_1(\overline{p}r) + H_2 Y_1(\overline{p}r) \tag{8.39}$$

Application of Ampère's law to our problem yields the following boundary conditions:

$$\overline{H}(r_0) = \frac{\overline{I}}{2\pi r_0} \text{ and } \overline{H}(0) = 0 \tag{8.40}$$

Taking into account that $J_1(0) = 0$, and that $Y_1(0) \to \infty$, we readily conclude from (8.39) and (8.40) that

$$H_2 = 0 \text{ and } H_1 = \frac{\overline{I}}{2\pi r_0 J_1(\overline{p}r_0)}$$

Substituting the above results into (8.39), we finally obtain the magnetic field inside the conductor

$$\overline{H}(r) = \frac{\overline{I}}{2\pi r_0}\frac{J_1(\overline{p}r)}{J_1(\overline{p}r_0)}, \text{ for } 0 \leq r \leq r_0 \tag{8.41}$$

As for the electric field, we have from (8.35), $\overline{E}(r) = j\omega\mu_0 \int \overline{H}(r)dr$, and noting that $\int J_1(\xi)d\xi = -J_0(\xi)$, we obtain

$$\overline{E}(r) = \frac{\overline{p}\overline{I}}{2\pi\sigma r_0}\frac{J_0(\overline{p}r)}{J_1(\overline{p}r_0)}, \text{ for } 0 \leq r \leq r_0 \tag{8.42}$$

Q_4 From $\overline{J} = \sigma\overline{E}$ we obtain for the current density

$$\overline{J}(r) = \frac{\overline{p}\overline{I}}{2\pi r_0}\frac{J_0(\overline{p}r)}{J_1(\overline{p}r_0)}, \text{ for } 0 \leq r \leq r_0 \tag{8.43}$$

For the case of very low frequencies, that is $|\overline{p}r| \ll 1$, or, put another way, when the conductor radius is much smaller than the penetration depth (8.38), $r_0 \ll \delta$, we may substitute the Bessel functions J_0 and J_1 in (8.43) by their Taylor expansions, retaining only the leading terms, $J_0(\overline{p}r) \simeq 1$ and $J_1(\overline{p}r_0) \simeq \overline{p}r_0/2$, and yielding

$$\overline{J} \simeq \frac{\overline{I}}{\pi r_0^2} = \frac{\overline{I}}{S} \tag{8.44}$$

This result is no surprise. It shows that for very low frequencies the current density is uniformly distributed over the conductor cross-section S.

Q_5 Particularizing (8.41) and (8.42) for $r = r_0$, recalling that $\overline{I}\overline{I}^* = 2I_{rms}^2$, and noting that $\vec{e}_z \times \vec{e}_\phi = -\vec{e}_r$, we obtain for the complex Poynting vector

$$\overline{\mathbf{S}}(r_0) = \frac{\overline{\mathbf{E}}(r_0) \times \overline{\mathbf{H}}^*(r_0)}{2} = -\overline{S}(r_0)\vec{e}_r, \text{ where } \overline{S}(r_0) = \frac{\overline{p}I_{rms}^2}{\sigma(2\pi r_0)^2}\frac{J_0(\overline{p}r_0)}{J_1(\overline{p}r_0)} \tag{8.45}$$

Q_6 Now we compute the complex power corresponding to the inward flux of $\overline{\mathbf{S}}(r_0)$ across the conductor's cylindrical surface of radius r_0 and unit length

$$\overline{P} = \int_{S_V} \overline{\mathbf{S}}(r_0) \cdot (-\vec{e}_r)\, dS = 2\pi r_0 \overline{S}(r_0) = \left(\frac{\overline{p}}{2\pi r_0 \sigma} \frac{J_0(\overline{p} r_0)}{J_1(\overline{p} r_0)} \right) I_{rms}^2 \tag{8.46}$$

According to (8.34), the term in the large parentheses is to be interpreted as the per-unit-length impedance of the conductor, hence

$$\overline{Z}(\omega) = R(\omega) + jX(\omega) = \frac{\overline{p}}{2\pi r_0 \sigma} \frac{J_0(\overline{p} r_0)}{J_1(\overline{p} r_0)} = R_{DC} \overline{p} r_0 \frac{J_0(\overline{p} r_0)}{2J_1(\overline{p} r_0)} \tag{8.47}$$

where $R_{DC} = 1/(\sigma \pi r_0^2)$ is the per-unit-length DC resistance of the conductor ($\omega = 0$).

Q_7 Here we analyze the limit cases of the low- and high-frequency regimes.

For low frequencies, $|\overline{p} r_0| \ll 1$, $r_0 \ll \delta$, we may substitute the Bessel functions J_0 and J_1 in (8.47) by their Taylor expansions, retaining only the first two terms

$$J_0(\overline{p} r_0) \simeq 1 - \left(\frac{\overline{p} r_0}{2} \right)^2, \quad J_1(\overline{p} r_0) \simeq \frac{\overline{p} r_0}{2} \left[1 - 2 \left(\frac{\overline{p} r_0}{4} \right)^2 \right]$$

On doing this, and recalling that $\overline{p}^2 = -j\omega\mu_0\sigma$, we find

$$\overline{Z}(\omega) = R_{DC} + j\omega R_{DC} \frac{\mu_0 \sigma r_0^2}{8} = R_{DC} + j\omega \underbrace{\left(\frac{\mu_0}{8\pi} \right)}_{L_{\text{inner}}} \tag{8.48}$$

While the real part of the impedance coincides with the R_{DC} contribution, the imaginary part, which is associated with the magnetic energy stored inside the conductor itself, is characterized by a per-unit-length internal inductance $L_{\text{inner}} = \mu_0/(8\pi)$. This result is not new – go to Chapter 4 and check Problem 4.15.1.

For high-frequencies, $|\overline{p} r_0| \gg 1$, $\delta \ll r_0$, we may substitute the Bessel functions J_0 and J_1 in (8.47) for their asymptotic behavior,

$$\lim_{\overline{p} r_0 \to \infty} \frac{J_0(\overline{p} r_0)}{J_1(\overline{p} r_0)} = j$$

from which we get

$$\overline{Z}(\omega) = j \frac{\overline{p}}{2\pi r_0 \sigma} = R_\delta + jR_\delta, \text{ with } R_\delta(\omega) = \frac{1}{(2\pi r_0 \delta)\sigma} \propto \sqrt{\omega} \tag{8.49a}$$

where use was made of

$$\overline{p} = \frac{1-j}{\delta}, \text{ with } \delta = \sqrt{\frac{2}{\omega \mu_0 \sigma}}$$

For high-frequency regimes the real and imaginary parts of the impedance tend to be equal and to increase proportionally to the square root of the frequency. For the physical interpretation of the per-unit-length resistance R_δ introduced in (8.49a), you can think of the conductor current as being concentrated near to the conductor's peripheral surface and flowing across a thin circular crown of thickness equal to the penetration depth δ. From this point of view, the conductor's effective area utilized by the current is $S_\delta = 2\pi r_0 \delta$.

The result in (8.49a) can be rewritten in order to reveal the high-frequency conductor's internal inductance

$$\overline{Z}(\omega) = R_\delta + j\omega L_{\text{inner}}, \text{ with } L_{\text{inner}}(\omega) = \frac{1}{2\pi r_0}\sqrt{\frac{\mu_0}{2\omega\sigma}} \propto \frac{1}{\sqrt{\omega}} \tag{8.49b}$$

As the frequency increases, the magnetic field is expelled from inside the conductor, becoming confined to a peripheral layer whose thickness gradually decreases. In the limiting case $\omega \to \infty$, the magnetic energy stored in the conductor goes to zero and, consequently, $L_{\text{inner}} \to 0$.

The general result in (8.47) is illustrated in Figure 8.9, where a graph of the typical evolution of the conductor's impedance against the frequency is shown.

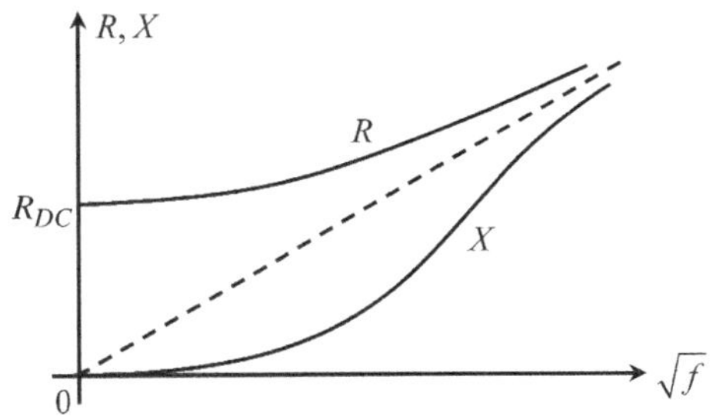

Figure 8.9 Graph of the conductor's internal resistance and reactance against frequency

8.8 Proposed Homework Problems

Problem 8.8.1

Consider a coaxial cable characterized by inner and outer radii r_1 and r_2. Assume that the cable conductors and the insulation medium are perfect. Denote by $u(t)$ and $i(t)$ the cable voltage and current at a given cross-section, $z =$ constant.

Q_1 Write the equations for the electric and magnetic fields inside the cable, $r_1 \le r \le r_2$.

Q_2 Determine the Poynting vector **S**.

Q_3 Determine the power p transmitted by the cable.

Answers

Q_1

$$\mathbf{E}(t) = \frac{u(t)}{r\ln(r_2/r_1)}\,\vec{e}_r,\ \mathbf{H}(t) = \frac{i(t)}{2\pi r}\,\vec{e}_\phi$$

Q_2

$$\mathbf{S} = \mathbf{E}\times\mathbf{H} = \frac{ui}{2\pi r^2\ln(r_2/r_1)}\,\vec{e}_z$$

Q_3

$$p = \int_{S_z} \mathbf{S}\cdot\vec{e}_z\,dS = \frac{ui}{2\pi\ln(r_2/r_1)}\int_{r_1}^{r_2}\frac{1}{r^2}\,2\pi r\,dr = ui$$

Problem 8.8.2

A uniform monochromatic plane wave propagating along the positive z axis is characterized, in free space, by the following electric field:

$$\mathbf{E}(z, t) = E_0\cos(\omega t - \beta z)\,\vec{e}_x + E_0\cos(\omega t - \beta z - \pi/2)\,\vec{e}_y$$

Q_1 Write $\overline{\mathbf{E}}(z)$ and determine $\overline{\mathbf{H}}(z)$ and $\mathbf{H}(z, t)$.

Q_2 Classify both fields as to their type of polarization and determine the corresponding rms values.

Q_3 Determine the complex Poynting vector.

Q_4 Assuming $E_0 = 5\,\text{V/m}$, find the time-averaged value of $\mathbf{S}$.

Answers

Q_1 $\overline{\mathbf{E}}(z) = E_0\,e^{-j\beta z}\,\vec{e}_x + E_0\,e^{-j(\beta z+\pi/2)}\,\vec{e}_y.$

$$\overline{\mathbf{H}}(z) = \frac{j}{\omega\mu_0}\text{curl}\,\overline{\mathbf{E}}(z) = H_0\,e^{-j(\beta z-\pi/2)}\,\vec{e}_x + H_0\,e^{-j\beta z}\,\vec{e}_y$$

where

$$H_0 = \frac{E_0}{R_w}\ \text{and}\ R_w = \frac{\omega\mu_0}{\beta} = \sqrt{\frac{\mu_0}{\varepsilon_0}}$$

$$\mathbf{H}(z, t) = \Re\left(\overline{\mathbf{H}}(z)\,e^{j\omega t}\right) = H_0\cos(\omega t - \beta z + \pi/2)\,\vec{e}_x + H_0\cos(\omega t - \beta z)\,\vec{e}_y.$$

Q_2 Both **E** and **H** are circularly polarized fields. The tip of both vectors rotates in the xy plane with angular velocity ω, **E** and **H** remaining orthogonal to each other.

$$E_{rms} = E_0, \quad H_{rms} = H_0.$$

Q_3

$$\overline{\mathbf{S}} = \frac{\overline{\mathbf{E}} \times \overline{\mathbf{H}}^*}{2} = E_0 H_0 \, \vec{e}_z$$

Q_4

$$(\mathbf{S})_{av} = \Re\left(\overline{\mathbf{S}}\right) = E_0 H_0 \, \vec{e}_z, \; E_0 H_0 = 66.3\,\mathrm{mW/m^2}.$$

Problem 8.8.3

An electric Hertz dipole of length $l = 50\,\mathrm{m}$ oriented along the z axis is excited by a sinusoidal current $i(t) = I\cos\omega t$, with $I = 100\,\mathrm{A}$ and frequency $f = 60\,\mathrm{kHz}$ (LF band/radio beacon applications). In a spherical coordinate system (r, θ, ϕ), the radiated electromagnetic field in free space (μ_0, ε_0) far away from the antenna is given by

$$\mathbf{E} = E_r(r, \theta, t)\,\vec{e}_r + E_\theta(r, \theta, t)\,\vec{e}_\theta, \quad \mathbf{H} = H_\phi(r, \theta, t)\,\vec{e}_\phi$$

where

$$E_r = \frac{IR_w l}{2\pi r^2}\cos\theta\cos(\omega t - \beta r), \quad E_\theta = \frac{IR_w l}{2\lambda r}\sin\theta\cos(\omega t - \beta r + \pi/2), \quad H_\phi = \frac{E_\theta}{R_w}$$

where $R_w = \sqrt{\mu_0/\varepsilon_0}$, $\beta = \omega\sqrt{\mu_0\varepsilon_0} = \omega/c = 2\pi/\lambda$, with $\lambda = c/f = 5\,\mathrm{km}$ denoting the wavelength.

Q_1 What is the type of polarization of the radiated fields?

Q_2 Show that there is no radiation along the dipole axis, $\theta = 0$.

Q_3 Determine the complex amplitudes of **E** and **H** in the dipole equatorial plane $(\theta = \pi/2)$, at a distance $r = 100\,\mathrm{km}$. Obtain the corresponding complex Poynting vector.

Q_4 Evaluate the time-averaged power P radiated by the antenna.

Q_5 Determine the dipole radiation resistance defined as $R_R = P/I_{rms}^2$.

Answers

Q_1 The electric field is elliptically polarized in meridian planes, that is planes defined by $\vec{e}_r$ and $\vec{e}_\theta$. The magnetic field is linearly polarized along $\vec{e}_\phi$.

Q_2 For $\theta = 0$ you have $E_\theta = H_\phi = 0$. Since $\mathbf{H} = 0$, the Poynting vector is also zero, $\mathbf{S} = 0$, which implies the absence of radiated power along z, that is along the antenna axis.

Q_3 For $\theta = \pi/2$:

$$\begin{cases} E_\theta = \dfrac{IR_w l}{2\lambda r}\cos(\omega t - \beta r + \pi/2) \\ E_r = 0 \\ H_\phi = \dfrac{Il}{2\lambda r}\cos(\omega t - \beta r + \pi/2) \end{cases}$$

$$\overline{\mathbf{E}} = j\frac{IR_w l}{2\lambda r}e^{-j\beta r}\vec{e}_\theta = 1.885\, e^{j\pi/2}\,\vec{e}_\theta \text{ mV/m}$$

$$\overline{\mathbf{H}} = j\frac{Il}{2\lambda r}e^{-j\beta r}\vec{e}_\phi = 5\, e^{j\pi/2}\,\vec{e}_\phi\ \mu\text{A/m}$$

Noting that $\vec{e}_\theta \times \vec{e}_\phi = \vec{e}_r$, you find

$$\overline{\mathbf{S}} = \frac{\overline{\mathbf{E}} \times \overline{\mathbf{H}}^*}{2} = 4.71\,\vec{e}_r \text{ nW/m}^2$$

Q_4 The complex amplitudes of **E** and **H**, which depend on r and θ, are

$$\overline{\mathbf{E}}(r,\theta) = \frac{IR_w l}{2\pi r^2}\cos\theta\, e^{-j\beta r}\vec{e}_r + j\frac{IR_w l}{2\lambda r}\sin\theta\, e^{-j\beta r}\vec{e}_\theta,\quad \overline{\mathbf{H}}(r,\theta) = j\frac{Il}{2\lambda r}\sin\theta\, e^{-j\beta r}\vec{e}_\phi$$

The corresponding complex Poynting vector has two components, one directed along $\vec{e}_\theta$ and the other along $\vec{e}_r$. The component $\overline{S}_\theta$ not only is of little importance for $r \gg \lambda$, but, above all, does not contribute to the radiation of energy. The radial component $\overline{S}_r$ is given by

$$\overline{S}_r = \frac{R_w}{8}\left(\frac{Il}{\lambda r}\sin\theta\right)^2$$

The power radiated outward by the antenna is obtained by integrating $\overline{\mathbf{S}}$ across a spherical surface S_V of radius $r \gg \lambda$, centered at the antenna feeding point:

$$\overline{P} = P = \int_{S_V} \overline{\mathbf{S}}\cdot\vec{e}_r\, dS = \int_{S_V} \overline{S}_r\, dS$$

where $dS = r^2 \sin\theta\, d\theta\, d\phi$, with $\theta \in [0,\ \pi]$ and $\phi \in [0,\ 2\pi]$. Therefore we have

$$P = \frac{R_w}{8}\left(\frac{Il}{\lambda}\right)^2 \int_0^{2\pi} d\phi \int_0^{\pi} \sin^3\theta\, d\theta = R_w \frac{\pi}{3}\left(\frac{Il}{\lambda}\right)^2 = 395\,\text{W}$$

Q_5

$$R_R = R_w\frac{2\pi}{3}\left(\frac{l}{\lambda}\right)^2 = 80\pi^2\left(\frac{l}{\lambda}\right)^2\ \Omega = 79\,\text{m}\Omega$$

Problem 8.8.4

As far as quasi-stationary regimes are concerned, a capacitor is a circuit component whose characteristic parameter is its capacitance. For high-frequency regimes, things are much more complicated. At certain frequencies a capacitor may behave as an inductor!

Consider a parallel-plate capacitor consisting of two circular metallic electrodes of radius r_0 separated by a perfect insulating dielectric medium of very small thickness h – see Figure 8.10. The capacitor current is a time-varying sinusoidal current $i(t) = I\cos(\omega t + \alpha_i)$.

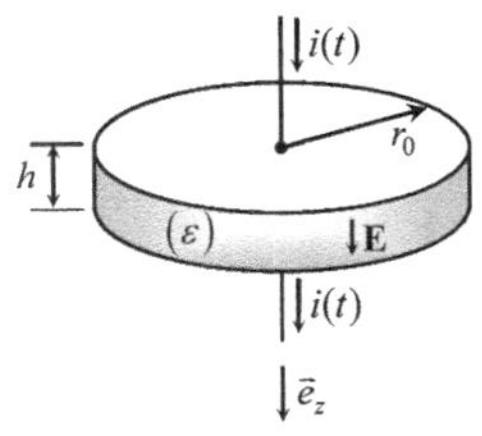

Figure 8.10 A parallel-plate disk-capacitor structure. For high-frequency regimes the device may exhibit an inductor-like behavior

Given the small thickness of the insulation medium, assume that all fields are independent of z. Assume, further, that the electric field $\mathbf{E}$ between the capacitor plates is vertically oriented along z. According to the geometry of the problem, a particular field solution to Maxwell's equations must be invariant under rotation operations around the z axis.

Take the permittivity and permeability of the insulation medium as ε and μ_0 respectively.

Questions

Q_1 By using a cylindrical coordinate reference frame, write the phasor-domain Maxwell's equations for the problem and obtain a solution for $\overline{\mathbf{H}}$ and $\overline{\mathbf{E}}$.

Q_2 Find the complex Poynting vector $\overline{\mathbf{S}}$ at any point belonging to the boundary cylindrical surface of radius $r = r_0$.

Q_3 Evaluate the complex power $\overline{P}$ corresponding to the inward flux of $\overline{\mathbf{S}}$ across the boundary cylindrical surface of radius r_0 and height h.

Q_4 Determine the frequency-dependent impedance $\overline{Z}(\omega)$ of the capacitor and discuss the resonance conditions of the device. Particularize $\overline{Z}(\omega)$ for very low- and very high-frequency regimes.

Q_5 Take $\varepsilon = 4\varepsilon_0$ and $r_0 = 15.92\,\text{mm}$. Find the first five resonance frequencies of the device.

(Hint: Follow the same line of thought employed in Application Example 8.7, substituting $\varepsilon\,\partial\mathbf{E}/\partial t$ for $\sigma\mathbf{E}$.)

Answers

Q_1

$$\overline{\mathbf{E}} = \overline{E}(r)\,\vec{e}_z, \quad \overline{\mathbf{H}} = \overline{H}(r)\,\vec{e}_\phi.$$

$$\frac{d\overline{E}}{dr} = j\omega\mu_0\overline{H}$$

$$\frac{d\overline{H}}{dr} + \frac{\overline{H}}{r} = j\omega\varepsilon\overline{E}$$

$$\frac{d^2\overline{E}}{dr^2} + \frac{1}{r}\frac{d\overline{E}}{dr} + \underbrace{\left(\omega\sqrt{\mu_0\varepsilon}\right)}_{\beta}{}^2\overline{E} = 0$$

$$\overline{E}(r) = E_1\, J_0(\beta r) + E_2\, Y_0(\beta r), \text{ where } \begin{cases} E_1 = \dfrac{\beta\overline{I}}{j\omega\varepsilon 2\pi r_0 J_1(\beta r_0)} \\ E_2 = 0 \end{cases}$$

$$\overline{E}(r) = -j\frac{\overline{I}}{2\pi r_0}\sqrt{\frac{\mu_0}{\varepsilon}}\frac{J_0(\beta r)}{J_1(\beta r_0)}$$

$$\overline{H}(r) = \frac{1}{j\omega\mu_0}\frac{d\overline{E}(r)}{dr} = \frac{\overline{I}}{2\pi r_0}\frac{J_1(\beta r)}{J_1(\beta r_0)}$$

Q_2

$$\overline{\mathbf{S}}(r_0) = \frac{\overline{\mathbf{E}}(r_0)\times\overline{\mathbf{H}}^*(r_0)}{2} = -\overline{S}(r_0)\vec{e}_r, \text{ where } \overline{S}(r_0) = \frac{I_{rms}^2}{j(2\pi r_0)^2}\sqrt{\frac{\mu_0}{\varepsilon}}\frac{J_0(\beta r_0)}{J_1(\beta r_0)}$$

Q_3

$$\overline{P} = \int_{S_V}\overline{\mathbf{S}}(r_0)\cdot(-\vec{e}_r)\,dS = 2\pi r_0 h\overline{S}(r_0) = \left(\frac{h}{j2\pi r_0}\sqrt{\frac{\mu_0}{\varepsilon}}\frac{J_0(\beta r_0)}{J_1(\beta r_0)}\right) I_{rms}^2$$

Q_4 According to (8.34), the term above, in the large parentheses, is to be interpreted as the impedance of the capacitor, which can be rewritten as

$$\overline{Z}(\omega) = jX(\omega) = \frac{1}{j\omega C_0}\beta r_0\frac{J_0(\beta r_0)}{2J_1(\beta r_0)} \tag{8.51}$$

where C_0 is the electrostatic capacitance, $C_0 = \left(\varepsilon\pi r_0^2\right)/h$, and β depends on the frequency, $\beta = \omega\sqrt{\mu_0\varepsilon}$.

From (8.51) you can see that the capacitor reactance $X(\omega)$ changes sign whenever the Bessel functions J_0 and J_1 go through zero. The first zeros of J_0 occur for $\beta r_0 \approx 2.405, 5.520, 8.654, \ldots$. The first zeros of J_1 occur for $\beta r_0 \approx 0$, 3.832, 7.016, 10.174, ...

In the range $0 < \beta r_0 < 2.405$ the parallel-plate device behaves as a capacitor, $X(\omega) < 0$. For $\beta r_0 = 2.405$ it behaves as a short circuit (resonance). In the range $2.405 < \beta r_0 < 3.832$

the device behaves as an inductor, $X(\omega) > 0$. For $\beta r_0 = 3.832$ it behaves as an open circuit (resonance). In the range $3.832 < \beta r_0 < 5.520$ the device again behaves as a capacitor, $X(\omega) < 0$. And so on.

This alternate behavior between capacitive and inductive character corresponds, respectively, to the dominance of electric energy or the dominance of magnetic energy stored in the insulation medium. (Note: Magnetic energy storage in the dielectric is associated with the magnetic field produced by displacement currents.)

Resonance situations occur for frequencies such that $(W_m)_{av} = (W_e)_{av}$.

The general result in (8.51) is illustrated in Figure 8.11, where a graph of the typical evolution of the capacitor reactance is plotted against βr_0.

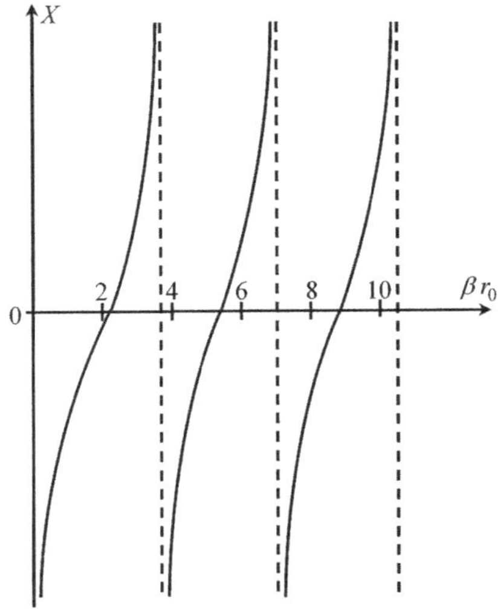

Figure 8.11 Graph of the reactance X of the disk capacitor against βr_0, where $\beta = \omega\sqrt{\mu_0 \varepsilon}$

For very low-frequency regimes, $\beta r_0 \to 0$, you have

$$\beta r_0 \frac{J_0(\beta r_0)}{2J_1(\beta r_0)} \to 1$$

hence

$$\overline{Z}(\omega) \approx \frac{1}{j\omega C_0}$$

For very high-frequency regimes, $\beta r_0 \gg 1$, you must consider the asymptotic behavior of the Bessel functions, that is

$$\frac{J_0(\beta r_0)}{J_1(\beta r_0)} \to \frac{1}{\tan(\beta r_0 - \pi/4)}$$

hence

$$\overline{Z}(\omega) \approx \frac{1}{j\omega C_0} \frac{\beta r_0/2}{\tan(\beta r_0 - \pi/4)}$$

Q_5 $f_1 = 4.81\,\text{GHz}$; $f_2 = 7.66\,\text{GHz}$; $f_3 = 11.04\,\text{GHz}$: $f_4 = 14.03\,\text{GHz}$: $f_5 = 17.31\,\text{GHz}$.

Problem 8.8.5

Consider a copper wire ($\mu = \mu_0$, $\sigma = 5.7 \times 10^7\,\text{S/m}$) of radius $r_0 = 1\,\text{mm}$ and length $l = 1\,\text{m}$. The wire carries a time-varying current given by $i(t) = I_0 + I_1 \cos(\omega_1 t)$, with $\omega_1 = 2\pi f_1$, $f_1 = 1\,\text{MHz}$.

Q_1 Making use of skin-effect results (see Application Example 8.7) determine the wire's resistance and inductance parameters for the DC component and for the AC component.

Q_2 Using the superposition principle, determine the longitudinal voltage drop $u(t)$ along the wire length.

Q_3 Show that $u(t)$ and $i(t)$ cannot, in any way, be related by an equation of the usual type $u = Ri + L\,di/dt$.

Answers

Q_1 From (8.48) you have

$$\omega = 0 \begin{cases} R_0 = \dfrac{l}{\sigma \pi r_0^2} = 5.6\,\text{m}\Omega \\ L_0 = \dfrac{\mu_0 l}{8\pi} = 50\,\text{nH} \end{cases}$$

For the AC component, the skin-effect penetration depth, (8.38), is

$$\delta = \sqrt{\frac{2}{\omega \mu_0 \sigma}} = 67\,\mu\text{m}$$

Since $\delta \ll r_0$, you can use the high-frequency results established in (8.49):

$$\omega = \omega_1 \begin{cases} R_1 = \dfrac{l}{\sigma 2\pi r_0 \delta} = R_0 \dfrac{r_0}{2\delta} = 41.7\,\text{m}\Omega \\ L_1 = \dfrac{R_1}{\omega_1} = 6.64\,\text{nH} \end{cases}$$

Q_2 For the DC component: $u_0 = R_0 I_0$.

For the AC component: $\overline{U}_1 = \sqrt{2} R_1 I_1 e^{j\pi/4} \rightarrow u_1(t) = \sqrt{2} R_1 I_1 \cos(\omega_1 t + \pi/4)$.

Hence

$$u(t) = R_0 I_0 + \sqrt{2} R_1 I_1 \cos(\omega_1 t + \pi/4) \tag{8.52}$$

Q_3 Using $i(t) = I_0 + I_1 \cos(\omega_1 t)$ in an equation like $u = Ri + L\,di/dt$ would yield

$$u(t) = RI_0 + RI_1 \cos(\omega_1 t) - \omega_1 L I_1 \sin(\omega_1 t)$$

or, which is the same,

$$u(t) = RI_0 + \sqrt{R^2 + (\omega_1 L)^2}\, I_1 \cos\left(\omega_1 t + \arctan\frac{\omega_1 L}{R}\right) \tag{8.53}$$

In order to ensure that (8.53) would coincide with the correct result in (8.52), it would be required that $R = R_0 = R_1 = \omega_1 L$, which is an impossible condition!

This clearly shows that, when analyzing skin-effect phenomena in the time domain, utilization of equations of the type $u = Ri + L\,di/dt$ is, in general, a mistake.

9

Transmission-Line Analysis

9.1 Introduction

Transmission-line analysis is an important topic in many electrical engineering areas that range from power line systems to telecommunications, including computer networks.

The simplest examples of a transmission line are the two-wire line and the coaxial cable. In both cases we are dealing with a system made of two conductors separated by a dielectric medium where waves propagate from one end to the other guided by the geometrical configuration of the line conductors.

Since typical transmission-line systems have longitudinal dimensions of the order of the working wavelength λ (or above), the standard approximations of quasi-stationary regimes no longer apply – lumped-parameters circuit approaches simply cannot be used. Some examples follow:

- Transmission power lines: length 1000 km; frequency 60 Hz; wavelength 5000 km.
- Telephone lines: length 100 km; frequency 3 kHz; wavelength 100 km.
- Transmitter–antenna line links: length 100 m; frequency 10 MHz; wavelength 30 m.
- Computer boards: length 20 cm; frequency 1 GHz; wavelength 30 cm.

Nonetheless, we will assume in our analysis that the transversal system dimensions are negligibly small compared to λ, that is transversal wave phenomena are absent.

Besides the simple cases of the two-wire line and the coaxial cable, transmission-line analysis is also of crucial importance in the study of multiconductor transmission-line structures; however, such a topic will only be tackled superficially due to the introductory nature of this textbook. Likewise, the topic of non-uniform lines – those where the transverse profile of the line changes along the line length – will only be briefly referred to.

Before we get into the mathematical details of the equations that govern transmission-line phenomena, let us introduce the subject from a qualitative point of view.

Consider a transmission line of length l connecting a generator to a load (Figure 9.1), where, for simplification purposes, the line conductors are assumed to be perfect conductors.

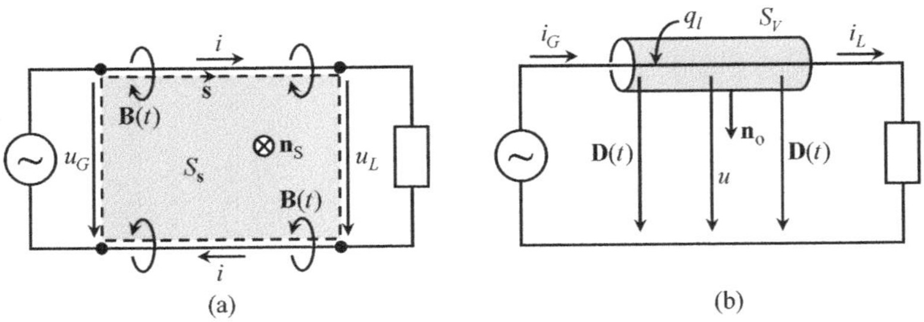

Figure 9.1 (a) Generator and load voltages are not the same, due to magnetic field distributed effects along the line. (b) Generator and load currents are not the same, due to electric field distributed effects along the line

You may recall that in the introductory matter to Part III we emphasized that the equalities $u_G(t) = u_L(t)$ and $i_G(t) = i_L(t)$ would not be valid for rapid time-varying field phenomena. Let us see why this happens.

As shown in Figure 9.1(a), conductor currents give rise to a distributed time-varying magnetic field along the line. By application of the induction law (Chapter 5) you find

$$-u_G(t) + u_L(t) = -\frac{d}{dt}\psi_l(t) = -\int_{S_s} \frac{\partial \mathbf{B}}{\partial t} \cdot \mathbf{n}_S \, dS \qquad (9.1)$$

where $\psi_l(t)$ denotes the flux linked with the rectangular path of longitudinal length l. Therefore you can see that the presence of distributed inductive effects is ultimately responsible for the inequality $u_G(t) \neq u_L(t)$.

Also, as shown in Figure 9.1(b), conductor voltages give rise to a distributed time-varying electric field along the line (displacement currents between conductors). By application of the charge continuity equation in integral form (Chapter 6) you find

$$-i_G(t) + i_L(t) = -\frac{d}{dt}q_l(t) = -\int_{S_V} \frac{\partial \mathbf{D}}{\partial t} \cdot \mathbf{n}_o \, dS \qquad (9.2)$$

where $q_l(t)$ denotes the total electric charge distributed along the top conductor of length l. Therefore you can see that the presence of distributed capacitive effects is ultimately responsible for the inequality $i_G(t) \neq i_L(t)$.

For slow time-varying regimes, where both the inductive and capacitive distributed effects are negligibly small, we have $u_G(t) = u_L(t)$ and $i_G(t) = i_L(t)$. For rapid time-varying regimes, the voltages and currents at the generator and load terminals do not coincide; nevertheless, a simple identity between their time integrals still exists.

Assume that the driving voltage is a natural time-limited function; that is, assume the line is inactive for $t = \pm\infty$. Try a time integration of (9.1) and (9.2) from $t = -\infty$ to $t = \infty$.

When you do that, you will find, from (9.1),

$$-\int_{-\infty}^{\infty} u_G(t)\, dt + \int_{-\infty}^{\infty} u_L(t)\, dt = \underbrace{\psi_l(-\infty)}_{0} - \underbrace{\psi_l(\infty)}_{0}$$

Hence,

$$\int_{-\infty}^{\infty} u_G(t)\, dt = \int_{-\infty}^{\infty} u_L(t)\, dt \tag{9.3}$$

Likewise, from (9.2),

$$-\int_{-\infty}^{\infty} i_G(t)\, dt + \int_{-\infty}^{\infty} i_L(t)\, dt = \underbrace{q_l(-\infty)}_{0} - \underbrace{q_l(\infty)}_{0}$$

Hence,

$$\int_{-\infty}^{\infty} i_G(t)\, dt = \int_{-\infty}^{\infty} i_L(t)\, dt \tag{9.4}$$

The conclusions in (9.3) and (9.4) are not just theoretical curiosities; they can be very helpful for checking how correct your results are when solving a given problem, whether you do it by hand or employ dedicated software tools.

The fact that voltages and currents at the accessible ends of the line are different, $u_G(t) \neq u_L(t)$, $i_G(t) \neq i_L(t)$, is a clear indication that line voltages and line currents will undergo a continuous evolution along the line's longitudinal coordinate, $u = u(z, t)$, $i = i(z, t)$. In the next section we will obtain the laws that govern such an evolution.

9.2 Time-Domain Transmission-Line Equations for Lossless Lines

Consider a two-conductor line immersed in a non-magnetic linear homogeneous dielectric medium, whose conductors run parallel to the longitudinal z axis.

If you assume that the current-carrying conductors are perfect, the electric field component E_z will be absent. This means that outside the conductors, the field lines of **E** and **H** exist in planes transversal to the z axis; in this case we say that we are dealing with transverse electromagnetic waves (TEM),

$$\mathbf{E} = E_x(z, t)\vec{e}_x + E_y(z, t)\vec{e}_y, \quad \mathbf{H} = H_x(z, t)\vec{e}_x + H_y(z, t)\vec{e}_y$$

In these circumstances we can unambiguously define line voltages and line currents as

$$u(z, t) = \int_{\overrightarrow{ba}} \mathbf{E}(z, t) \cdot d\mathbf{s}$$

$$i(z, t) = \int_{S_1} \mathbf{J}_1(z, t) \cdot \vec{e}_z\, dS = \int_{S_2} \mathbf{J}_2(z, t) \cdot (-\vec{e}_z)\, dS = \oint_{\mathrm{s}} \mathbf{H}(z, t) \cdot d\mathbf{s}$$

where, as shown in Figure 9.2, the arbitrary open path $\overrightarrow{ba}$, the arbitrary closed path s and the conductor cross-sections S_1 and S_2 all belong to the same transversal plane $z =$ constant.

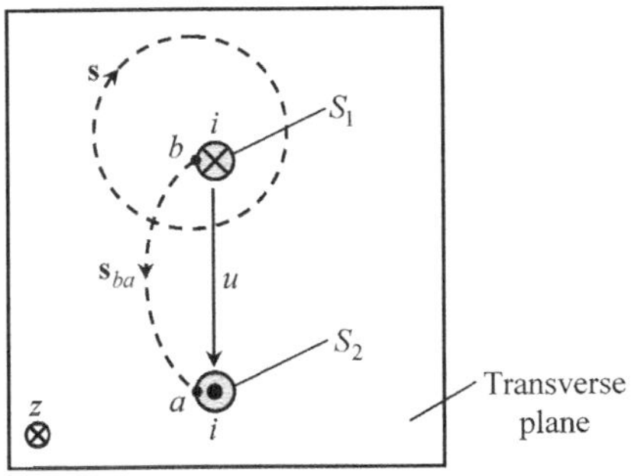

Figure 9.2 Line voltages and line currents are unambiguously defined in the transverse plane of the transmission line. **E**-field integration along an arbitrary open path from b to a yields the same result for the voltage u. Similarly, **H**-field integration along an arbitrary closed path encircling the upper conductor yields the same result for the current i

However, as you move from a z plane to another z plane both u and i will change. Our next goal is to determine the equations that govern those changes.

Consider the application of the induction law to a very small line section of length $\Delta z \to 0$ (Figure 9.3).

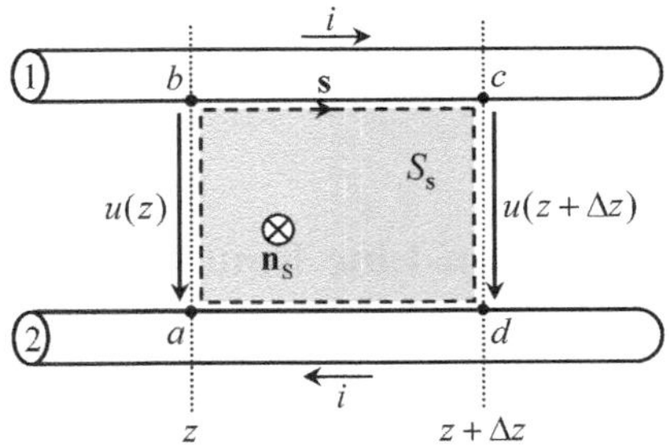

Figure 9.3 Application of the induction law to a very small line section of length $\Delta z \to 0$

The rectangular circulation path $\mathbf{s} = \overrightarrow{abcda}$ includes four contributions: one on the plane z, another on the plane $z + \Delta z$ and two others along the conductors' surfaces (where $E_z = 0$). The magnetic flux $\psi_{\Delta z}$ linked to the circulation path originated by the line currents can be evaluated by taking into account the per-unit-length external inductance L_e of the line

$$\begin{cases} \oint_{\mathbf{S}} \mathbf{E} \cdot d\mathbf{s} = \int_{\overrightarrow{ab}} \mathbf{E} \cdot d\mathbf{s} + \int_{\overrightarrow{bc}} \mathbf{E} \cdot d\mathbf{s} + \int_{\overrightarrow{cd}} \mathbf{E} \cdot d\mathbf{s} + \int_{\overrightarrow{da}} \mathbf{E} \cdot d\mathbf{s} = -u(z,t) + 0 + u(z+\Delta z, t) + 0 \\ -\dfrac{d}{dt} \int_{S_s} \mathbf{B} \cdot \mathbf{n}_S \, dS = -\dfrac{d}{dt} \psi_{\Delta z} = -L_e \Delta z \dfrac{\partial}{\partial t} i(z,t) \end{cases}$$

Equating the results above, dividing by Δz and taking the limit as $\Delta z \to 0$ we obtain

$$\lim_{\Delta z \to 0} \frac{u(z+\Delta z, t) - u(z,t)}{\Delta z} = \frac{\partial}{\partial z} u(z,t) = -L_e \frac{\partial}{\partial t} i(z,t)$$

Next, consider the application of the integral form of the charge continuity equation to a very small line section of length $\Delta z \to 0$; conductor 1 is enclosed by the integration surface S_V (Figure 9.4).

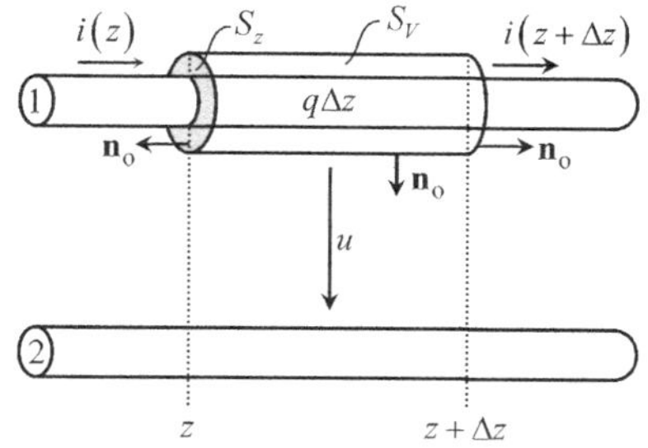

Figure 9.4 Application of the integral form of the charge continuity equation to a very small line section of length $\Delta z \to 0$

The current intensities through S_V are $-i(z, t)$ and $i(z + \Delta z, t)$, on the left and right cross-sections, respectively. The electric charge $q\Delta z$ inside S_V (originating the electric field between line conductors) can be evaluated by taking into account the per-unit-length capacitance C of the line

$$\begin{cases} \int_{S_V} \mathbf{J} \cdot \mathbf{n}_o \, dS = \int_{S_z} \mathbf{J} \cdot (-\vec{e}_z) \, dS + \int_{S_{z+\Delta z}} \mathbf{J} \cdot \vec{e}_z \, dS = -i(z, t) + i(z + \Delta z, t) \\ -\int_V \rho dV = -q\Delta z = -C\Delta z \, u(z, t) \end{cases}$$

Equating the results above, dividing by Δz and taking the limit as $\Delta z \to 0$ we obtain

$$\lim_{\Delta z \to 0} \frac{i(z + \Delta z, t) - i(z, t)}{\Delta z} = \frac{\partial}{\partial z} i(z, t) = -C \frac{\partial}{\partial t} u(z, t)$$

Summarizing, the time-domain transmission-line equations for lossless lines are

$$\frac{\partial}{\partial z} u(z, t) = -L_e \frac{\partial}{\partial t} i(z, t) \tag{9.5}$$

$$\frac{\partial}{\partial z} i(z, t) = -C \frac{\partial}{\partial t} u(z, t) \tag{9.6}$$

The line voltage rate of change in space is proportional to the line current rate of change in time; the link between those two rates is established via the per-unit-length inductance of the line. The line current rate of change in space is proportional to the line voltage rate of change in time; the link between those two rates is established via the per-unit-length capacitance of the line.

9.2.1 Wave Parameters, Propagation Velocity, Characteristic Wave Resistance

If you take the z derivative of (9.5) and then use (9.6) or, the other way around, if you take the z derivative of (9.6) and then use (9.5), you will obtain the already familiar one-dimensional wave equation, which applies indistinctly to line voltages and line currents

$$\frac{\partial^2}{\partial z^2}\begin{Bmatrix} u(z,t) \\ i(z,t) \end{Bmatrix} - L_e C \frac{\partial^2}{\partial t^2}\begin{Bmatrix} u(z,t) \\ i(z,t) \end{Bmatrix} = 0 \tag{9.7}$$

where the product $L_e C$ is the inverse of the squared wave propagation velocity, that is

$$v^2 = \frac{1}{L_e C} \tag{9.8}$$

from which two solutions can be found, $v = \pm 1/\sqrt{L_e C}$. While one solution describes a propagating wave along the positive z direction (the so-called incident wave), the other solution describes a propagating wave along the negative z direction (the so-called reflected wave). This means that, in general, the line voltage and line current solutions are a superposition of two counter-propagating waves, with velocity $v = 1/\sqrt{L_e C}$, which we can write in the form

$$u(z,t) = u_i(z - vt) + u_r(z + vt) \tag{9.9a}$$

$$i(z,t) = i_i(z - vt) + i_r(z + vt) = \frac{u_i(z - vt)}{(R_w)_i} + \frac{u_r(z + vt)}{(R_w)_r} \tag{9.9b}$$

where subscripts i and r are remainders for the incident and reflected wave solutions, respectively.

From (9.7), you can see that voltage and current waves are similar except for a scale factor with the physical dimensions of a resistance. For the incident wave, the scaling factor $(R_w)_i = u_i/i_i$ is the so-called characteristic resistance of the incident wave. The same thing applies, identically, to the reflected wave, with $(R_w)_r = u_r/i_r$.

Let us now determine the characteristic resistances concerning both waves.

By making $\xi = z - vt$, $\zeta = z + vt$, and taking into account that

$$\frac{\partial u}{\partial z} = \frac{\partial (u_i + u_r)}{\partial z} = \overbrace{\left(\frac{du_i}{d\xi}\right)}^{u_i'}\frac{\partial \xi}{\partial z} + \overbrace{\left(\frac{du_r}{d\zeta}\right)}^{u_r'}\frac{\partial \zeta}{\partial z} = u_i' + u_r'$$

$$\frac{\partial i}{\partial t} = \frac{\partial (i_i + i_r)}{\partial t} = \frac{di_i}{d\xi}\frac{\partial \xi}{\partial t} + \frac{di_r}{d\zeta}\frac{\partial \zeta}{\partial z} = \frac{u_i'}{(R_w)_i}(-v) + \frac{u_r'}{(R_w)_r}(v)$$

substituting (9.9) into (9.5) we find

$$u_i' + u_r' = \frac{L_e v}{(R_w)_i}u_i' + \frac{-L_e v}{(R_w)_r}u_r'$$

from which you see that the characteristic wave resistances are

$$(R_w)_i = -(R_w)_r = R_w = L_e v = \frac{1}{Cv} = \sqrt{\frac{L_e}{C}} \tag{9.10}$$

In the same way that two counter-propagating waves have reciprocal velocities ($\pm v$), the same thing happens with their characteristic resistances ($\pm R_w$).

According to this conclusion, we rewrite the transmission-line wave equation solutions in the final form

$$u(z, t) = u_i(z - vt) + u_r(z + vt) \tag{9.11a}$$

$$i(z, t) = i_i(z - vt) + i_r(z + vt) = \frac{u_i(z - vt)}{R_w} + \frac{u_r(z + vt)}{-R_w} \tag{9.11b}$$

In Chapter 8 we saw that the velocity of electromagnetic waves in a lossless homogeneous medium was determined solely by the intrinsic properties of the medium, that is by the parameters μ and ε. For non-magnetic media we have found

$$v = \frac{1}{\sqrt{\mu\varepsilon}} = \frac{1}{\sqrt{\mu_0\varepsilon_0\varepsilon_r}} = \frac{c}{\sqrt{\varepsilon_r}} \tag{9.12}$$

We can check that this result still applies to waves guided by lossless homogeneous transmission lines. In fact, if you make use of the per-unit-length capacitance and inductance parameters established in Chapters 2 and 4, concerning the coaxial cable and the two-thin-wire line configuration, you will obtain v as given in (9.12) – see Table 9.1.

Table 9.1 Transmission-line parameters

	Coaxial cable	Two-thin-wire line
Per-unit-length capacitance C (F/m)	$\dfrac{2\pi\varepsilon}{\ln(r_2/r_1)}$	$\dfrac{\pi\varepsilon}{\ln(2d/r)}$
Per-unit-length inductance L_e (H/m)	$\dfrac{\mu_0}{2\pi}\ln(r_2/r_1)$	$\dfrac{\mu_0}{\pi}\ln(2d/r)$
Propagation velocity $v = \dfrac{1}{\sqrt{L_e C}}$ (m/s)	$\dfrac{1}{\sqrt{\mu_0\varepsilon}}$	$\dfrac{1}{\sqrt{\mu_0\varepsilon}}$
Characteristic resistance $R_w = \sqrt{\dfrac{L_e}{C}}$ (Ω)	$\sqrt{\dfrac{\mu_0}{\varepsilon}}\dfrac{\ln(r_2/r_1)}{2\pi}$	$\sqrt{\dfrac{\mu_0}{\varepsilon}}\dfrac{\ln(2d/r)}{\pi}$

However, the transmission-line characteristic wave resistance R_w is not to be confused with the free-space characteristic resistance we spoke about in Chapter 8; while the latter is a relationship between electric and magnetic field intensities, the former relates the voltage and the current of a given wave. Transmission-line characteristic wave resistances depend not only on the dielectric medium properties, but also on the transverse profile of the line itself – see Table 9.1.

When reading Table 9.1, you should bear in mind that, for the coaxial cable configuration, r_1 and r_2 denote the radius of the inner conductor and the inner radius of the outer conductor, respectively. For the two-thin-wire line, r denotes the radius of both conductors, and $2d$ denotes the axial separation between them.

9.2.2 Pulse Propagation, Pulse Reflection

Consider a transmission line of length l, and assume that at the input end of the line ($z = 0$) a triangular pulse voltage of short duration ΔT is provided by an ideal generator (Figure 9.5(a)).

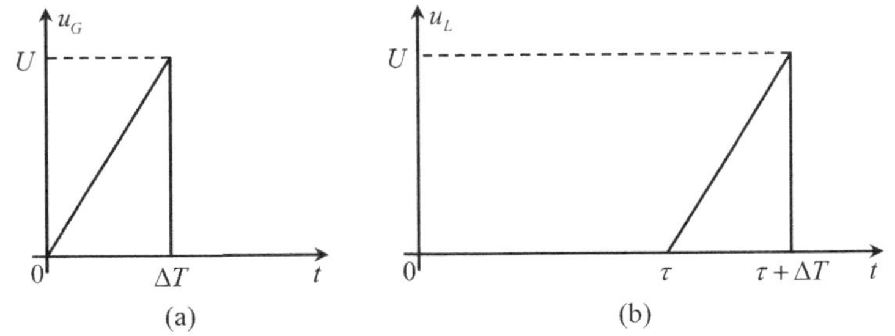

Figure 9.5 (a) Generator voltage shape. (a) Load voltage shape for the particular case of a matched line; the delay time τ is defined as $\tau = l/v$

The time delay between the emission of the pulse and its arrival at the load terminal is $\tau = l/v$ (with $\Delta T < \tau$). If you take a photo of the line voltage and line current along the line length at an instant $t < \tau$ you will only observe the propagating incident wave (Figure 9.6).

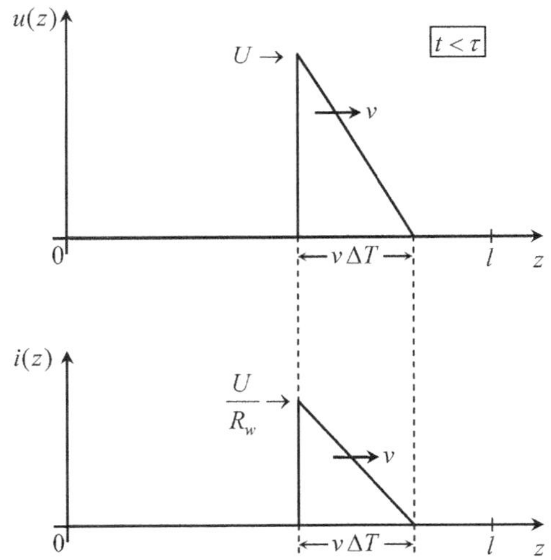

Figure 9.6 Snapshot, at instant t ($t < \tau$), of the line voltage and line current (incident wave) observed along z

A parenthetical remark: at this point some of you may be thinking that there is a mistake in the line drawing shown in Figure 9.6, for the shape of the triangular voltage pulse is the reverse of the triangular pulse shape in Figure 9.5(a). Well, you are thinking wrong; there is no mistake in Figure 9.6. The thing is that Figure 9.5(a) and Figure 9.6 are different representations of the pulse; while the first is a representation in time, the second is a representation in space. From Figure 9.5(a) you can see that the wavefront ($t = 0$) starts with zero voltage and only later in time ($t = \Delta T$) is the peak value U reached; as a consequence, when the wave travels in space, the pulse tail is the peak value.

As the incident wave reaches $z = l$ part of its energy is absorbed by the load and another part is reflected back towards the generator (reflected wave).

You may be wondering whether the reflection process can be prevented in some way. The answer is yes. If the load is properly chosen it can happen that the incident pulse is totally absorbed by the load; in such a case we say that the line is perfectly matched. In this case the voltage at the load terminals is simply a delayed image of the generator voltage (Figure 9.5(b)).

Let us see how you should choose the load R_L in order to get a matched line.

If you enforce the goal condition $u_r = 0$ in (9.11), and particularize the result to $z = l$, you will find

$$u(l, t) = u_i(l - vt); \quad i(l, t) = \frac{u_i(l - vt)}{R_w}$$

Noting that $R_L = u(l, t)/i(l, t)$, you readily conclude

$$\text{Matched line} : R_L = R_w \tag{9.13}$$

Whenever a lossless line is terminated by a resistor whose resistance is equal to the characteristic resistance of the line, reflected waves will be absent. This result provides you with a physical interpretation for the concept of characteristic wave resistance.

Take for instance a 50 Ω coaxial cable. Such a specification has nothing to do with the resistance of the cable conductors; the 50 Ω information concerns the cable's characteristic resistance, and tells you that if you want to avoid reflected waves inside the cable then you must terminate it with a 50 Ω resistor.

Next, let us examine the reflection processes that take place in a transmission line when the line is a mismatched one. For exemplification purposes, we keep considering that at $z = 0$ the voltage pulse is the one shown in Figure 9.5(a), but now we assume that the load terminals are left open, that is $i(l, t) = 0$.

As before in Figure 9.6, if you take a photo of the line voltage and line current along the line length at an instant $t < \tau$ you will observe only the propagating incident wave.

Since the load is an open circuit, it cannot absorb energy and a reflected wave is produced such that, at $z = l$,

$$i = 0 = i_i + i_r \quad \rightarrow \quad i_r = -i_i \text{ and } u_r = -R_w i_r = +R_w i_i = u_i \tag{9.14}$$

Now, if you take a photo of the line voltage and line current along the line length at an instant $\tau < t < 2\tau$ you will observe only the propagating reflected wave where the current pulse has changed its polarity – see Figure 9.7.

When the wave reaches the generator terminal, the generator voltage is zero (remember that the pulse duration ΔT is smaller than τ); put another way, the incoming wave sees a

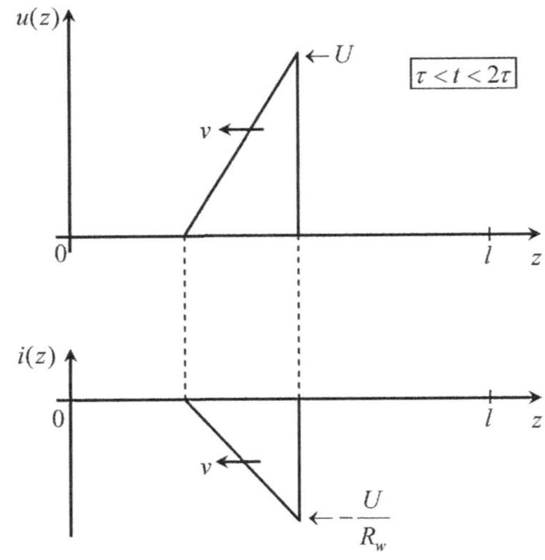

Figure 9.7 Snapshot, at instant t ($\tau < t < 2\tau$), of the line voltage and line current (reflected wave) observed along z

short circuit at $z = 0$. A new reflection phenomenon takes place and a new incident wave is originated, such that

$$u = 0 = u_i + u_r \quad \rightarrow \quad u_i = -u_r \text{ and } i_i = u_i/R_w \tag{9.15}$$

Again, if you take a photo of the line voltage and line current along the line length at an instant $2\tau < t < 3\tau$ you will observe only the propagating incident wave, where the voltage pulse has changed its polarity – see Figure 9.8.

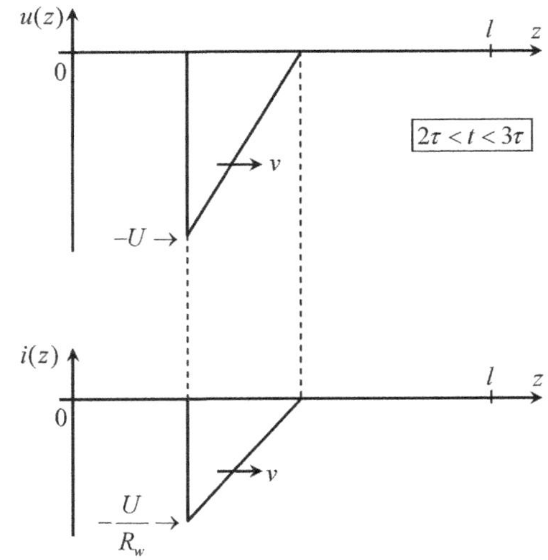

Figure 9.8 Snapshot, at instant t ($2\tau < t < 3\tau$), of the line voltage and line current (incident wave) observed along z

Since the line is lossless and its terminals do not include any dissipative components, reflection phenomena at both ends will occur in succession, and indefinitely.

Before ending this section, an important remark is in order. You have seen in the preceding analysis that current pulse waves travel along the line at a velocity $v = 1/\sqrt{\mu_0 \varepsilon}$, which typically is of the order of 10^8 m/s. Does this mean that the conductors' free electrons are moving at that same speed?

Absolutely not! Electron speeds inside good conductors are of the order of 1 m/s (Chapter 3). So, where is the catch?

Have you ever witnessed the so-called Mexican wave that fans usually perform in football stadiums? They get up and sit down sector after sector in a synchronized way producing a revolving wave around the stadium. Football fans are not running at the wave velocity, indeed they barely move. Well, the same thing happens with the conductor charges. They barely move. When the guided electromagnetic wave travels along the line, the free charges are pulled from their rest position to the conductor surface. After the wave has passed, the free charges return to their rest position.

9.3 Application Example (Parallel-Plate Transmission Line)

Consider a transmission line consisting of two parallel-plate perfect conductors of length l and width w. The space between the plates is filled with a non-magnetic dielectric material of very small thickness δ and permittivity $\varepsilon = \varepsilon_r \varepsilon_0$.

As shown in Figure 9.9, the line is fed at $z = 0$ by a voltage source described by its internal resistance R_G plus an ideal generator. The latter produces a single rectangular voltage pulse of amplitude U and duration ΔT. At the opposite end, $z = l$, the line is terminated by a resistor R_L.

Data: $l = 20\,\text{cm}$, $w = 25.13\,\text{mm}$, $\delta = 1\,\text{mm}$, $\varepsilon_r = 2.25$, $U = 4\,\text{V}$, $\Delta T = 1\,\text{ns}$, $R_L = 30\,\Omega$.

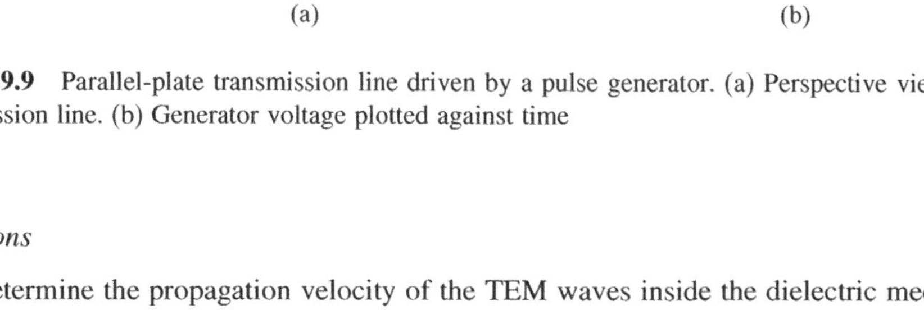

Figure 9.9 Parallel-plate transmission line driven by a pulse generator. (a) Perspective view of the transmission line. (b) Generator voltage plotted against time

Questions

Q_1 Determine the propagation velocity of the TEM waves inside the dielectric medium, v.

Q_2 Determine the per-unit-length capacitance and inductance of the line, C and L_e.

Q_3 Determine the characteristic wave resistance of the line, R_w.

Q$_4$ Assume that $R_G = R_w$.
Obtain the time shape of the voltages $u_0(t)$ and $u_L(t)$.
Check that

$$\int_{-\infty}^{\infty} u_0(t)\, dt = \int_{-\infty}^{\infty} u_L(t)\, dt$$

Evaluate the energy brought into play by the generator as well as the total energy dissipated in the system.

Q$_5$ Repeat the calculations in Q$_4$, but considering that $R_G = 0$.

Solutions

Q$_1$ $v = 1/\sqrt{\mu_0 \varepsilon} = c/\sqrt{\varepsilon_r} = 2 \times 10^8\,\text{m/s}$.

Q$_2$ The electrostatic capacitance of the configuration of length l is $C_l = \varepsilon w l/\delta$. The per-unit-length capacitance of the line is $C = C_l/l = \varepsilon w/\delta = 500\,\text{pF/m}$.

$$v^2 = \frac{1}{L_e C} \to L_e = \frac{1}{v^2 C} = \frac{\mu_0 \delta}{w} = 50\,\text{nH/m}$$

Q$_3$

$$R_w = \sqrt{\frac{L_e}{C}} = \sqrt{\frac{\mu_0}{\varepsilon}} \times \frac{\delta}{w} = 10\,\Omega$$

Q$_4$ We start by noting that the time delay coincides with the pulse duration

$$\tau = \frac{l}{v} = 1\text{ns} = \Delta T$$

When the process begins, only the incident wave is present, $u = u_i$, $i = i_i = u_i/R_w$.
At $z = 0$ we have

$$u_0 = u_G - R_G i_0 \text{ or } u_i = u_G - R_G \frac{u_i}{R_w}$$

Since $R_G = R_w$, we find $u_0 = u_i = u_G/2 = 2\,\text{V}$ and $i_0 = i_i = u_G/(2R_w) = 200\,\text{mA}$.
At $t = \tau$, when the incident wave reaches the load, reflection occurs. The reflected wave is determined using the boundary condition at $z = l$, that is

$$u_L = u_i + u_r,\ i_L = \frac{u_i - u_r}{R_w},\ u_L = R_L i_L \tag{9.16}$$

from which we find

$$u_r = \Gamma u_i = \underbrace{\left(\frac{R_L - R_w}{R_L + R_w}\right)}_{\Gamma} u_i, \quad \Gamma = 0.5 \tag{9.17}$$

where the dimensionless factor Γ is the so-called load reflection coefficient. The reflected wave voltage is a positive pulse with half the magnitude of the incident one, $u_r = 1\,\mathrm{V}$. The reflected wave current is a negative pulse, $i_r = -u_r/R_w = -100\,\mathrm{mA}$.

From (9.16) and (9.17), we obtain the load voltage,

$$u_L = (1+\Gamma)\ u_i = (1+\Gamma)\frac{u_G}{2} = 3\ \mathrm{V}$$

The reflected wave travels along the line towards the generator ($\tau < t < 2\tau$) and when it gets there it finds a matched load and the process ceases (note that the generator voltage is zero for $t > \tau$). The voltage at $z = 0$ is simply determined by the arriving wave u_r, that is $u_0 = 1\,\mathrm{V}$, the corresponding current is $i_0 = -100\,\mathrm{mA}$.

Figure 9.10 summarizes the preceding results, showing the time shape of both $u_0(t)$ and $u_L(t)$:

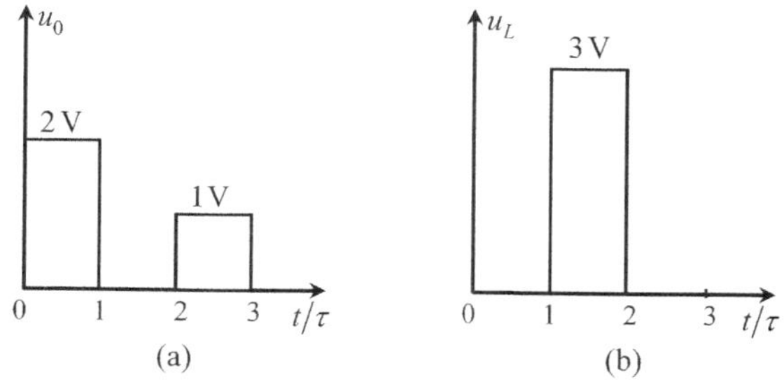

Figure 9.10 Voltages observed at the input and output ends of the line when $R_G = R_w$. (a) Input voltage $u_0(t)$. (b) Output voltage $u_L(t)$

$$\int_{-\infty}^{\infty} u_0(t)\ dt = \int_{0}^{\tau} u_0(t)\ dt + \int_{2\tau}^{3\tau} u_0(t)\ dt = 3\ \mathrm{nWb};\ \int_{-\infty}^{\infty} u_L(t)\ dt = \int_{\tau}^{2\tau} u_L(t)\ dt = 3\ \mathrm{nWb}$$

$$W_{R_G} = \int_{-\infty}^{\infty} R_G i_0^2\ dt = \int_{0}^{\tau} R_G i_0^2\ dt + \int_{2\tau}^{3\tau} R_G i_0^2\ dt = 0.5\ \mathrm{nJ}$$

$$W_{R_L} = \int_{-\infty}^{\infty} \frac{u_L^2}{R_L}\ dt = \int_{\tau}^{2\tau} \frac{u_L^2}{R_L}\ dt = 0.3\ \mathrm{nJ}$$

$$W_G = \int_{-\infty}^{\infty} u_G i_0\ dt = \int_{0}^{\tau} u_G i_0\ dt = 0.8\ \mathrm{nJ}$$

Note that $W_G = W_{R_G} + W_{R_L}$.

Q_5 With $R_G = 0$ we have $u_0(t) = u_G(t)$.

The initially launched incident wave is $u_i = u_G = U = 4\,\mathrm{V}$, $i_i = u_i/R_w = 400\,\mathrm{mA}$.

When the incident wave reaches the load at $t = \tau$, reflection occurs dictated by the same equations as in (9.16) and (9.17), from which we obtain $u_r = \Gamma u_i = \Gamma U = 2\,\mathrm{V}$ and $u_L = (1+\Gamma)U = 6\,\mathrm{V}$.

The reflected wave travels along the line towards the generator ($\tau < t < 2\tau$) and when it gets there it finds a short-circuit condition, $u = 0$, hence

$$u = 0 = u_i + u_r \quad \rightarrow \quad u_i = -u_r$$

A new reflection occurs, and a new incident voltage wave consisting of a negative pulse is produced, $u_i = -u_r = -\Gamma U = -2\,\text{V}$.

For $t = 3\tau$, the negative pulse arrives at the load, where it is reflected back. The new reflected voltage wave is obtained from (9.17), $u_r = \Gamma u_i = -\Gamma^2 U = -1\,\text{V}$. The load voltage is determined from $u_L = (1+\Gamma)u_i = -(1+\Gamma)\Gamma U = -3\,\text{V}$.

The reflected wave travels along the line towards the generator ($3\tau < t < 4\tau$) and when it gets there it again finds a short-circuit condition; consequently, a new incident wave is produced such that $u_i = -u_r = \Gamma^2 U = 1\,\text{V}$.

This wave reaches the load at $t = 5\tau$, and produces a new reflection, the load voltage being given by $u_L = (1+\Gamma)u_i = (1+\Gamma)\Gamma^2 U = 1.5\,\text{V}$.

The process of successive reflections at both ends of the line lasts indefinitely with u_L tending to zero as time passes.

For clarity, we summarize next the results obtained for the load voltage:

For $\tau < t < 2\tau$: $u_L = (1+\Gamma)U = 6\,\text{V}$.
For $3\tau < t < 4\tau$: $u_L = -\Gamma(1+\Gamma)U = -3\,\text{V}$.
For $5\tau < t < 6\tau$: $u_L = \Gamma^2(1+\Gamma)U = 1.5\,\text{V}$.

In general: For $(2n-1)\tau < t < 2n\tau$: $u_L = (-\Gamma)^{n-1}(1+\Gamma)U$, with $n \geq 1$.

Figure 9.11 shows the time shape of both $u_0(t)$ and $u_L(t)$:

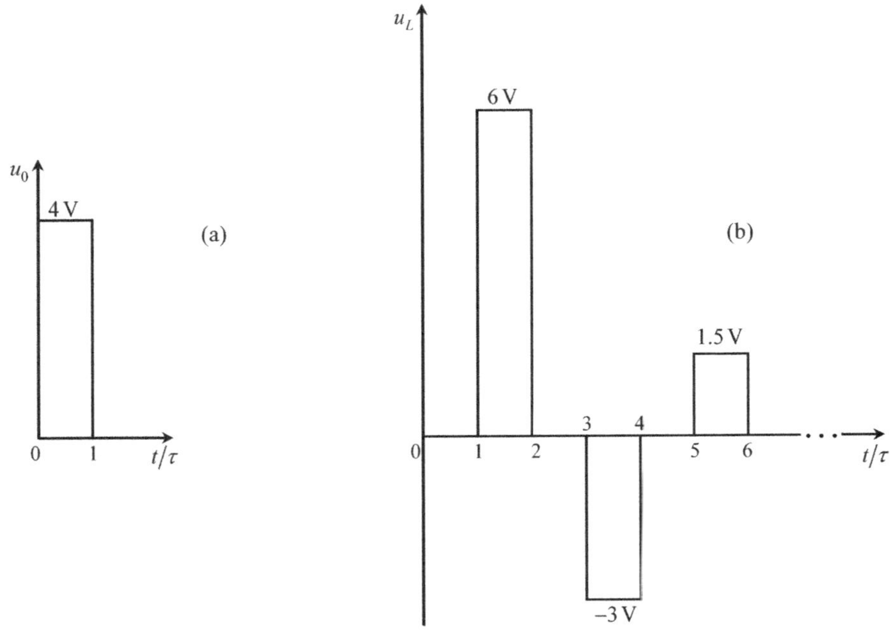

Figure 9.11 Voltages observed at the input and output ends of the line when $R_G = 0$. (a) Input voltage $u_0(t)$. (b) Output voltage $u_L(t)$

$$\int_{-\infty}^{\infty} u_0(t)\, dt = \int_{0}^{\tau} U\, dt = U\tau = 4\,\text{nWb}$$

$$\int_{-\infty}^{\infty} u_L(t)\, dt = \sum_{n=1}^{\infty} \int_{(2n-1)\tau}^{2n\tau} u_L\, dt = (1+\Gamma)U\tau \times \sum_{n=1}^{\infty} (-\Gamma)^{n-1}$$

Taking into account that the geometric series $\sum_{n=1}^{\infty} (-\Gamma)^{n-1}$ converges to $1/(1+\Gamma)$, we conclude that $\int_{-\infty}^{\infty} u_L(t)dt = U\tau$, as expected:

$$W_G = \int_{-\infty}^{\infty} u_G i_0\, dt = \int_{0}^{\tau} u_G i_0\, dt = \frac{U^2}{R_w}\tau = 1.6\,\text{nJ}$$

$$W_{R_L} = \int_{-\infty}^{\infty} \frac{u_L^2}{R_L}\, dt = \frac{1}{R_L}\sum_{n=1}^{\infty} \int_{(2n-1)\tau}^{2n\tau} u_L^2\, dt = \frac{(1+\Gamma)^2 U^2}{R_L}\tau \times \sum_{n=1}^{\infty} \Gamma^{2(n-1)}$$

Taking into account that the geometric series $\sum_{n=1}^{\infty} \Gamma^{2(n-1)}$ converges to $1/(1-\Gamma^2)$, and noting, from (9.17), that

$$R_L = R_w \frac{1+\Gamma}{1-\Gamma}$$

we conclude

$$W_{R_L} = \frac{(1+\Gamma)^2 U^2}{(1-\Gamma^2) R_L}\tau = \frac{U^2}{R_w}\tau = W_G$$

9.4 Frequency-Domain Transmission-Line Equations for Lossy Lines

Actual transmission-line structures are made of imperfect conductors and imperfect dielectric media, which are a cause of undesirable losses. Regrettably, the perturbations introduced by loss mechanisms cannot be easily taken into account directly in the time-domain transmission-line equations. To help you understand the difficulty, it may suffice to recall what you learnt in Section 8.7 and Problem 8.8.5 about the skin-effect phenomenon in imperfect conductors. In Section 8.7 we established, in the frequency domain, an equation for the per-unit-length impedance of a circular cylindrical conductor, $\overline{Z}(\omega) = R(\omega) + j\omega L_i(\omega)$. Nonetheless, in Problem 8.8.5, we showed that such an equation does not translate into the time domain as $u = Ri + L_i\, di/dt$ (that would only be possible if R and L_i were frequency-independent parameters, which, clearly, is not the case).

Therefore, our approach to the analysis of lossy lines is conducted entirely in the frequency domain, where line voltages and line currents are sinusoidal functions described by the corresponding phasors $\overline{U}(z)$ and $\overline{I}(z)$, that is

$$u(z,t) = \Re\left(\overline{U}(z)\, e^{j\omega t}\right), \quad i(z,t) = \Re\left(\overline{I}(z)\, e^{j\omega t}\right)$$

A word of advice: since the assumption of perfect conductors has been dropped, a longitudinal component for the electric field does now exist. The electromagnetic field structure is no longer of the TEM type, and some ambiguity arises in the definition of line voltages. However, for good conductors, the longitudinal component of the **E** field is, in general, much smaller than the transverse component. This allows us to keep using the same formalism we established for lossless lines. But, in any case, you must realize that this approach is in fact an approximation (quasi-TEM approach).

9.4.1 Per-Unit-Length Longitudinal Impedance, Per-Unit-Length Transverse Admittance

Let us return to the (lossless) transmission-line equation in (9.5), which was obtained considering the application of the induction law: $\partial u/\partial z = -L_e \partial i/\partial t$. The corresponding frequency-domain equation is

$$\frac{d}{dz}\overline{U}(z) = -j\omega L_e \overline{I}(z) \tag{9.18}$$

In order to take into account the skin-effect phenomena in the two imperfect conductors of the line, two additional contributions should be added to the right-hand side of this equation. Those contributions are associated with the longitudinal voltage drops along line conductors 1 and 2,

$$\int_{\overrightarrow{bc}} \mathbf{E} \cdot d\mathbf{s} \text{ and } \int_{\overrightarrow{da}} \mathbf{E} \cdot d\mathbf{s}$$

which were null in the case of perfect conductors (see again Figure 9.3).

Therefore, if we add the per-unit-length conductor impedances $\overline{Z}_1(\omega) = R_1(\omega) + j\omega L_1(\omega)$ and $\overline{Z}_2(\omega) = R_2(\omega) + j\omega L_2(\omega)$ as perturbations to the main term $j\omega L_e$ in (9.18), we obtain the generalized equation

$$\frac{d}{dz}\overline{U}(z) = -\{\underbrace{(R_1(\omega) + R_2(\omega))}_{R} + j\omega\underbrace{(L_1(\omega) + L_2(\omega) + L_e)}_{L}\}\overline{I}(z) \tag{9.19}$$

This result can be written compactly if we introduce, as is usual, the so-called longitudinal impedance per unit length of the line $\overline{Z}_l$ (units: Ω/m)

$$\overline{Z}_l = R(\omega) + j\omega L(\omega) = Z_l \, e^{j(\pi/2 - \delta_Z)} \tag{9.20}$$

where δ_Z is the longitudinal loss angle.

Now, let us return to the (lossless) transmission-line equation in (9.6), which was obtained considering the application of the charge continuity equation: $\partial i/\partial z = -C\partial u/\partial t$. The corresponding frequency-domain equation is

$$\frac{d}{dz}\overline{I}(z) = -j\omega C\, \overline{U}(z) \tag{9.21}$$

Dielectric losses can occur because of two distinct mechanisms: one is the presence of transverse conduction currents across the imperfect insulation medium ($\sigma_{\text{dielectric}} \neq 0$), the other is the periodic polarization of the dielectric material – a phenomenon that can be accounted for by replacing the permittivity ε by a complex permittivity $\overline{\varepsilon} = \varepsilon'(\omega) - j\varepsilon''(\omega)$; see Section 8.4.

The first contribution can be included in the right-hand side of (9.21) by adding a perturbation term $G_\sigma \overline{U}$. The second contribution can be incorporated into (9.21) by replacing C by a complex per-unit-length capacitance $\overline{C} = C' - jC''$, where, typically, $C'' \ll C'$ and $C' \simeq C$. On doing this we obtain the generalized equation

$$\frac{d}{dz}\overline{I}(z) = -\{\underbrace{(G_\sigma + \omega C''\omega)}_{G} + j\omega C\}\overline{U}(z) \tag{9.22}$$

This result can be written compactly if we introduce, as is usual, the so-called transverse admittance per unit length of the line $\overline{Y}_t$ (units: S/m)

$$\overline{Y}_t = G(\omega) + j\omega C = Y_t\ e^{j(\pi/2-\delta_Y)} \tag{9.23}$$

where δ_Y is the transversal loss angle.

In short, the frequency-domain transmission-line equations for the steady-state harmonic analysis of lossy lines read as

$$\frac{d}{dz}\overline{U}(z) = -\overline{Z}_l(\omega)\overline{I}(z) \tag{9.24}$$

$$\frac{d}{dz}\overline{I}(z) = -\overline{Y}_t(\omega)\overline{U}(z) \tag{9.25}$$

9.4.2 *Propagation Constant, Phase Velocity, Characteristic Wave Impedance*

Let us seek the solutions to the pair of equations in (9.24) and (9.25).

Taking the z derivative of (9.24) and substituting $-\overline{Y}_t(\omega)\overline{U}(z)$ for $d\overline{I}/dz$ yields

$$\frac{d^2\overline{U}(z)}{dz^2} - \overline{Z}_l\overline{Y}_t\overline{U}(z) = 0 \tag{9.26}$$

This is a homogeneous linear differential equation with constant coefficients of order 2.

Note that we have already dealt with equations of this type in the analysis of transient regimes (Chapter 7); the only difference is that, now, the equation is on z (not on t). Since this a familiar subject we can skip some details, and at once write

$$\text{Characteristic equation}: s^2 - \overline{Z}_l\overline{Y}_t = 0$$

$$\text{Characteristic roots}: s_{1,2} = \pm\overline{\gamma}, \quad \overline{\gamma} = \sqrt{\overline{Z}_l\overline{Y}_t} \tag{9.27}$$

$$\text{Line voltage solution}: \overline{U}(z) = \overline{U}_i e^{-\overline{\gamma}z} + \overline{U}_r e^{+\overline{\gamma}z} \tag{9.28}$$

In (9.27) the complex constant $\overline{\gamma}$ is designated a propagation constant

$$\overline{\gamma}(\omega) = \sqrt{\overline{Z}_l\overline{Y}_t} = \sqrt{Z_l Y_t}\ e^{j(\pi-\delta_Z-\delta_Y)/2} = \alpha + j\beta \tag{9.29}$$

Since both the loss angles δ_Z and δ_Y are confined to the range $[0,\ \pi/2]$, the propagation constant must necessarily belong to the first quadrant of the complex plane and, consequently, α and β are positive numbers; α is the so-called attenuation constant (units: Np/m), and β is the so-called phase constant (units: rad/m).

As shown in (9.28), the voltage solution is made up of two contributions, the first associated to the incident wave and the second to the reflected wave. The constants $\overline{U}_i$ and $\overline{U}_r$ are to be determined upon imposition of the pertinent boundary conditions at the ends of the line.

The solution for the line current $\overline{I}(z)$ is determined from (9.24),

$$\overline{I}(z) = -\frac{1}{\overline{Z}_l}\frac{d\overline{U}(z)}{dz}$$

yielding

$$\text{Line current solution:}\quad \overline{I}(z) = \frac{\overline{U}_i e^{-\overline{\gamma}z} - \overline{U}_r e^{+\overline{\gamma}z}}{\overline{Z}_w} \tag{9.30}$$

where $\overline{Z}_w$ is the characteristic wave impedance of the line (units: Ω) – a concept that, in the frequency domain, can be viewed as the generalization of the characteristic wave resistance introduced in Section 9.2.1 for lossless line analysis,

$$\overline{Z}_w(\omega) = \sqrt{\frac{\overline{Z}_l}{\overline{Y}_t}} = Z_w\ e^{j\theta_w},\ \ Z_w = \sqrt{\frac{Z_l}{Y_t}},\ \ \theta_w = \frac{\delta_Y - \delta_Z}{2} \tag{9.31}$$

Since both the loss angles δ_Z and δ_Y are confined to the range $[0,\ \pi/2]$, the characteristic wave impedance must belong to the first or eighth octant of the complex plane, that is $\theta_w \in [-\pi/4,\ \pi/4]$ and $\Re\left(\overline{Z}_w\right) > 0$.

As in our prior discussion on lossless lines, here too we can state that in order to avoid the presence of the reflected wave, $\overline{U}_r = 0$ (matched line case), all you have to do is to terminate the line by a load impedance equal to the characteristic wave impedance, $\overline{Z}_L = \overline{Z}_w$. Nevertheless, you should bear in mind that $\overline{Z}_w$ is, now, a frequency-dependent parameter and, therefore, if the working frequency changes you will most likely fail with perfect matching (unless you properly readjust the load impedance value).

To better grasp the significance of the wave parameters $\overline{\gamma}$ and $\overline{Z}_w$ introduced in this section, we are going to obtain space–time descriptions of $u(z, t)$ and $i(z, t)$. To simplify matters we assume that the working frequency and the line termination are such that the line is perfectly matched, that is

$$\overline{U}(z) = \overline{U}_i\ e^{-(\alpha+j\beta)z},\ \text{with}\ \overline{U}_i = U\ e^{j\phi_u}$$

$$\overline{I}(z) = \frac{\overline{U}_i\ e^{-(\alpha+j\beta)z}}{Z_w\ e^{j\theta_w}}$$

Taking into account that $u(z, t) = \Re\left(\overline{U}(z)\ e^{j\omega t}\right)$ and $i(z, t) = \Re\left(\overline{I}(z)\ e^{j\omega t}\right)$, we get

$$u(z, t) = U\ e^{-\alpha z}\cos(\omega t - \beta z + \phi_u) \tag{9.32a}$$

$$i(z, t) = I\ e^{-\alpha z}\cos(\omega t - \beta z + \phi_i) \tag{9.32b}$$

where $I = U/Z_w$ and $\phi_i = \phi_u - \theta_w$.

Since the preceding expressions are very similar, let us pay attention to only one of them, for example the line voltage $u(z, t)$.

The phase under the cosine function depends on z and t, $\phi(z, t) = \omega t - \beta z + \phi_u$. At which speed v should you run along the line in order to always observe the same phase value?

By making $z = z_0 + vt$ you find the condition

$$\omega t - \beta v t - \beta z_0 + \phi_u = \text{time-invariant constant}$$

from which you readily get

$$v = \frac{\omega}{\beta} \tag{9.33}$$

The velocity defined in (9.33) is the so-called phase velocity. It coincides with the propagation velocity defined in (9.8) and (9.12) in the case of lossless lines.

Next, consider that, at a given instant t_0, you take a photo of the voltage distribution along the line. You will observe an oscillating function on z with decaying amplitude – see Figure 9.12.

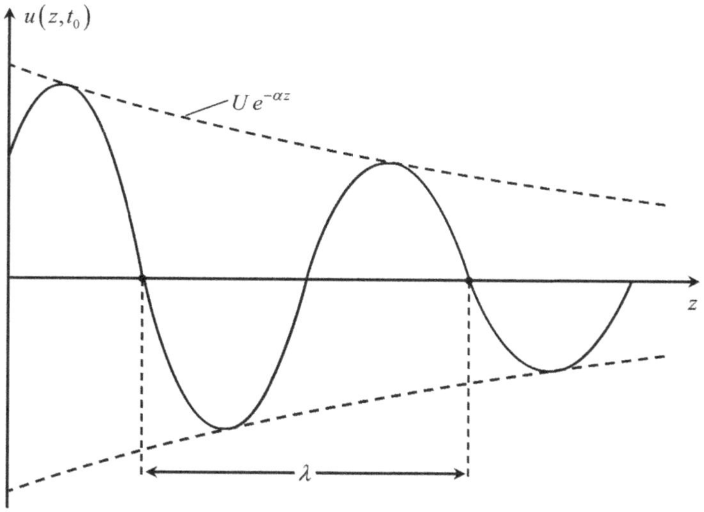

Figure 9.12 Snapshot, taken at $t = t_0$, of the line voltage distribution along the axial coordinate z, for the case of a lossy line submitted to a sinusoidal steady-state regime

The rate of decay is determined by the attenuation constant α. On the other hand, the periodicity of the oscillating function is determined by the phase constant β. In fact, the space period λ (wavelength) is such that the product $\beta\lambda$ must equal 2π, that is

$$\lambda = \frac{2\pi}{\beta} = \frac{2\pi v}{\omega} = \frac{v}{f}$$

which, manifestly, is no surprise.

To end this section, imagine that you place an oscilloscope at $z = z_0$ to observe the time evolution of $u(z_0, t)$. The image you will obtain on the screen is shown in Figure 9.13; it is a pure sinusoidal function with a time period $T = \omega/(2\pi) = 1/f$, which displays no decay at all. Pay attention to the fact that, according to (9.32), the attenuation constant affects the function evolution on z, but not on t. However, if you place another oscilloscope at $z_1 > z_0$ you will observe another sinusoidal time function with smaller magnitude (see again Figure 9.13).

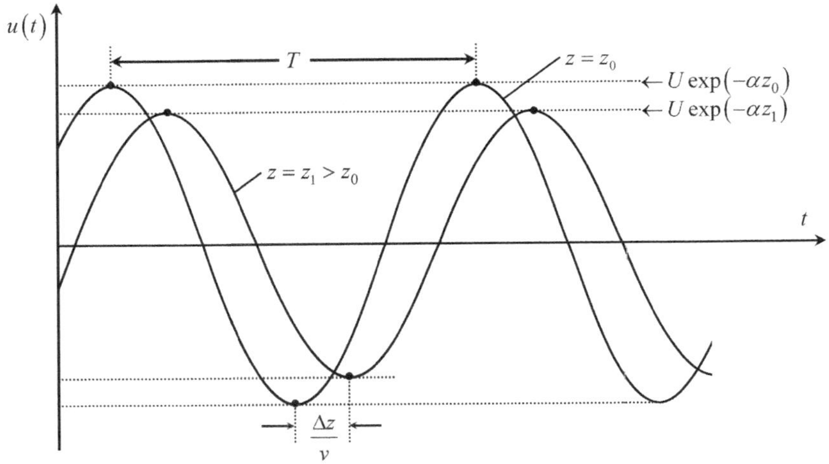

Figure 9.13 Oscilloscope readings of the time evolution of the line voltage at two different places $z = z_0$ and $z = z_1 = z_0 + \Delta z$

9.4.3 Transfer Matrix, Non-Uniform Line Analysis

In many applications a detailed knowledge of the voltage and current distributions along the line is of little concern. Quite often the line is simply treated as a black box, a two-port network, where only the voltages and currents at the accessible ends really matter – see Figure 9.14.

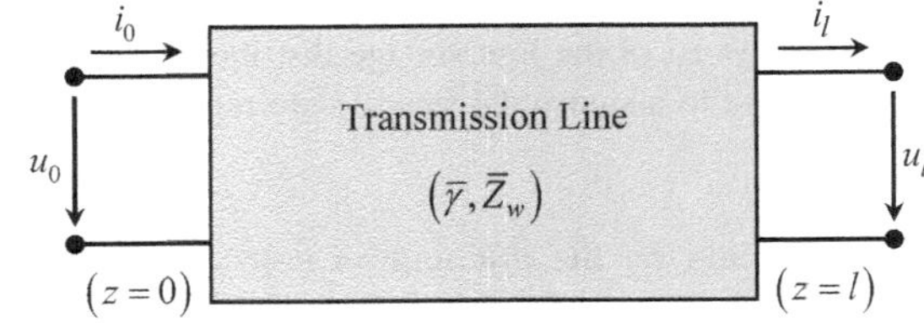

Figure 9.14 A two-port, black box representation of a transmission line of length l

Consider a transmission-line section of length l where $\overline{U}_0$ and $\overline{I}_0$ are the input quantities, and $\overline{U}_l$ and $\overline{I}_l$ are the output quantities. Our next goal is to establish a matrix relationship among those quantities, that is

$$\begin{bmatrix} \overline{U}_0 \\ \overline{I}_0 \end{bmatrix} = [T] \begin{bmatrix} \overline{U}_l \\ \overline{I}_l \end{bmatrix} \tag{9.34}$$

where $[T]$ is the transfer matrix (also termed the chain matrix).

Making use of (9.28) and (9.30) we get

$$z = l: \begin{cases} \overline{U}_l = \overline{U}_i e^{-\overline{\gamma} l} + \overline{U}_r e^{+\overline{\gamma} l} \\ \overline{Z}_w \overline{I}_l = \overline{U}_i e^{-\overline{\gamma} l} - \overline{U}_r e^{+\overline{\gamma} l} \end{cases} \rightarrow \begin{cases} \overline{U}_i = \frac{1}{2}(\overline{U}_l + \overline{Z}_w \overline{I}) e^{+\overline{\gamma} l} \\ \overline{U}_r = \frac{1}{2}(\overline{U}_l + \overline{Z}_w \overline{I}) e^{-\overline{\gamma} l} \end{cases}$$

$$z = 0: \begin{cases} \overline{U}_0 = \overline{U}_i + \overline{U}_r = \cosh(\overline{\gamma} l) \overline{U}_l + \sinh(\overline{\gamma} l) \overline{Z}_w \overline{I}_l \\ \overline{Z}_w \overline{I}_0 = \overline{U}_i - \overline{U}_r = \sinh(\overline{\gamma} l) \overline{U}_l + \cosh(\overline{\gamma} l) \overline{Z}_w \overline{I}_l \end{cases}$$

from which the transfer matrix is determined

$$[T] = \begin{bmatrix} \cosh(\overline{\gamma} l) & \sinh(\overline{\gamma} l) \overline{Z}_w \\ \overline{Z}_w^{-1} \sinh(\overline{\gamma} l) & \cosh(\overline{\gamma} l) \end{bmatrix} \tag{9.35}$$

The transfer matrix is a precious tool for handling non-uniform line analysis. Imagine that you have a transmission-line structure whose transverse profile changes along z (see the example in Figure 9.15). You may, conceptually, break down the structure into a chain of N small line sections, of appropriate lengths, so that each and every section can be considered a uniform line.

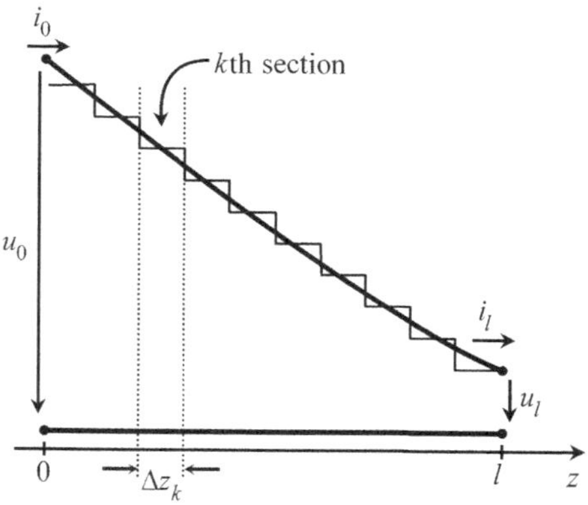

Figure 9.15 Segmentation technique ordinarily used for the analysis of non-uniform lines

Once you have evaluated the transfer matrices $[T_1], \ldots, [T_k], \ldots, [T_N]$ pertaining to the individual line sections, you can finally determine the global transfer matrix $[T]$ of the non-uniform structure through

$$\begin{bmatrix} \overline{U}_0 \\ \overline{I}_0 \end{bmatrix} = \underbrace{[T_1]\ [T_2]\ [T_3] \ldots [T_N]}_{[T]} \begin{bmatrix} \overline{U}_l \\ \overline{I}_l \end{bmatrix}$$

The accuracy of this method obviously depends on the number of line sections you have utilized. The thinner the segmentation, the better the results.

9.5 Frequency-Domain Transmission-Line Equations for Lossless Lines

In a lossy line, the attenuation and phase constants increase with the frequency, but while $\alpha \propto \sqrt{f}$, we have $\beta \propto f$. This means that the loss angles

$$\delta_Z + \delta_Y = 2\arctan(\alpha/\beta)$$

ordinarily tend to zero as the frequency increases. Therefore, for very high-frequency regimes, it makes sense to consider the approximations $\overline{Z}_l \to j\omega L_e$ and $\overline{Y}_t \to j\omega C$, which in turn imply

$$\overline{\gamma}(\omega) = \sqrt{\overline{Z}_l \overline{Y}_t} = j\omega\sqrt{L_e C} = 0 + j\beta$$

$$\overline{Z}_w = \sqrt{\frac{\overline{Z}_l}{\overline{Y}_t}} = \sqrt{\frac{L_e}{C}} = R_w + j0$$

This means that we are back again to the framework of lossless-line analysis. Consequently, the transmission-line solutions established in (9.28) and (9.30) greatly simplify:

$$\overline{U}(z) = \overline{U}_i e^{-j\beta z} + \overline{U}_r e^{+j\beta z} \tag{9.36a}$$

$$\overline{I}(z) = \frac{\overline{U}_i e^{-j\beta z} - \overline{U}_r e^{+j\beta z}}{R_w} \tag{9.36b}$$

where $\overline{U}_i$ and $\overline{U}_r$ denote, respectively, the complex amplitudes of the incident and reflected voltage waves at the line sending end, $z = 0$.

9.5.1 Terminated Line, Load Reflection Coefficient, Line Input Impedance

Consider a transmission line of length l driven by a sinusoidal voltage generator at one end and terminated at the opposite end by a passive load impedance $\overline{Z}_L$.

For the analysis of terminated lines it is frequent, and convenient, to introduce a longitudinal y axis with origin at the load, antiparallel to z – see Figure 9.16.

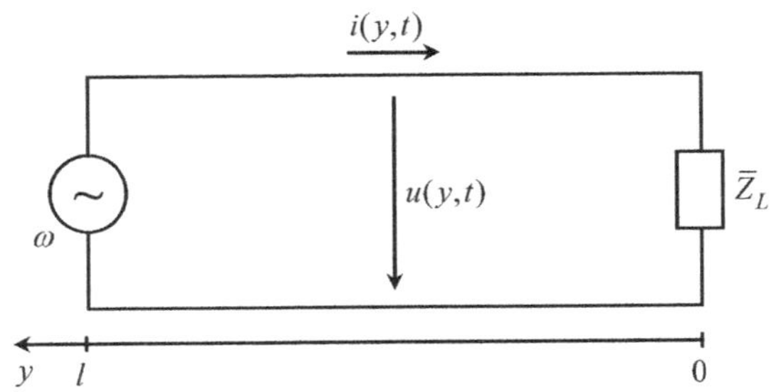

Figure 9.16 Terminated transmission line showing the longitudinal y axis starting at the load terminals

Taking into account the change of variable, $y = l - z$, the solution in (9.36) reads as

$$\overline{U}(y) = \overline{U}'_i \; e^{+j\beta y} + \overline{U}'_r \; e^{-j\beta y} \tag{9.37a}$$

$$\overline{I}(y) = \frac{\overline{U}'_i \; e^{+j\beta y} - \overline{U}'_r \; e^{-j\beta y}}{R_w} \tag{9.37b}$$

where $\overline{U}'_i = \overline{U}_i \; e^{-j\beta l}$ and $\overline{U}'_r = \overline{U}_r \; e^{+j\beta l}$ respectively denote the incident wave voltage and reflected wave voltage phasors, measured at the load terminals. A complex dimensionless number, the load reflection coefficient $\overline{\Gamma}$, is defined as the ratio of those wave phasors, that is

$$\overline{\Gamma} = \frac{\overline{U}'_r}{\overline{U}'_i} \tag{9.38}$$

Taking into account the definition of $\overline{\Gamma}$, we can rewrite (9.37) as

$$\overline{U}(y) = \overline{U}'_i \; e^{+j\beta y} \times \overbrace{(1 + \overline{\Gamma} \; e^{-j2\beta y})}^{\overline{V}_U} \tag{9.39a}$$

$$\overline{I}(y) = \frac{\overline{U}'_i \; e^{+j\beta y}}{R_w} \times \underbrace{(1 - \overline{\Gamma} \; e^{-j2\beta y})}_{\overline{V}_I} \tag{9.39b}$$

On the right-hand side of (9.39) the leading factor is to be interpreted as the incident wave contribution; the factors on the right, $\overline{V}_U$ and $\overline{V}_I$, are complex vectors where the influence of the terminating load is included via $\overline{\Gamma}$. Those two auxiliary vectors are such that

$$\sphericalangle \overline{V}_U, \overline{V}_I = \sphericalangle \overline{U}, \overline{I} = \varphi(y); \qquad \frac{\overline{V}_U + \overline{V}_I}{2} = 1 \tag{9.40}$$

If the equations in (9.39) are particularized to the load terminals, $y = 0$, we get

$$\overline{U}_L = \overline{U}'_i (1 + \overline{\Gamma}); \; \overline{I}_L = \frac{\overline{U}'_i}{R_w} (1 - \overline{\Gamma})$$

Dividing the two equations above, we find

$$\overline{Z}_L = R_w \frac{1 + \overline{\Gamma}}{1 - \overline{\Gamma}} \tag{9.41}$$

Solving (9.41) for $\overline{\Gamma}$, we obtain

$$\overline{\Gamma} = \Gamma \; e^{j\theta} = \frac{\overline{Z}_L - R_w}{\overline{Z}_L + R_w} \tag{9.42}$$

For passive loads (consisting of R, L, C arrangements), where $\Re\left(\overline{Z}_L\right) \geq 0$, we have $\Gamma \leq 1$ and $0 \leq \theta \leq 2\pi$. The physical reason behind the fact that Γ cannot exceed unity is that the energy reflected by the load cannot exceed the energy carried by the incident wave.

Total reflection ($\Gamma = 1$) can only occur in the case of non-dissipative loads, that is

$$\text{Short circuit}: \overline{Z}_L = 0 \rightarrow \overline{\Gamma} = 1\ e^{j\pi}$$

$$\text{Open circuit}: \overline{Z}_L = \infty \rightarrow \overline{\Gamma} = 1$$

$$\text{Reactive load}: \overline{Z}_L = jX_L \rightarrow \overline{\Gamma} = 1\ e^{j[\pi - 2\arctan(X_L/R_w)]}$$

Total absorption ($\Gamma = 0$) happens when the line is perfectly matched, $\overline{Z}_L = R_w$.

The line input impedance $\overline{Z}_{in}$, defined as the ratio $\overline{U}/\overline{I}$ at $y = l$, depends on the frequency, on the line length and on the line load. If the equations in (9.39) are particularized to the input end, we get

$$\overline{U}_{in} = \overline{U}_i' e^{j\beta l}(1 + \overline{\Gamma}\ e^{-j2\beta l}) \text{ and } \overline{I}_{in} = \left[\overline{U}_i'\ e^{j\beta l}(1 - \overline{\Gamma}\ e^{-j2\beta l})\right]/R_w$$

Hence

$$\overline{Z}_{in} = \frac{\overline{U}_{in}}{\overline{I}_{in}} = R_w \frac{1 + \overline{\Gamma}\ e^{-j2\beta l}}{1 - \overline{\Gamma}\ e^{-j2\beta l}} \tag{9.43}$$

9.5.2 *Matched Line, Open Line, Short-Circuited Line*

We now particularize the results from Section 9.5.1 to three particularly important cases.

9.5.2.1 Matched Line ($\overline{Z}_L = R_w$, $\overline{\Gamma} = 0$)

In the frequency domain:

$$\begin{cases} \overline{U}(y) = \overline{U}_i' e^{j\beta y} \\ \overline{I}(y) = \overline{U}_i' e^{j\beta y}/R_w \end{cases} \tag{9.44}$$

where $\overline{U}_i' = U'\ e^{j\phi_u'}$.

The input impedance is

$$\overline{Z}_{in} = \frac{\overline{U}_{y=l}}{\overline{I}_{y=l}} = R_w$$

In the time domain:

$$\begin{cases} u(y,t) = U' \cos(\omega t + \beta y + \phi_u') \\ i(y,t) = \dfrac{U'}{R_w} \cos(\omega t + \beta y + \phi_u') \end{cases} \tag{9.45}$$

In a matched line, the input impedance $\overline{Z}_{in}$ is equal to the characteristic wave resistance; it depends neither on the frequency nor on the line length. Since the reflected wave is absent, the propagation phenomena along the line are entirely described by the incident wave component. In this case we say we are dealing with a purely traveling wave. If you take two photos of $u(y)$, at $t = t_1$ and $t = t_2 > t_1$, you will observe two sinusoidal functions of the same magnitude but shifted by a distance $\Delta y = v(t_2 - t_1)$ – see Figure 9.17.

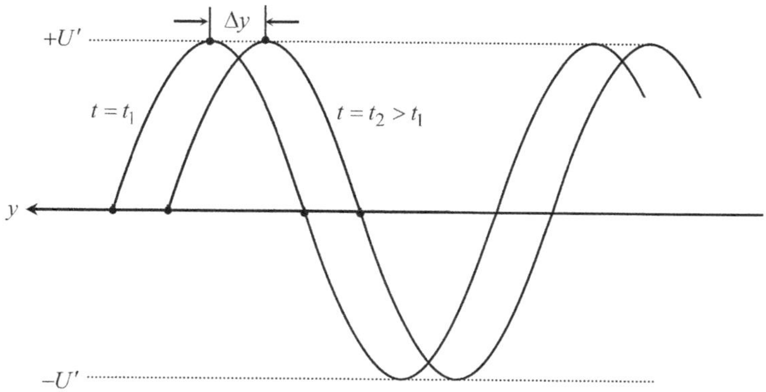

Figure 9.17 Snapshots, taken at $t = t_1$ and $t = t_2 > t_1$, of the line voltage distribution along the axial coordinate y, for the case of a matched lossless line (traveling wave). $\Delta y = v(t_2 - t_1)$

9.5.2.2 Open Line ($\overline{Z}_L = \infty$, $\overline{\Gamma} = 1$)

In the frequency domain:

$$\begin{cases} \overline{U}(y) = \overline{U}'_i \left(e^{+j\beta y} + e^{-j\beta y}\right) \\ \overline{I}(y) = \dfrac{\overline{U}'_i}{R_w} \left(e^{+j\beta y} - e^{-j\beta y}\right) \end{cases} \tag{9.46a}$$

or, which is the same thing,

$$\begin{cases} \overline{U}(y) = 2\overline{U}'_i \cos(\beta y) \\ \overline{I}(y) = \dfrac{2j\overline{U}'_i}{R_w} \sin(\beta y) \end{cases} \tag{9.46b}$$

where $\overline{U}'_i = U' e^{j\phi'_u}$.

The input impedance is

$$\overline{Z}_{in} = \frac{\overline{U}_{y=l}}{\overline{I}_{y=l}} = \frac{-jR_w}{\tan(\beta l)} \tag{9.47}$$

In the time domain:

$$\begin{cases} u(y,t) = 2U' \cos(\omega t + \phi'_u) \cos(\beta y) \\ i(y,t) = 2\dfrac{U'}{R_w} \cos(\omega t + \phi'_u + \pi/2) \sin(\beta y) \end{cases} \tag{9.48}$$

In an open-line situation the input impedance is purely reactive, its sign depending on the value of the product $\beta l = 2\pi l/\lambda$. For instance, if $l = \lambda/4$ you will find $\beta l = \pi/2$ and the input impedance is zero; the generator at the sending end of an open line faces a short-circuit situation at its own terminals! On the contrary, if $l = \lambda/2$ the generator will face an open-circuit situation. See Figure 9.18.

For those of you who are studying transmission lines for the first time, the preceding results must be shocking and unexpected. However, you must realize that transmission-line

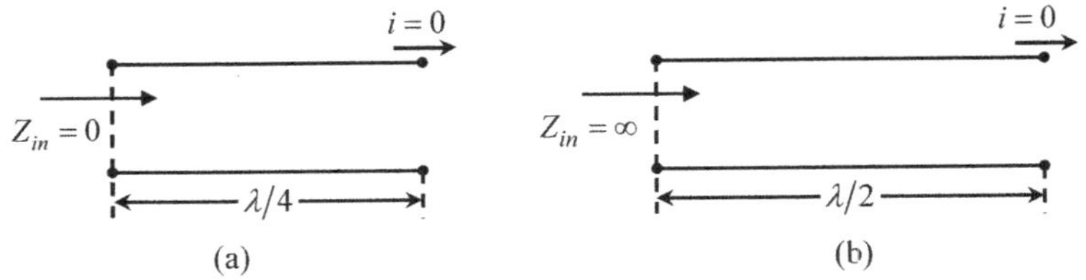

Figure 9.18 Lossless transmission line terminated by an open circuit. (a) The input impedance is zero when $l = \lambda/4$. (b) The input impedance is infinite when $l = \lambda/2$

behavior is completely inexplicable if you insist on using the ordinary tools and concepts from circuit analysis (which pertain to slow time-varying field phenomena, and are not valid here).

As you can see from (9.46a), the line voltage and line current solutions are the result of two counter-propagating waves, both of the same magnitude. These two waves interact with each other giving rise to constructive or destructive interference at certain critical points along the y axis. If you compare the time-domain solutions in (9.48) and (9.45) you will see that in the latter the space and time variables are linearly combined under the cosine function (which is typical of traveling waves); however, in (9.48) the space and time variables appear separated under different trigonometric functions. When two counter-propagating traveling waves of the same magnitude add together they give rise to a purely stationary wave, that is a wave that pulses but does not move. In fact, if you take a series of photos of $i(y)$ at different time instants you will observe the result shown in Figure 9.19.

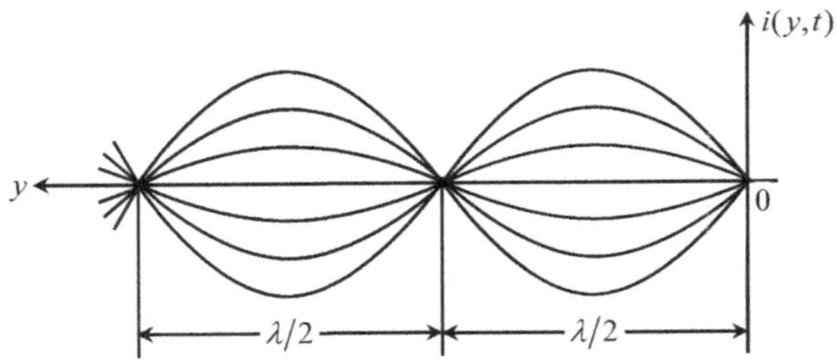

Figure 9.19 Successive snapshots, taken at different time instants, of the line current distribution along the axial coordinate y, for the case of an open lossless line (stationary wave)

In addition you may note, from (9.48), that at certain instants of time the line current is zero everywhere along the line length, the same thing happening to the line voltage, but at other instants:

$$\begin{cases} i(y,t) = 0; \text{ for } \omega t = (k-1)\dfrac{\pi}{2} - \phi'_u \\ u(y,t) = 0; \text{ for } \omega t = k\dfrac{\pi}{2} - \phi'_u \end{cases}$$

with k an odd integer.

9.5.2.3 Short-Circuited Line ($\overline{Z}_L = 0,\ \overline{\Gamma} = -1$)

This case can be considered the dual of the preceding situation, where the roles of u and i are interchanged.

In the frequency domain:

$$\begin{cases} \overline{U}(y) = \overline{U}_i' \left(e^{+j\beta y} - e^{-j\beta y}\right) \\ \overline{I}(y) = \dfrac{\overline{U}_i'}{R_w} \left(e^{+j\beta y} + e^{-j\beta y}\right) \end{cases} \tag{9.49a}$$

or, which is the same thing,

$$\begin{cases} \overline{U}(z) = 2j\overline{U}_i' \sin(\beta y) \\ \overline{I}(y) = \dfrac{2\overline{U}_i'}{R_w} \cos(\beta y) \end{cases} \tag{9.49b}$$

where $\overline{U}_i' = U' e^{j\phi_u'}$.

The input impedance is

$$\overline{Z}_{in} = \frac{\overline{U}_{y=l}}{\overline{I}_{y=l}} = jR_w \tan(\beta l) \tag{9.50}$$

In the time domain:

$$\begin{cases} u(y, t) = 2U' \cos(\omega t + \phi_u' + \pi/2) \sin(\beta y) \\ i(y, t) = 2\dfrac{U'}{R_w} \cos(\omega t + \phi_u') \cos(\beta y) \end{cases} \tag{9.51}$$

In a short-circuited line the input impedance is purely reactive, its sign depending on the value of the product $\beta l = 2\pi l/\lambda$. For instance, if $l = \lambda/4$ you will find $\beta l = \pi/2$ and the input impedance goes to ∞; the generator at the sending end of a short-circuited line faces an open-circuit situation at its own terminals! On the contrary, if $l = \lambda/2$ the generator will face a short-circuit situation. See Figure 9.20.

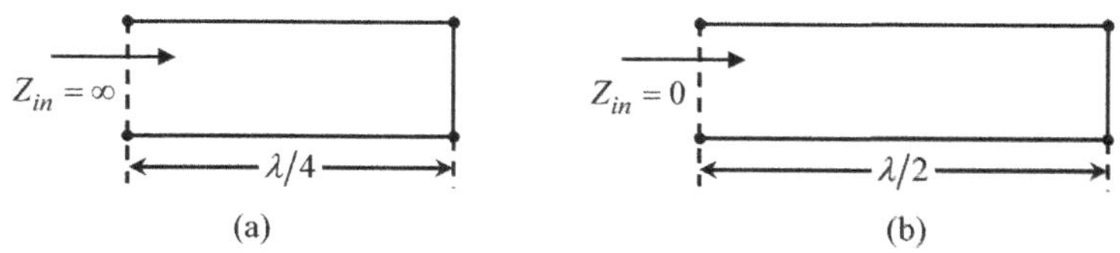

Figure 9.20 Lossless transmission line terminated by a short circuit. (a) The input impedance is infinite when $l = \lambda/4$. (b) The input impedance is zero when $l = \lambda/2$

Again, as in the case of an open line, here too the line voltage and line current are described by purely stationary waves. Apart from a vertical scale factor, the graph for $i(y)$ in Figure 9.19 could be used here to reproduce the evolution $u(y)$ at several instants of time.

9.5.3 Standing Wave Pattern, Standing Wave Ratio, Active Power

In the case of a matched line, if you record the rms values of the line voltage and line current distributions along the length of the line you will obtain constant readings. In fact, from (9.44), you have

$$\begin{cases} U_{rms}(y) = \dfrac{|\overline{U}(y)|}{\sqrt{2}} = \dfrac{U_i}{\sqrt{2}} = (U_i)_{rms} = \text{constant} \\ I_{rms}(y) = \dfrac{|\overline{I}(y)|}{\sqrt{2}} = \dfrac{(U_i)_{rms}}{R_w} = \text{constant} \end{cases}$$

However, when the line is mismatched, the superposition of the incident and reflected waves gives rise to more or less pronounced fluctuations in both $U_{rms}(y)$ and $I_{rms}(y)$. The larger the magnitude of the reflection coefficient Γ, the more important the fluctuation. In fact, the complex vectors $\overline{V}_U$ and $\overline{V}_I$ introduced in (9.39), which vary with y, have their minimum and maximum magnitudes given by $1-\Gamma$ and $1+\Gamma$, the overall fluctuation being 2Γ. In addition, note that when $|\overline{V}_U|$ is at a maximum, $|\overline{V}_I|$ is at a minimum, and vice versa – see Figure 9.21.

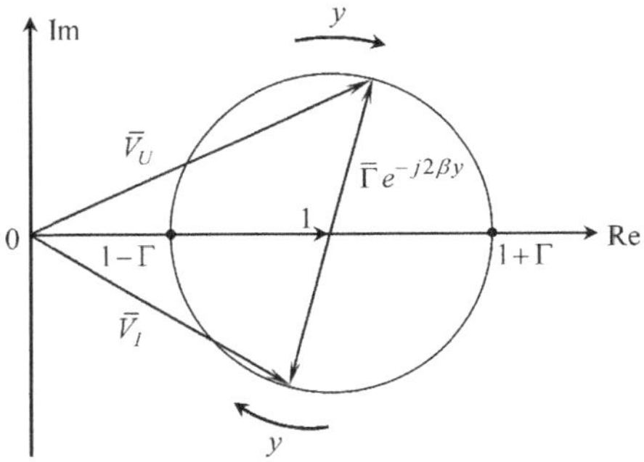

Figure 9.21 Complex plane vector diagram used for analyzing the evolution of $\overline{V}_U$ and $\overline{V}_I$ as one moves away from the load terminals (increasing y)

The evolution of $U_{rms}(y)$ and $I_{rms}(y)$ can be retrieved from the geometrical vector diagram in Figure 9.21 where, as you move away from the load towards the generator terminals, the tips of $\overline{V}_U(y)$ and $\overline{V}_I(y)$ move clockwise, synchronously, along a circumference of radius Γ.

An analytical alternative for obtaining the evolution of $U_{rms}(y)$ and $I_{rms}(y)$ is also available. Go back to the general equations in (9.39), take their absolute value and divide by $\sqrt{2}$. This enables you to write U_{rms} and I_{rms} as follows:

$$U_{rms}(y) = (U_i)_{rms}\left|1+\Gamma\, e^{j(\theta-2\beta y)}\right| = (U_i)_{rms}\sqrt{1+\Gamma^2+2\Gamma\cos(2\beta y-\theta)} \tag{9.52a}$$

$$I_{rms}(y) = \frac{(U_i)_{rms}}{R_w}\left|1-\Gamma\, e^{j(\theta-2\beta y)}\right| = \frac{(U_i)_{rms}}{R_w}\sqrt{1+\Gamma^2-2\Gamma\cos(2\beta y-\theta)} \tag{9.52b}$$

where $\theta = \sphericalangle\overline{\Gamma}$.

Typical standing wave patterns representative of $U_{rms}(y)$ and $I_{rms}(y)$ are depicted in Figure 9.22.

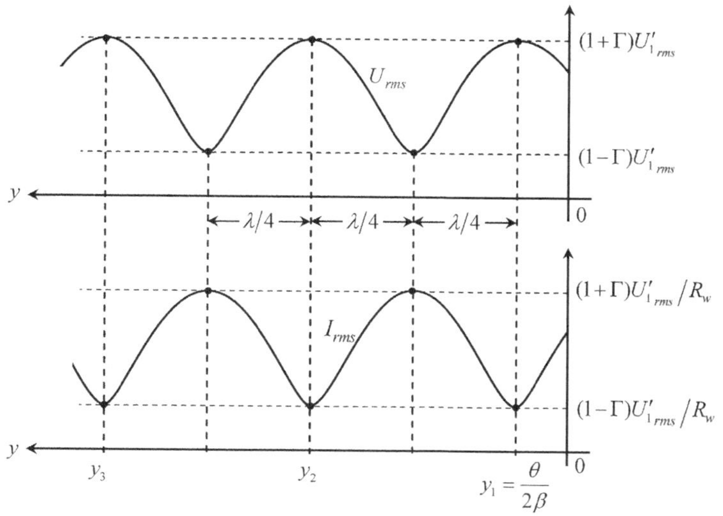

Figure 9.22 Standing wave patterns representative of $U_{rms}(y)$ and $I_{rms}(y)$

The main features of the standing wave patterns are summarized next:

- The patterns are periodic, with period $\lambda/2$.
- Consecutive maxima and minima are separated by a distance equal to $\lambda/4$.
- A point of maximum voltage is a point of minimum current, and vice versa.
- Voltage maxima and minima are given by

$$(U_{rms})_{\max} = (U_i)_{rms} \times (1+\Gamma); \quad (U_{rms})_{\min} = (U_i)_{rms} \times (1-\Gamma) \tag{9.53}$$

- Current maxima and minima are given by

$$(I_{rms})_{\max} = \frac{(U_i)_{rms}}{R_w} \times (1+\Gamma); \quad (I_{rms})_{\min} = \frac{(U_i)_{rms}}{R_w} \times (1-\Gamma) \tag{9.54}$$

- The abscissas y_k where the line voltage is at a maximum are given by

$$y_k = \frac{\theta}{2\beta} + k\frac{\lambda}{2}, \text{ with } k = 0, 1, 2, \ldots \tag{9.55}$$

- From (9.53) and (9.54) you have

$$\frac{(U_{rms})_{\max}}{(I_{rms})_{\max}} = \frac{(U_{rms})_{\min}}{(I_{rms})_{\min}} = R_w \tag{9.56}$$

- The voltage and current standing wave patterns are essentially identical. They just differ in two details: a translation of $\lambda/4$ in the horizontal axis, and a scale factor of R_w in the vertical axis.

In order to quantify the magnitude of the fluctuations observed in the standing wave patterns, one usually defines the so-called standing wave ratio, SWR, a dimensionless parameter that can vary from 1 to ∞:

$$\text{SWR} = \frac{(U_{rms})_{\max}}{(U_{rms})_{\min}} = \frac{(I_{rms})_{\max}}{(I_{rms})_{\min}} = \frac{1+\Gamma}{1-\Gamma} \tag{9.57}$$

For matched loads, where $\Gamma = 0$, you have an SWR of 1 (fluctuations do not exist). For loads where total reflection occurs, $\Gamma = 1$, you have an SWR of ∞ (fluctuations are as large as possible, and $(U_{rms})_{\min} = (I_{rms})_{\min} = 0$).

Note especially that, in (9.57), the absolute value of $\overline{\Gamma}$ is used, not the complex $\overline{\Gamma}$ itself.

The active power delivered by a lossless line is constant along y because, by definition, loss mechanisms along the line are absent. You may remember that the active power was defined in Chapter 7 through $P = (p(t))_{av} = U_{rms} I_{rms} \cos\varphi$. However, in a mismatched line, the quantities U_{rms}, I_{rms} and φ – all of them – vary with y. So, the obvious question is how can you evaluate P in as simple a manner as possible?

The key for this question can be found in (9.40).

Whenever $\overline{V}_U(y)$ and $\overline{V}_I(y)$ happen to be parallel vectors you have $\varphi = 0$. But, from Figure 9.21, you will see that such a situation corresponds exactly to extremal points in the standing wave patterns. Therefore, the answer you are looking for is

$$P = (U_{rms})_{\max} (I_{rms})_{\min} = (U_{rms})_{\min} (I_{rms})_{\max} = (1-\Gamma^2)\frac{(U_i)_{rms}^2}{R_w}$$

9.5.4 The Low-Frequency Limit Case, Short Lines

The transmission-line results developed in the foregoing sections should agree with the familiar results from circuit analysis when low frequencies are considered. To confirm that this statement is true, let us examine two illustrative situations.

For an open-line situation ($\overline{Z}_L = \infty$) we saw in (9.47) that the input impedance observed at the generator terminals was given by $\overline{Z}_{in} = -jR_w/\tan(\beta l)$. For low-frequency regimes we have $\beta l \ll 1$, and consequently $\tan(\beta l) \simeq \beta l$. Therefore, taking into account the definitions of R_w and β, we find

$$\overline{Z}_{in} = \frac{-jR_w}{\tan(\beta l)} \simeq \frac{-jR_w}{\beta l} = -j\frac{\sqrt{L_e/C}}{\omega\sqrt{L_e C}l} = \frac{1}{j\omega(Cl)}$$

from which you can see that the input impedance of an insulated two-conductor system of length l immersed in a dielectric medium is nothing but the impedance of a capacitor whose capacitance is Cl, as expected from a circuit theory approach.

Likewise, for a short-circuited line ($\overline{Z}_L = 0$) we saw in (9.50) that the input impedance observed at the generator terminals was given by $\overline{Z}_{in} = jR_w \tan(\beta l)$. For low-frequency regimes we have $\beta l \ll 1$, and $\tan(\beta l) \simeq \beta l$. Therefore, taking into account the definitions of R_w and β, we find

$$\overline{Z}_{in} = jR_w \tan(\beta l) \simeq jR_w\, \beta l = j\sqrt{\frac{L_e}{C}}\omega\sqrt{L_e C}l = j\omega(L_e l)$$

from which you can see that the input impedance of a two-conductor system of length l short-circuited at one end is nothing but the impedance of a rectangular single-loop inductor whose inductance is $L_e l$, as expected from a circuit theory approach.

9.6 Application Example (Line-Matching Techniques)

The problem of mismatched loads can be handled by using several techniques. Here we address two commonly used solutions, the quarter-wave transformer and the single matching stub.

Consider an aerial lossless transmission line characterized by $R_w = 150\ \Omega$. The line, of length $l = 3\,\text{m}$, is terminated by a resistive load $R_L = 50\,\Omega$. The line is driven by an ideal generator whose voltage is $u_G(t) = \sqrt{2}\,(U_G)_{rms}\cos(\omega t)$, where $(U_G)_{rms} = 200\,\text{V}$, $f = 100\,\text{MHz}$.

Questions

Q_1 Determine the following parameters: v, λ, $\overline{\Gamma}$ and SWR. Determine the complex amplitude $\overline{U}'_i$ of the incident wave voltage at the load terminals. Obtain the load voltage $\overline{U}_L$. Draw the standing wave pattern of the rms voltage along the line. Find the active power delivered to the load.

Q_2 Consider that the aerial line of length l is made of two distinct sections (Figure 9.23), the first of length $l - d$ with $R_w = 150\,\Omega$, and the second of length $d = \lambda/4$ with characteristic resistance R_{w0} (this second section is called a quarter-wave transformer).

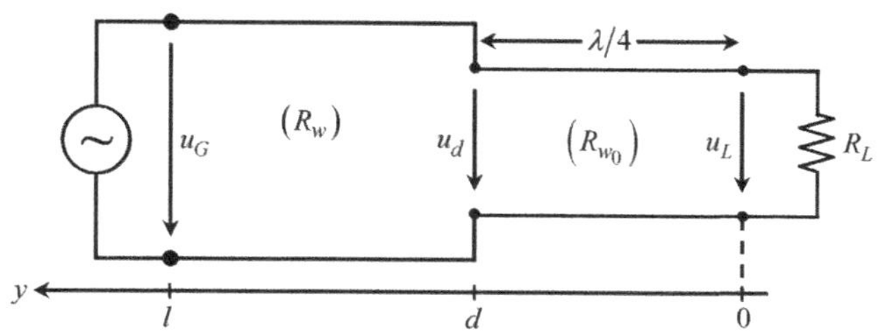

Figure 9.23 Line-matching technique using a quarter-wave line transformer

The purpose of the quarter-wave transformer is to ensure that at the transverse plane $y = d$ the line impedance is equal to R_w, therefore ensuring that the first line section is perfectly matched.

Regarding the second line section, determine d, R_{w0}, $\overline{\Gamma}_0$ and SWR_0. Obtain $\overline{U}_d$ and $\overline{U}_L$. Draw the standing wave pattern of the rms voltage along the line. Find the active power delivered to the load.

Q_3 As shown in Figure 9.24, the aerial line of length l includes a stub positioned at $y = d$. The stub consists of a short-circuited line section of adjustable length h (with the same characteristics of the original line). Parameters d and h are the unknowns of the problem. The minimal distance d is chosen so that the line admittance to the right of $y = d$ is of the type $\overline{Y}_{y=d^+} = R_w^{-1} + jS$. The minimal distance h is chosen so that the input admittance of the auxiliary stub is $\overline{Y}_{st} = -jS$. The parallel association of these two admittances results in an equivalent impedance $\overline{Z}_{y=d} = (\overline{Y}_{y=d^+} + \overline{Y}_{st})^{-1}$ equal to R_w, thus ensuring that the line section fed by the generator is perfectly matched.

Determine d, S and h. Obtain $\overline{U}_d$ and $\overline{U}_L$. Draw the standing wave pattern of the rms voltage along the line. Find the active power delivered to the load.

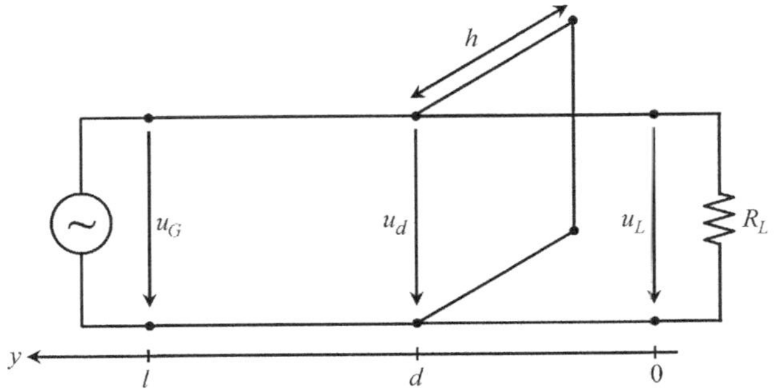

Figure 9.24 Line-matching technique using a parallel-connected stub

Solutions

Q$_1$ $v = 1/\sqrt{\mu_0\varepsilon_0} = 3\times 10^8\,\text{m/s}$; $\lambda = v/f = 3\,\text{m}$;

$$\overline{\Gamma} = \frac{R_L - R_w}{R_L + R_w} = 0.5\,e^{j\pi}; \quad \text{SWR} = \frac{1+\Gamma}{1-\Gamma} = 3$$

From $\overline{U}_G = \overline{U}_i' e^{j\beta l}(1 + \overline{\Gamma} e^{-j2\beta l})$, noting that $\beta l = 2\pi$, we get

$$\overline{U}_i' = \frac{\overline{U}_G}{1+\overline{\Gamma}} = \sqrt{2}\,400\,\text{V}$$

From $\overline{U}_L = \overline{U}_i'(1+\overline{\Gamma})$, we get $\overline{U}_L = \overline{U}_G$ (note that the line is one wavelength long).
From $(U_{rms})_{\max} = (U_i')_{rms}(1+\Gamma)$, we get $(U_{rms})_{\max} = 600\,\text{V}$.
From $(U_{rms})_{\min} = (U_i')_{rms}(1-\Gamma)$, we get $(U_{rms})_{\min} = 200\,\text{V}$.
From $P = (1-\Gamma^2)\,(U_i')^2_{rms}/R_w$, we get $P = 800\,\text{W}$.
See the sketch of the voltage standing wave pattern in Figure 9.25(a).

Q$_2$ $d = \lambda/4 = 75\,\text{cm}$. Taking into account that for a quarter-wave line $\beta d = \pi/2$, we can write, from (9.39),

$$\overline{U}_d = j\overline{U}_i'(1-\overline{\Gamma}_0) \quad \text{and} \quad \overline{I}_d = j\overline{U}_i'(1+\overline{\Gamma}_0)/R_{w_0}$$

Dividing both equations and enforcing $\overline{U}_d/\overline{I}_d = R_w$, we obtain

$$R_w = R_{w_0}\frac{1-\overline{\Gamma}_0}{1+\overline{\Gamma}_0}$$

Making use of

$$\overline{\Gamma}_0 = \frac{R_L - R_{w_0}}{R_L + R_{w_0}}$$

we finally conclude

$$R_w = R_{w_0} \frac{R_{w_0}}{R_L} \quad \rightarrow \quad R_{w_0} = \sqrt{R_L R_w} = 86.6\ \Omega$$

from which we evaluate

$$\overline{\Gamma}_0 = 0.268\ e^{j\pi} \text{ and } \mathrm{SWR}_0 = \frac{1+\Gamma_0}{1-\Gamma_0} = 1.73$$

Since the first line section is matched, the voltage $\overline{U}_d$ is equal to $\overline{U}_G$ apart from a phase delay $\beta(l-d) = 3\pi/2$, that is

$$\overline{U}_d = \sqrt{2}\,200\,e^{-j3\pi/2} = j\sqrt{2}\,200\,\mathrm{V}.$$

Knowledge of $\overline{U}_d$ permits the determination of

$$\overline{U}_i' = \frac{\overline{U}_d}{j(1-\overline{\Gamma}_0)} = \sqrt{2}\ 157.73\ \mathrm{V}$$

which finally allows for the evaluation of the load voltage,

$$\overline{U}_L = \overline{U}_i'(1+\overline{\Gamma}_0) = \sqrt{2}\,115.47\,\mathrm{V}.$$

You may note that

$$\frac{(U_d)_{rms}}{(U_L)_{rms}} = \mathrm{SWR}_0$$

The active power can be determined via

$$P = \frac{(U_G)^2_{rms}}{R_w} \text{ or } P = \frac{(U_L)^2_{rms}}{R_L}$$

yielding $P = 266.7\,\mathrm{W}$.

See the sketch of the voltage standing wave pattern in Figure 9.25(b).

Q_3 The input admittance of the end section of length d is obtained from (9.43) as

$$\overline{Y}_{y=d^+} = R_w^{-1} \frac{1-\overline{\Gamma}\,e^{-j2\beta d}}{1+\overline{\Gamma}\,e^{-j2\beta d}}$$

Taking into account, from Q_1, that $\overline{\Gamma} = -0.5$, we can write

$$R_w^{-1} + jS = \overline{Y}_{y=d^+} = R_w^{-1} \frac{2+e^{-jx}}{2-e^{-jx}}, \text{ where } x = 2\beta d = \frac{4\pi d}{\lambda}$$

After some algebraic manipulation we find

$$1+jSR_w = \frac{3-j4\sin x}{5-4\cos x} \rightarrow \begin{cases} 1 = \dfrac{3}{5-4\cos x} \\ SR_w = \dfrac{-4\sin x}{5-4\cos x} \end{cases} \rightarrow \begin{cases} x = \pi/3 \\ SR_w = -1.155 \end{cases}$$

In conclusion, $d = \lambda/12 = 25\,\text{cm}$ and $S = -7.7\,\text{mS}$.

The input admittance of the stub is determined from (9.50),

$$\overline{Y}_{st} = -jS = \frac{-j}{R_w \tan(\beta h)}$$

from which we get $\tan(\beta h) = (SR_w)^{-1} = -0.866$, yielding $h = 116\,\text{cm}$ (in your calculations do not forget that the angle βh is expressed in radians, not degrees!).

Since the first line section is matched, the voltage $\overline{U}_d$ is equal to $\overline{U}_G$ apart from a phase delay $\beta(l-d) = 11\pi/6$, that is $\overline{U}_d = \sqrt{2}\,200\, e^{+j\pi/6}\,\text{V}$.

For the end line section to the right of the stub insertion we utilize

$$\overline{U}(y) = \overline{U}_i' e^{+j\beta y}\left(1+\overline{\Gamma} e^{-j2\beta y}\right)$$

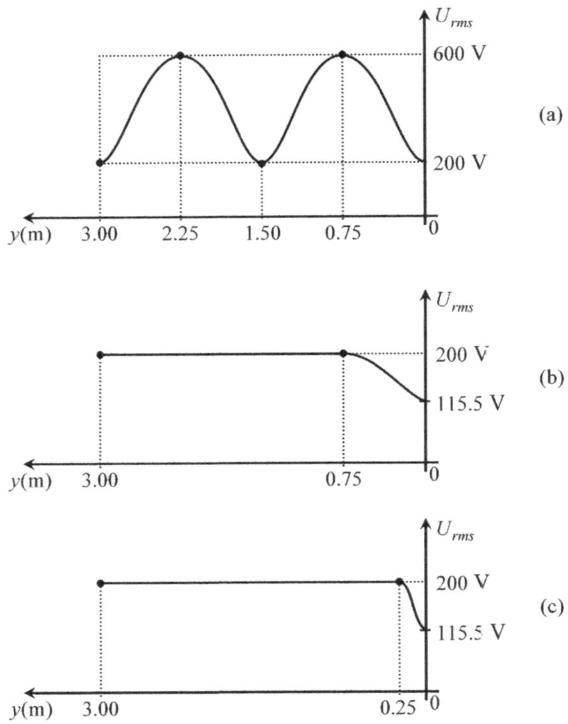

Figure 9.25 Standing wave voltage patterns. (a) Without the use of mitigation matching techniques. (b) Using a quarter-wave line transformer. (c) Using a parallel-connected stub

with the boundary condition $\overline{U}_{y=d} = \sqrt{2}\,200\,e^{+j\pi/6}\,\text{V}$, from which we find

$$\overline{U}_i' = \frac{\sqrt{2}\,200\,e^{j\pi/6}}{e^{j\pi/6}\left(1 - 0.5\,e^{-j\pi/3}\right)} = \sqrt{2}\;230.95\;e^{-j\pi/6}\;\text{V}$$

Finally, at the load terminals, we obtain $\overline{U}_L = \overline{U}_i'\left(1 + \overline{\Gamma}\right) = \sqrt{2}\,115.47\;e^{-j\pi/6}\,\text{V}$.

The active power can be determined via

$$P = \frac{(U_G)_{rms}^2}{R_w} \text{ or } P = \frac{(U_L)_{rms}^2}{R_L}$$

yielding $P = 266.7\,\text{W}$.

See the sketch of the voltage standing wave pattern in Figure 9.25(c).

9.7 Multiconductor Transmission Lines

In the foregoing sections we paid attention to the analysis of two-conductor transmission lines; however, many transmission-line structures include more than just two conductors. As a subject of the utmost interest to electrical engineers, multiconductor transmission lines can be found almost everywhere, with applications in overhead power lines, railway systems, printed circuit boards, flatpack or ribbon cables for electronic systems interconnection, as well as several microwave structures, to name just a few.

In this section we address the topic of multiconductor transmission lines (MTLs) at an introductory level.

Consider a uniform transmission-line system made of $N+1$ conductors parallel to the longitudinal z axis, where conductor (0) is taken as the reference conductor – see Figure 9.26.

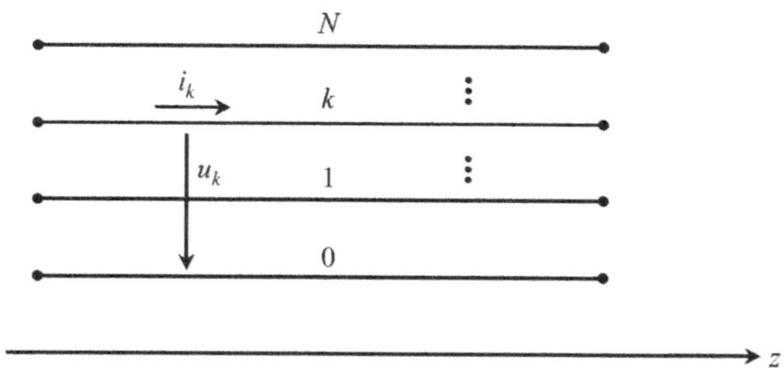

Figure 9.26 Multiconductor transmission-line structure with $N+1$ coupled conductors

For $N > 1$, the generalization of the frequency-domain two-conductor line equations in (9.24) and (9.25) take the matrix form

$$\frac{d}{dz}\left[\overline{U}(z)\right] = -\left[\overline{Z}_l\right]\left[\overline{I}(z)\right]; \quad \frac{d}{dz}\left[\overline{I}(z)\right] = -\left[\overline{Y}_t\right]\left[\overline{U}(z)\right] \tag{9.58}$$

where $\left[\overline{U}(z)\right]$ and $\left[\overline{I}(z)\right]$ are complex column vectors, of size N, gathering the complex amplitudes of the line voltages and line currents

$$\left[\overline{U}(z)\right]=\begin{bmatrix}\overline{U}_1(z)\\ \overline{U}_2(z)\\ \vdots\\ \overline{U}_N(z)\end{bmatrix};\quad \left[\overline{I}(z)\right]=\begin{bmatrix}\overline{I}_1(z)\\ \overline{I}_2(z)\\ \vdots\\ \overline{I}_N(z)\end{bmatrix}$$

and $\left[\overline{Z}_l\right]$ and $\left[\overline{Y}_t\right]$ respectively denote the longitudinal impedance and transverse admittance matrices per unit length of the line.

The coupled differential equations in (9.58) can be decoupled by introducing an appropriate change of variables

$$\left[\overline{U}\right]=[T]\left[\overline{U}_m\right],\quad \left[\overline{I}\right]=[W]\left[\overline{I}_m\right] \tag{9.59}$$

where $\left[\overline{U}_m\right]$ and $\left[\overline{I}_m\right]$ are arrays for the so-called modal voltages and modal currents, respectively. Substituting (9.59) into (9.58), we find

$$\frac{d}{dz}\left[\overline{U}_m(z)\right]=-\underbrace{\left([T]^{-1}\left[\overline{Z}_l\right][W]\right)}_{\left[\overline{Z}_m\right]}\left[\overline{I}_m(z)\right] \tag{9.60a}$$

$$\frac{d}{dz}\left[\overline{I}_m(z)\right]=-\underbrace{\left([W]^{-1}\left[\overline{Y}_t\right][T]\right)}_{\left[\overline{Y}_m\right]}\left[\overline{U}_m(z)\right] \tag{9.60b}$$

where $\left[\overline{Z}_m\right]$ and $\left[\overline{Y}_m\right]$ are the per-unit-length longitudinal impedance and transverse admittance matrices in modal coordinates, respectively.

In order to ensure that the equations in (9.60) define a decoupled set, both $\left[\overline{Z}_m\right]$ and $\left[\overline{Y}_m\right]$ must be diagonal matrices, that is

$$\frac{d}{dz}\begin{bmatrix}\overline{U}_{1_m}\\ \vdots\\ \overline{U}_{k_m}\\ \vdots\\ \overline{U}_{N_m}\end{bmatrix}=-\begin{bmatrix}\overline{Z}_{1_m} & & & & \\ & \ddots & & & \\ & & \overline{Z}_{k_m} & & \\ & & & \ddots & \\ & & & & \overline{Z}_{N_m}\end{bmatrix}\begin{bmatrix}\overline{I}_{1_m}\\ \vdots\\ \overline{I}_{k_m}\\ \vdots\\ \overline{I}_{N_m}\end{bmatrix} \tag{9.61a}$$

$$\frac{d}{dz}\begin{bmatrix}\overline{I}_{1_m}\\ \vdots\\ \overline{I}_{k_m}\\ \vdots\\ \overline{I}_{N_m}\end{bmatrix}=-\begin{bmatrix}\overline{Y}_{1_m} & & & & \\ & \ddots & & & \\ & & \overline{Y}_{k_m} & & \\ & & & \ddots & \\ & & & & \overline{Y}_{N_m}\end{bmatrix}\begin{bmatrix}\overline{U}_{1_m}\\ \vdots\\ \overline{U}_{k_m}\\ \vdots\\ \overline{U}_{N_m}\end{bmatrix} \tag{9.61b}$$

or, which is the same,

$$\text{For } k=1 \text{ to } N: \begin{cases} \dfrac{d}{dz}\overline{U}_{k_m} = -\overline{Z}_{k_m}\,\overline{I}_{k_m} \\ \dfrac{d}{dz}\overline{I}_{k_m} = -\overline{Y}_{k_m}\,\overline{U}_{k_m} \end{cases} \tag{9.61c}$$

The equations in (9.61c) are the familiar two-conductor transmission-line equations, whose solution we have already established in (9.28) and (9.30)

$$\overline{U}_{k_m}(z) = \left(\overline{U}_i\right)_k e^{-\overline{\gamma}_k z} + \left(\overline{U}_r\right)_k e^{+\overline{\gamma}_k z}; \quad \overline{I}_{k_m}(z) = \frac{\left(\overline{U}_i\right)_k e^{-\overline{\gamma}_k z} - \left(\overline{U}_r\right)_k e^{+\overline{\gamma}_k z}}{\overline{Z}_{w_k}}$$

where $\overline{\gamma}_k = \sqrt{\overline{Z}_{k_m}\overline{Y}_{k_m}}$ and $\overline{Z}_{w_k} = \sqrt{\overline{Z}_{k_m}/\overline{Y}_{k_m}}$ respectively denote the propagation constant, and the characteristic wave impedance for the kth mode.

Once the modal voltages and currents $\left[\overline{U}_m\right]$ and $\left[\overline{I}_m\right]$ have been found, we reuse (9.59) to determine the natural voltages and currents in the multiconductor line.

So far, the procedure described above is straightforward. However, we still have to deal with the problem of the determination of the transformation matrices $[T]$ and $[W]$ – a major point in this formalism.

Taking into account the definition of the diagonal matrices $\left[\overline{Z}_m\right]$ and $\left[\overline{Y}_m\right]$ in (9.60), and bearing in mind that diagonal matrices always commute, we find

$$\left[\overline{Z}_m\right]\left[\overline{Y}_m\right] = \left[\overline{\gamma}^2\right] = [T]^{-1}\left[\overline{Z}_l\overline{Y}_t\right][T] \tag{9.62a}$$

$$\left[\overline{Y}_m\right]\left[\overline{Z}_m\right] = \left[\overline{\gamma}^2\right] = [W]^{-1}\left[\overline{Y}_t\overline{Z}_l\right][W] \tag{9.62b}$$

where $\left[\overline{\gamma}^2\right]$ is a diagonal matrix that gathers the squared propagation constants of the N independent propagation modes.

From (9.62a) you can see that the transformation matrix $[T]$ is determined by solving an eigenvalue/eigenvector problem concerning the $\left[\overline{Z}_l\overline{Y}_t\right]$ matrix product; that is,

$$\left[\overline{Z}_l\overline{Y}_t\right]\,[T] = [T]\,\left[\overline{\gamma}^2\right] \quad \text{or} \quad \left(\left[\overline{Z}_l\,\overline{Y}_t\right] - \overline{\gamma}_k^2[1]\right)\,[t_k] = 0 \tag{9.63}$$

where $[t_k]$, the column k of $[T]$, represents the kth eigenvector of $\left[\overline{Z}_l\,\overline{Y}_t\right]$ associated to the eigenvalue $\overline{\gamma}_k^2$. Matrix $[1]$ is the identity matrix.

The numerical task involved in solving (9.63) is a heavy one, particularly if N is large. Fortunately, standard software packages exist for finding the eigenvalues and eigenvectors of general complex matrices. Note that the procedure in (9.63) allows you not only you to find $[T]$ but also to obtain the modal propagation constants.

For the determination of the transformation matrix $[W]$ we may adopt diverse strategies. For example, we may start by arbitrarily defining $\left[\overline{Z}_m\right]$ as a non-singular diagonal matrix. Then we determine $[\overline{Y}_m]$ by using $[\overline{Y}_m] = [\overline{Z}_m]^{-1}[\overline{\gamma}^2]$. Finally, from (9.60), we obtain, indifferently,

$$[W] = \begin{cases} [\overline{Z}_l]^{-1}[T]\,[\overline{Z}_m] \\ [\overline{Y}_t][T]\,[\overline{Y}_m]^{-1} \end{cases} \tag{9.64}$$

Alternatively, let us take the transpose of the matrix equation in (9.62b):

$$[\overline{Z}_m][\overline{Y}_m] = [\overline{\gamma}^2] = [W]^T[\overline{Z}_l\overline{Y}_t]\,[W]^{-1T} \tag{9.65}$$

where account is taken of the symmetry of $[\overline{Z}_l]$ and $[\overline{Y}_t]$. The comparison between (9.65) and (9.62a) indicates one possible choice:

$$[W]^T = [T]^{-1} \tag{9.66}$$

Final remarks are as follows:

- For lossy lines, extraordinary cases exist where the problem posed in (9.63) of transforming $[\overline{Z}_l\,\overline{Y}_t]$ into diagonal form cannot be solved at all. However, for typical MTL configurations, those abnormal situations are extremely rare.
- The solution of (9.63) does not uniquely define $[T]$. If you find one solution for $[T]$ and multiply it on the right by an arbitrary non-singular diagonal matrix, the resulting new matrix will also be a solution of (9.63).
- The degrees of freedom contained in the construction of $[T]$ and $[W]$ imply some sort of arbitrariness in the definition of $[\overline{Z}_m]$, $[\overline{Y}_m]$ and $[\overline{Z}_{w_m}]$; this remark should put you on your guard as far as the physical significance of those matrices is concerned.
- For lossy lines, the transformation $[T]$ is in general a complex frequency-dependent matrix.
- For lossless lines, where $[\overline{Z}_l] = j\omega[L]$ and $[\overline{Y}_t] = j\omega[C]$, a real frequency-independent transformation matrix $[T]$ can always be found.
- For lossless homogeneous lines, where the dielectric medium around the conductors is characterized by ε and μ, the eigenvalues of the product matrix $[\overline{Z}_l\,\overline{Y}_t]$ are all equal, that is $\overline{\gamma}_k = j\omega\sqrt{\mu\varepsilon}$, for $k = 1$ to N.

9.8 Application Example (Even and Odd Modes)

Consider a lossless inhomogeneous MTL structure consisting of two identical dielectric-coated cylindrical conductors in close contact seated on a reference conducting plane. The structure is operating at $\omega = 1$ Grad/s. The ratio of the radii r/r_0 is 0.8. The insulation coating material is characterized by $\varepsilon = 4\varepsilon_0$, $\mu = \mu_0$. As shown in Figure 9.27, the system under analysis displays a vertical plane of symmetry; the corresponding per-unit-length capacitance and inductance matrices have been evaluated as

$$[C] = \begin{bmatrix} 225.62 & -69.80 \\ -69.80 & 225.62 \end{bmatrix} \text{pF/m}, \quad [L] = \begin{bmatrix} 126.51 & 35.26 \\ 35.26 & 126.51 \end{bmatrix} \text{nH/m}$$

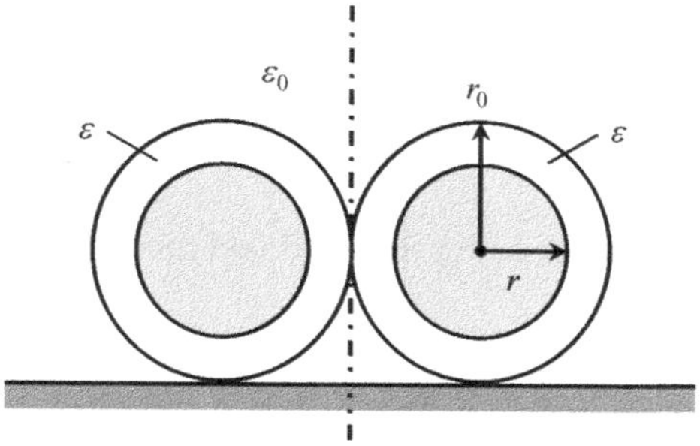

Figure 9.27 Cross-sectional view of an MTL structure consisting of two identical dielectric-coated cylindrical conductors in close contact seated on a reference ground plane

Questions

Q_1 Determine the matrix products $[\overline{Z}_l\overline{Y}_t]$ and $[\overline{Y}_t\overline{Z}_l]$.

Q_2 Determine the eigenvectors (transformation matrices) and eigenvalues of both matrices. Evaluate the phase velocities of the two propagation modes.

Q_3 Determine the modal matrices $[\overline{Z}_m]$ and $[\overline{Y}_m]$, and then evaluate the characteristic wave impedances of the two propagation modes.

Q_4 Assume that the receiving end of the structure is terminated in a matched load. Determine the solution for the complex amplitudes of the line voltages and currents along the line.

Q_5 Determine the characteristic impedance matrix describing the matched load.

Solutions

Q_1

$$[\overline{Z}_l\overline{Y}_t] = [\overline{Y}_t\overline{Z}_l] = (j\omega)^2[L][C] = (j\omega)^2[C][L] = \begin{bmatrix} -26.08 & 0.875 \\ 0.875 & -26.08 \end{bmatrix} \text{ m}^{-2}$$

Q_2

$$[T] = [W] = \frac{1}{\sqrt{2}}\begin{bmatrix} 1 & 1 \\ 1 & -1 \end{bmatrix}$$

The eigenvector

$$[t_1] = [w_1] = \frac{1}{\sqrt{2}}\begin{bmatrix} 1 \\ 1 \end{bmatrix}$$

defines the excitation of mode 1, the so-called even mode (or common mode), where $\overline{U}_1 = \overline{U}_2$ and $\overline{I}_1 = \overline{I}_2$.

The eigenvector

$$[t_2] = [w_2] = \frac{1}{\sqrt{2}}\begin{bmatrix} 1 \\ -1 \end{bmatrix}$$

defines the excitation of mode 2, the so-called odd mode (or differential mode), where $\overline{U}_1 = -\overline{U}_2$ and $\overline{I}_1 = -\overline{I}_2$.

Regarding the eigenvalues, we find

$$\left[\overline{\gamma}^2\right] = \begin{bmatrix} -25.21 & 0 \\ 0 & -26.96 \end{bmatrix} \text{ m}^{-2}$$

From $\overline{\gamma}_k = j\beta_k$ we obtain $\beta_1 = 5.021\,\text{rad/m}$ and $\beta_2 = 5.192\,\text{rad/m}$.
From $\beta_k = \omega/v_k$ we obtain $v_1 = 1.992 \times 10^8\,\text{m/s}$ and $v_2 = 1.926 \times 10^8\,\text{m/s}$.

Note that the normalization factor $1/\sqrt{2}$ which appears in both $[T]$ and $[W]$ is arbitrary; it has been chosen with the single purpose of getting unit norm eigenvectors.

Q_3

$$\left[\overline{Z}_m\right] = j\omega\,[T]^{-1}\,[L]\,[W] = \begin{bmatrix} 161.8 & 0 \\ 0 & 91.25 \end{bmatrix} \Omega/\text{m}$$

$$\left[\overline{Y}_m\right] = j\omega\,[W]^{-1}\,[C]\,[T] = \begin{bmatrix} 155.8 & 0 \\ 0 & 295.4 \end{bmatrix} \text{mS/m}$$

From $\overline{Z}_{w_k} = \sqrt{\overline{Z}_{k_m}/\overline{Y}_{k_m}}$ we obtain $\overline{Z}_{w_1} = R_{w_1} = 32.22\,\Omega$ and $\overline{Z}_{w_2} = R_{w_2} = 17.58\,\Omega$.

Q_4 Voltage solution:

$$\left[\overline{U}\right] = [T]\begin{bmatrix} \left(\overline{U}_i\right)_1 e^{-j\beta_1 z} \\ \left(\overline{U}_i\right)_2 e^{-j\beta_2 z} \end{bmatrix} \rightarrow \begin{cases} \overline{U}_1(z) = \frac{1}{\sqrt{2}}\left(\left(\overline{U}_i\right)_1 e^{-j\beta_1 z} + \left(\overline{U}_i\right)_2 e^{-j\beta_2 z}\right) \\ \overline{U}_2(z) = \frac{1}{\sqrt{2}}\left(\left(\overline{U}_i\right)_1 e^{-j\beta_1 z} - \left(\overline{U}_i\right)_2 e^{-j\beta_2 z}\right) \end{cases}$$

Current solution:

$$\left[\overline{I}\right] = [W]\begin{bmatrix} \dfrac{\left(\overline{U}_i\right)_1}{R_{w_1}} e^{-j\beta_1 z} \\ \dfrac{\left(\overline{U}_i\right)_2}{R_{w_2}} e^{-j\beta_2 z} \end{bmatrix} \rightarrow \begin{cases} \overline{I}_1(z) = \dfrac{1}{\sqrt{2}}\left(\dfrac{\left(\overline{U}_i\right)_1}{R_{w_1}} e^{-j\beta_1 z} + \dfrac{\left(\overline{U}_i\right)_2}{R_{w_2}} e^{-j\beta_2 z}\right) \\ \overline{I}_2(z) = \dfrac{1}{\sqrt{2}}\left(\dfrac{\left(\overline{U}_i\right)_1}{R_{w_1}} e^{-j\beta_1 z} - \dfrac{\left(\overline{U}_i\right)_2}{R_{w_2}} e^{-j\beta_2 z}\right) \end{cases}$$

Q_5 When a transmission-line structure is terminated in a matched load (as we have assumed in Q_4), the relationship between $\left[\overline{U}(z)\right]$ and $\left[\overline{I}(z)\right]$ is independent of z. Therefore, we can use the results derived in Q_4, particularized to $z = 0$, to get

$$\begin{cases} \sqrt{2}\,\overline{U}_1(0) = \left(\overline{U}_i\right)_1 + \left(\overline{U}_i\right)_2 \\ \sqrt{2}\,\overline{U}_2(0) = \left(\overline{U}_i\right)_1 - \left(\overline{U}_i\right)_2 \end{cases} \quad \text{and} \quad \begin{cases} \sqrt{2}\,\overline{I}_1(0) = \left(\overline{U}_i\right)_1/R_{w_1} + \left(\overline{U}_i\right)_2/R_{w_2} \\ \sqrt{2}\,\overline{I}_2(0) = \left(\overline{U}_i\right)_1/R_{w_1} - \left(\overline{U}_i\right)_2/R_{w_2} \end{cases}$$

Eliminating $(\overline{U}_i)_1$ and $(\overline{U}_i)_2$, we find

$$\begin{bmatrix} \overline{U}_1 \\ \overline{U}_2 \end{bmatrix} = [\overline{Z}_w] \begin{bmatrix} \overline{I}_1 \\ \overline{I}_2 \end{bmatrix}$$

where

$$[\overline{Z}_w] = \frac{1}{2} \begin{bmatrix} (R_{w_1} + R_{w_2}) & (R_{w_1} - R_{w_2}) \\ (R_{w_1} - R_{w_2}) & (R_{w_1} + R_{w_2}) \end{bmatrix} = \begin{bmatrix} 24.90 & 7.32 \\ 7.32 & 24.90 \end{bmatrix} \Omega$$

You can check that this impedance matrix corresponds to the line termination shown in Figure 9.28.

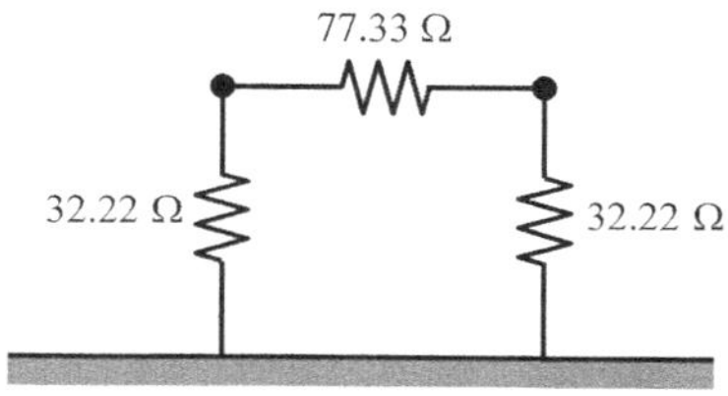

Figure 9.28 Matched termination for the MTL structure depicted in Figure 9.27

9.9 Proposed Homework Problems

Problem 9.9.1

Consider a lossless transmission line, 300 m long, characterized by $v = 3 \times 10^8$ m/s and $R_w = 60\,\Omega$. As shown in Figure 9.29, the line is excited at $z = 0$ by a step voltage $U_0 = 120$ V. A resistor R terminates the line at $z = l$.

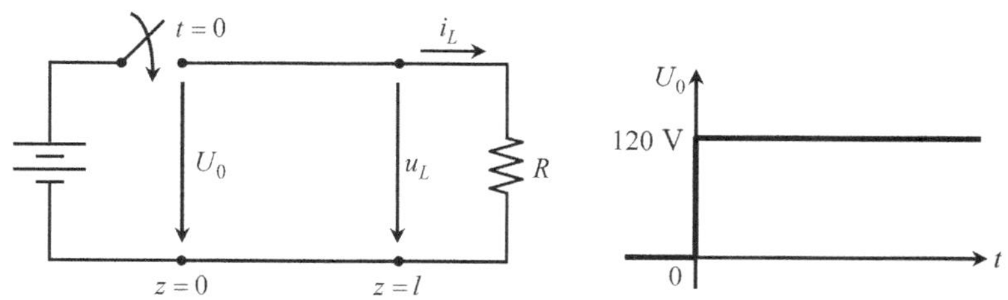

Figure 9.29 A step voltage is applied to the input end of a transmission line loaded by a resistor

Q_1 Determine the per-unit-length L and C parameters of the line. Determine the time delay τ between the two ends of the line.

Q_2 Assuming that $R = 0$, find the evolution of $i_L(t)$.

Q_3 Assuming that $R = \infty$, find the evolution of $u_L(t)$.

Q_4 Assuming that $R = 20\,\Omega$, find the evolution of $u_L(t)$.

Answers

Q_1

$$L = \frac{R_w}{v} = 200.0\,\text{nH/m}; \quad C = \frac{1}{vR_w} = 55.55\,\text{pF/m}; \quad \tau = l/v = 1\,\mu\text{s}$$

Q_2

$$i_L(t) = \begin{cases} 0 & ;\ \text{for } 0 < t < \tau \\ 4\text{ A} & ;\ \text{for } \tau < t < 3\tau \\ 8\text{ A} & ;\ \text{for } 3\tau < t < 5\tau \\ \vdots & \qquad \vdots \\ 2k\text{ A} & ;\ \text{for } (k-1)\tau < t < (k+1)\tau \end{cases}$$

(with k an even number). See Figure 9.30(a).

Q_3

$$u_L(t) = \begin{cases} 0 & ;\ \text{for } 0 < t < \tau \\ 240\text{ V} & ;\ \text{for } \tau < t < 3\tau \\ 0 & ;\ \text{for } 3\tau < t < 5\tau \\ 240\text{ V} & ;\ \text{for } 5\tau < t < 7\tau \\ 0 & ;\ \text{for } 7\tau < t < 9\tau \\ \vdots & \qquad \vdots \end{cases}$$

See Figure 9.30(b).

Q_4

$$u_L(t) = \begin{cases} 0 & ;\ \text{for } 0 < t < \tau \\ 60\text{ V} & ;\ \text{for } \tau < t < 3\tau \\ 90\text{ V} & ;\ \text{for } 3\tau < t < 5\tau \\ 105\text{ V} & ;\ \text{for } 5\tau < t < 7\tau \\ \vdots & \qquad \vdots \\ 120\text{ V} & \qquad t = \infty \end{cases}$$

See Figure 9.30(c).

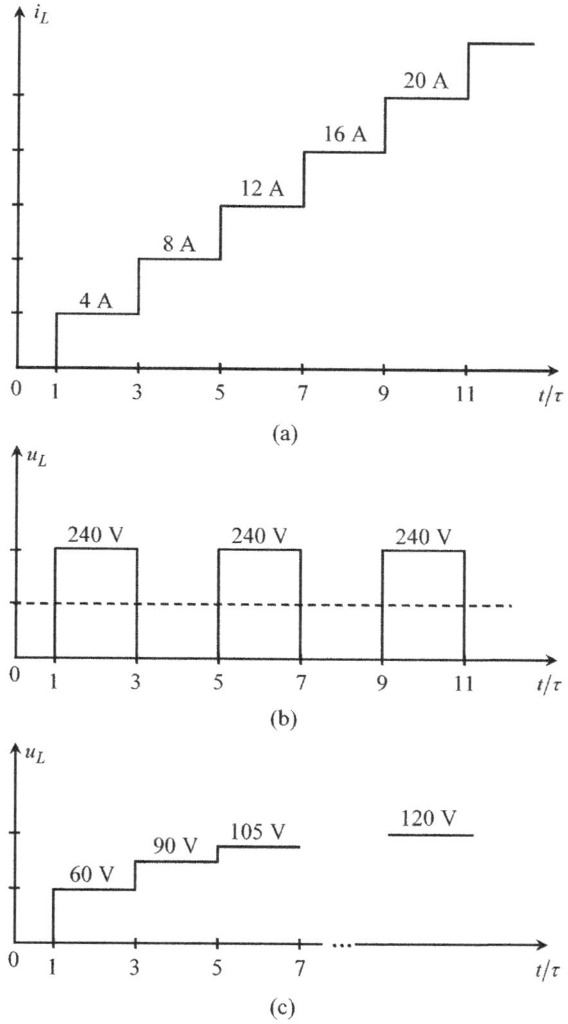

Figure 9.30 Load analysis. (a) Plot of the load current against time for the case $R = 0$. (b) Plot of the load voltage against time for the case $R = \infty$. (b) Plot of the load voltage against time for the case $R = 20\,\Omega$

Problem 9.9.2

Consider a coaxial cable of length $l = 1\,\text{m}$, whose per-unit-length parameters are known, $L_e = 0.25\,\mu\text{H/m}$ and $C = 100\,\text{pF/m}$. The operating frequency is $f = 100\,\text{MHz}$. The dielectric medium is assumed to be lossless, but the line conductors not. Assume that the perturbation arising from skin effect phenomena on the line conductors is such that the corresponding per-unit-length impedance correction is $\overline{Z}_1 + \overline{Z}_2 = \overline{Z}_{cond} = 10\,e^{j40°}\,\Omega/\text{m}$.

Q_1 Determine the per-unit-length longitudinal impedance and transverse admittance of the line, $\overline{Z}_l$ and $\overline{Y}_t$.

Q_2 Determine the phase velocity v and the attenuation constant α.

Q_3 Determine the load impedance $\overline{Z}_L$ necessary for perfect matching. For this case determine the relationship between the active powers at the generator and load terminals; that is, find $\eta = P_L/P_G$.

Answers

Q_1 $\overline{Z}_l = j\omega L_e + \overline{Z}_{cond} = 163.6\, e^{j87.32°}\ \Omega/\text{m};\ \overline{Y}_t = j\omega C = j62.83\,\text{mS/m}.$

Q_2 $\overline{\gamma} = \alpha + j\beta = \sqrt{\overline{Z}_l \overline{Y}_t} = 3.206\, e^{j88.66°} = 0.075 + j3.205\,\text{m}^{-1}.$

$\alpha = 75\,\text{mNp/m};\ v = \omega/\beta = 1.96 \times 10^8\,\text{m/s}.$

Q_3

$$\overline{Z}_L = \overline{Z}_w = Z_w\, e^{j\theta_w} = \sqrt{\frac{\overline{Z}_l}{\overline{Y}_t}} = 51.03\, e^{-j1.34°} = 51 - j1.2\,\Omega$$

$$\begin{cases} \overline{U}(z) = \overline{U}_G\, e^{-(\alpha+j\beta)z} \\ \overline{I}(z) = \dfrac{\overline{U}_G}{Z_w\, e^{j\theta_w}}\, e^{-(\alpha+j\beta)z} \end{cases} \rightarrow P(z) = U_{rms}(z)\ I_{rms}(z)\ \cos(\theta_w) = \frac{(U_G)^2_{rms}}{Z_w} \cos(\theta_\omega)\, e^{-2\alpha z}$$

$$\eta = \frac{P_L}{P_G} = \frac{P_{z=l}}{P_{z=0}} = e^{-2\alpha l} = 0.86$$

Problem 9.9.3

A transfer matrix representation of a two-conductor lossy line of length l was obtained in (9.35) in Section 9.4.3:

$$\begin{bmatrix} \overline{U}_0 \\ \overline{I}_0 \end{bmatrix} = \underbrace{\begin{bmatrix} \cosh(\overline{\gamma} l) & \sinh(\overline{\gamma} l)\, \overline{Z}_w \\ \overline{Z}_w^{-1} \sinh(\overline{\gamma} l) & \cosh(\overline{\gamma} l) \end{bmatrix}}_{[T]} \begin{bmatrix} \overline{U}_l \\ \overline{I}_l \end{bmatrix}$$

Consider the equivalent 'T' circuit shown in Figure 9.31, whose component blocks are two identical impedances $\overline{Z}_x$ and one admittance $\overline{Y}_x$.

Q_1 Determine the transfer matrix of the equivalent circuit.

Q_2 By comparing the result obtained above with the transfer matrix of the line section, determine both $\overline{Z}_x$ and $\overline{Y}_x$.

Q_3 Assume that the transmission line is operated at a low-frequency regime; that is, assume that $l \ll \lambda$. Find the corresponding approximations for $\overline{Z}_x$ and $\overline{Y}_x$, and redraw the equivalent circuit.

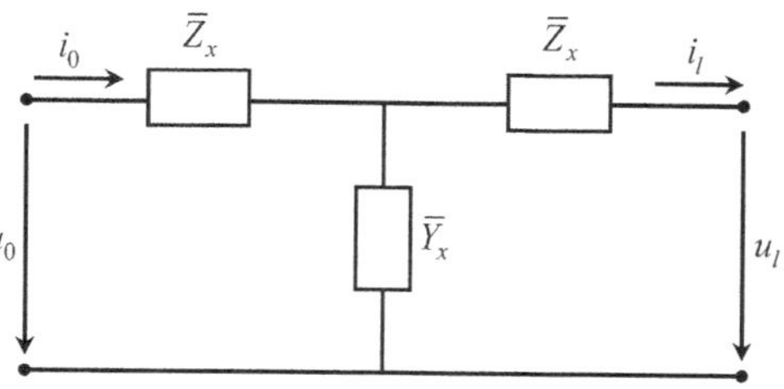

Figure 9.31 An equivalent model, with the shape of a 'T' circuit, aimed at representing a lossy line section of length l

Answers

Q_1

$$[T] = \begin{bmatrix} 1+\overline{Z}_x\overline{Y}_x & \overline{Z}_x(2+\overline{Z}_x\overline{Y}_x) \\ \overline{Y}_x & 1+\overline{Z}_x\overline{Y}_x \end{bmatrix}$$

Q_2

$$\overline{Y}_x = \overline{Z}_w^{-1}\sinh(\overline{\gamma}l); \quad \overline{Z}_x = \overline{Z}_w \tanh\frac{\overline{\gamma}l}{2}$$

Q_3 For $|\overline{\gamma}l| \ll 1$ we have

$$\sinh(\overline{\gamma}l) \simeq \overline{\gamma}l = \sqrt{\overline{Z}_l\overline{Y}_t}\, l \quad \text{and} \quad \tanh\frac{\overline{\gamma}l}{2} \simeq \frac{\overline{\gamma}l}{2} = \sqrt{\overline{Z}_l\overline{Y}_t}\,\frac{l}{2}$$

$$\overline{Y}_x = \overline{Z}_w^{-1}\sinh(\overline{\gamma}l) \simeq \sqrt{\frac{\overline{Y}_t}{\overline{Z}_l}} \times \sqrt{\overline{Z}_l\overline{Y}_t}\, l = \overline{Y}_t l = Gl + j\omega(Cl)$$

$$\overline{Z}_x = \overline{Z}_w \tanh\frac{\overline{\gamma}l}{2} \simeq \sqrt{\frac{\overline{Z}_l}{\overline{Y}_t}} \times \sqrt{\overline{Z}_l\overline{Y}_t}\,\frac{l}{2} = \frac{\overline{Z}_l l}{2} = \frac{Rl}{2} + j\omega\left(\frac{Ll}{2}\right)$$

The corresponding simplified equivalent circuit is shown in Figure 9.32.

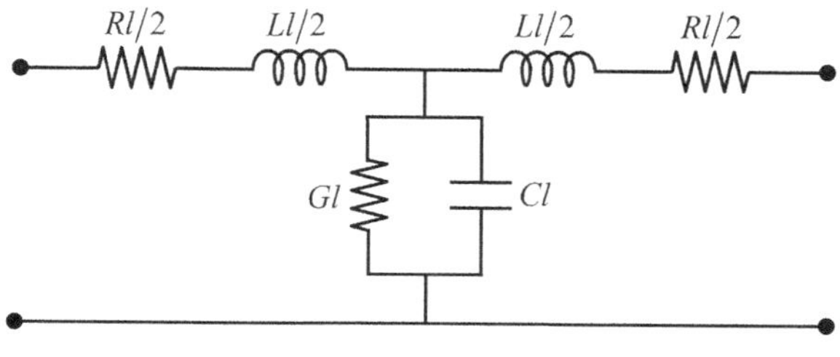

Figure 9.32 Identification of the constituent components of the 'T' circuit in Figure 9.31, for the special case of low-frequency regimes

Problem 9.9.4

A power distribution line of length $l = 100\,\text{km}$ is fed by an AC generator whose voltage is given by $u_G(t) = \sqrt{2}\,(U_G)_{rms} \cos \omega t$, where $(U_G)_{rms} = 20\,\text{kV}$, and $f = 50\,\text{Hz}$.

The per-unit-length parameters of the line are known: $R = 0.15\,\Omega/\text{km}$, $L = 2\,\text{mH/km}$, $G = 0$, $C = 6\,\text{nF/km}$.

Q_1 Determine the propagation constant $\overline{\gamma}$ and the wavelength λ. Check that $l \ll \lambda$.

Q_2 Taking into account the results obtained in Problem 9.9.3, show that $\left|\overline{Y}_x^{-1}\right| \gg \left|\overline{Z}_x\right|$.

Q_3 Assume that the receiving end of the line is short-circuited. Determine $\overline{I}_G$.

Q_4 Assume that the receiving end is left open. Determine $\overline{I}_G$.

Answers

Q_1 $\overline{\gamma} = \alpha + j\beta = \sqrt{(R + j\omega L)(j\omega C)} = (0.129 + j1.096) \times 10^{-6}\,\text{m}^{-1}$.

$\lambda = 2\pi/\beta = 5733\,\text{km} \gg l$.

Q_2 $\overline{Y}_x^{-1} = -j5.3\,\text{k}\Omega$; $\overline{Z}_x = 7.5 + j31.4 = 32.3\,e^{j76.57^\circ}\,\Omega$.

$\left|\overline{Y}_x^{-1}\right| \gg \left|\overline{Z}_x\right|$.

Q_3 By making use of the equivalent 'T' circuit in Figure 9.32 you get

$$\overline{I}_G \approx \frac{\overline{U}_G}{2\overline{Z}_x} = \frac{\overline{U}_G}{(R + j\omega L)\,l} = \sqrt{2}\;310\,e^{-j76.57^\circ}\ \text{A}$$

Q_4 By making use of the equivalent 'T' circuit in Figure 9.32 you get

$$\overline{I}_G \approx \overline{Y}_x \overline{U}_G = j\omega C l \overline{U}_G = \sqrt{2}\;3.77\,e^{j90^\circ}\ \text{A}$$

Problem 9.9.5

Consider a homogeneous lossless line, 50 cm long, whose dielectric medium is characterized by $\mu = \mu_0$ and $\varepsilon = 2.25\varepsilon_0$. An ideal generator is positioned at $y = l$, the corresponding voltage being given by $u_G(t) = \sqrt{2}\,(U_G)_{rms} \cos \omega t$, where $(U_G)_{rms} = 60\,\text{V}$, and $f = 0.1\,\text{GHz}$.

The characteristic wave resistance of the line is $R_w = 60\,\Omega$. As shown in Figure 9.33, the line is terminated by a load impedance $\overline{Z}_L$.

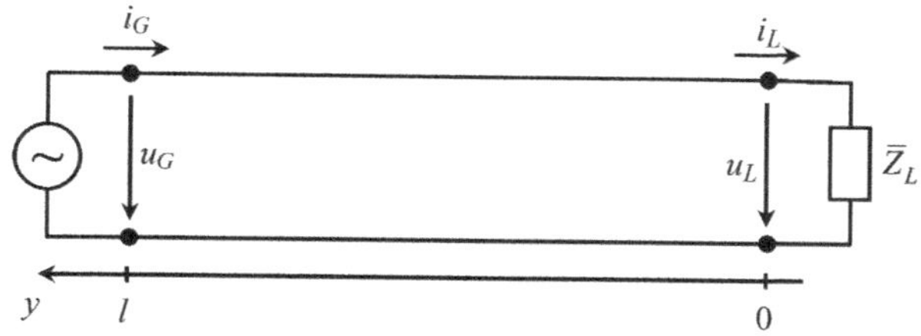

Figure 9.33 A homogeneous uniform lossless line terminated by a load impedance $\overline{Z}_L$

Q$_1$ Determine the propagation velocity v as well as the associated wavelength λ. Evaluate the per-unit-length L and C parameters.

Q$_2$ The load reflection coefficient is $\overline{\Gamma} = -1/3$. Determine the load impedance.

Q$_3$ Find the complex amplitude of the incident wave voltage at the load terminals, and determine the representative phasors for the voltage and current at $y = 0$ and $y = l$.

Q$_4$ Draw the standing wave patterns of $U_{rms}(y)$ and $I_{rms}(y)$ and, based on them, find SWR and the active power P transmitted along the line.

Q$_5$ Assume that the dielectric permittivity quadruples its original value. How would the parameters L, C, R_w, v and λ change? How would the new standing wave patterns look? Does the active power change?

Answers

Q$_1$ $v = 1/\sqrt{\mu_0 \varepsilon} = 2 \times 10^8\,\text{m/s}$; $\lambda = v/f = 2\,\text{m}$ (note that $l = \lambda/4$).

$$L = \frac{R_w}{v} = 300\,\text{nH/m};\ C = \frac{1}{vR_w} = 83.33\,\text{pF/m}$$

Q$_2$

$$\overline{Z}_L = R_w \frac{1+\overline{\Gamma}}{1-\overline{\Gamma}} = R_L = 30\,\Omega$$

Q$_3$

$$\begin{cases} \overline{U}_G = \overline{U}_i' e^{j\beta l}(1 + \overline{\Gamma} e^{-j2\beta l}) \\ \overline{I}_G = \dfrac{\overline{U}_i'}{R_w} e^{j\beta l}(1 - \overline{\Gamma} e^{-j2\beta l}) \end{cases} \quad \text{where } \overline{\Gamma} = -\frac{1}{3}, \quad \beta l = \frac{2\pi l}{\lambda} = \frac{\pi}{2}, \quad \overline{U}_G = \sqrt{2}\,60\ \text{V}$$

$$\overline{U}_i' = -j\tfrac{3}{4}\overline{U}_G = -j\sqrt{2}\,45\ \text{V}, \quad \overline{I}_G = \sqrt{2}\,0.5\,\text{A}.$$

$$\begin{cases} \overline{U}_L = \overline{U}_i'\,(1 + \overline{\Gamma}) = -j\sqrt{2}\,30\ \text{V} \\ \overline{I}_L = \dfrac{\overline{U}_i'}{R_w}\,(1 - \overline{\Gamma}) = -j\sqrt{2}\,1\,\text{A} \end{cases}$$

Q_4 The standing wave patterns are depicted in Figure 9.34.

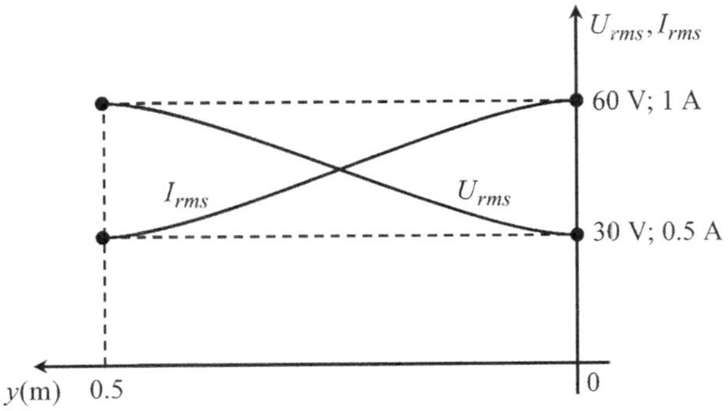

Figure 9.34 Voltage and current standing wave patterns (note that $l = \lambda/4$)

The SWR is 2; $P = (U_G)_{rms}(I_G)_{rms} = (U_L)_{rms}(I_L)_{rms} = 30\,\mathrm{W}$.

Q_5 L remains unaltered. C quadruples.

R_w, v and λ reduce to half of their original values (note that the new $R_w = 30\,\Omega = R_L$). Now the line is perfectly matched. The new standing wave patterns are horizontal lines, $U_{rms}(y) = 60\,\mathrm{V}$ and $I_{rms}(y) = 2\,\mathrm{A}$.

The active power quadruples, $P = 120\,\mathrm{W}$.

Problem 9.9.6

Consider a lossless transmission line of length l subjected to a time-harmonic regime. The line is left open at both ends. The generator that drives the line is positioned at an arbitrary location $y = y_G$ – see Figure 9.35.

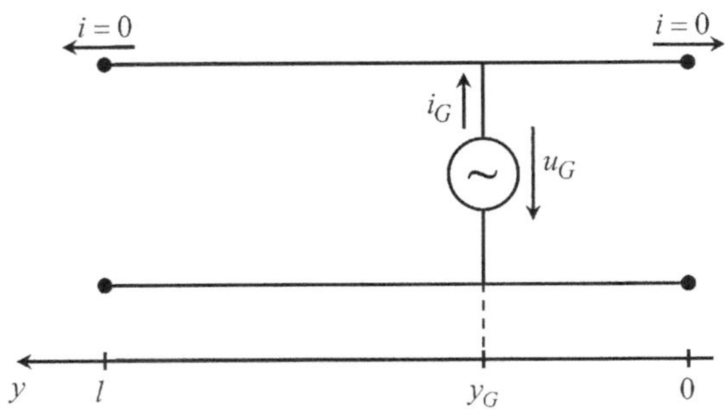

Figure 9.35 A two-conductor transmission line, left open at both ends, is excited by an ideal voltage generator placed at $y = y_G$

Q_1 Determine the complex amplitude of the generator current $\overline{I}_G$ as a function of its position.

Q_2 Consider the particular situation where the line length is an integer multiple of one half-wavelength, $l = n\lambda/2$. Find $\overline{I}_G$.

Answers

Q_1 Input admittance to the right of the generator:

$$\left(\overline{Y}_{in}\right)_R = j\frac{\tan(\beta y_G)}{R_w}$$

Input admittance to the left of the generator:

$$\left(\overline{Y}_{in}\right)_L = j\frac{\tan\left(\beta(l - y_G)\right)}{R_w}$$

$$\overline{I}_G = \overline{U}_G\left(\left(\overline{Y}_{in}\right)_R + \left(\overline{Y}_{in}\right)_L\right) = \frac{j\overline{U}_G}{R_w}\left(\tan(\beta y_G) + \tan(\beta(l - y_G))\right)$$

Q_2 From trigonometry: $\tan(\beta y) + \tan(\beta(l - y)) = \tan(\beta l) \times (1 - \tan(\beta y) \times \tan(\beta(l - y)))$. Therefore

$$\overline{I}_G = \frac{j\overline{U}_G}{R_w}\tan(\beta l) \times (1 - \tan(\beta y_G) \times \tan(\beta(l - y_G)))$$

For $l = n\lambda/2$ you have $\beta l = n\pi$, $\tan(\beta l) = 0$, $\overline{I}_G = 0$.

Problem 9.9.7

Consider a non-uniform transmission-line structure consisting of a chain connection of two uniform lossless coaxial cable sections, as shown in Figure 9.36. The first cable section, of length $l_1 = 10\,\text{m}$, is characterized by $R_{w_1} = 75\,\Omega$ and $v_1 = 2 \times 10^8\,\text{m/s}$. The second cable section, of length $l_2 = 7.5\,\text{m}$, is characterized by $R_{w_2} = 50\,\Omega$ and $v_2 = 1 \times 10^8\,\text{m/s}$. The working frequency is $f = 10\,\text{MHz}$. The load is an antenna that radiates 400 W and whose input impedance is $R_L = 100\,\Omega$.

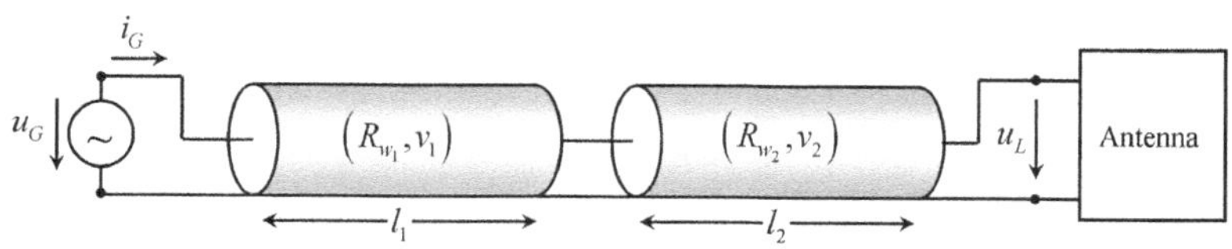

Figure 9.36 An antenna fed by the chain connection of two distinct coaxial cables

Q_1 Determine the transfer matrices describing both cable sections.

Q_2 Determine the transfer matrix of the global non-uniform transmission-line structure.

Q_3 Assuming that $\overline{U}_L = U_L e^{j0}$, determine $\overline{U}_G$ and $\overline{I}_G$.

Answers

Q_1

$$\lambda_1 = v_1/f = 20\,\text{m};\quad \beta_1 = 2\pi/\lambda_1 = \pi/10\,\text{rad/m};\quad \beta_1 l_1 = \pi.$$
$$\lambda_2 = v_2/f = 10\,\text{m};\quad \beta_2 = 2\pi/\lambda_2 = \pi/5\,\text{rad/m};\quad \beta_2 l_2 = 3\pi/2.$$

$$[T_1] = \begin{bmatrix} \cos(\beta_1 l_1) & jR_{w_1}\sin(\beta_1 l_1) \\ jR_{w_1}^{-1}\sin(\beta_1 l_1) & \cos(\beta_1 l_1) \end{bmatrix} = \begin{bmatrix} -1 & 0 \\ 0 & -1 \end{bmatrix}$$

$$[T_2] = \begin{bmatrix} \cos(\beta_2 l_2) & jR_{w_2}\sin(\beta_2 l_2) \\ jR_{w_2}^{-1}\sin(\beta_2 l_2) & \cos(\beta_2 l_2) \end{bmatrix} = \begin{bmatrix} 0 & -j50\,\Omega \\ -j0.02\,\text{S} & 0 \end{bmatrix}$$

Q_2

$$[T] = [T_1][T_2] = \begin{bmatrix} 0 & j50\,\Omega \\ j0.02\,\text{S} & 0 \end{bmatrix}$$

Q_3

$$\overline{U}_L = \sqrt{2}\,200\,\text{V};\quad \overline{I}_L = \sqrt{2}\,2\,\text{A}.$$

$$\begin{bmatrix} \overline{U}_G \\ \overline{I}_G \end{bmatrix} = [T]\begin{bmatrix} \overline{U}_L \\ \overline{I}_L \end{bmatrix} = j\sqrt{2}\begin{bmatrix} 100\,\text{V} \\ 4\,\text{A} \end{bmatrix}$$

Problem 9.9.8

Consider the connections shown in Figure 9.37, where a 100 MHz voltage generator simultaneously drives two lossless coaxial cables of length $l = 75\,\text{cm}$. The cables are geometrically identical, but they differ in their dielectric media properties, $\varepsilon_1 = 4\varepsilon_0$ and $\varepsilon_2 = 9\varepsilon_0$, $\mu_1 = \mu_2 = \mu_0$. The generator voltage is $u_G(t) = \sqrt{2}\,(U_G)_{rms}\cos(\omega t)$, with $(U_G)_{rms} = 48\,\text{V}$.

Q_1 For each cable, determine the propagation velocities v_1 and v_2 as well as the corresponding values of the wavelength λ_1 and λ_2.

Q_2 The per-unit-length inductance common to both cables is $L = 400\,\text{nH/m}$. Find the per-unit-length capacitances C_1 and C_2 of the cables, as well as the corresponding characteristic wave resistances R_{w_1} and R_{w_2}.

Q_3 Cable 1 is matched to its load. Cable 2 is short-circuited at $y = 0$. Determine R. Determine the load reflection coefficients $\overline{\Gamma}_1$ and $\overline{\Gamma}_2$ for both cables. Find the complex amplitudes of all of the voltages and currents marked in Figure 9.37.

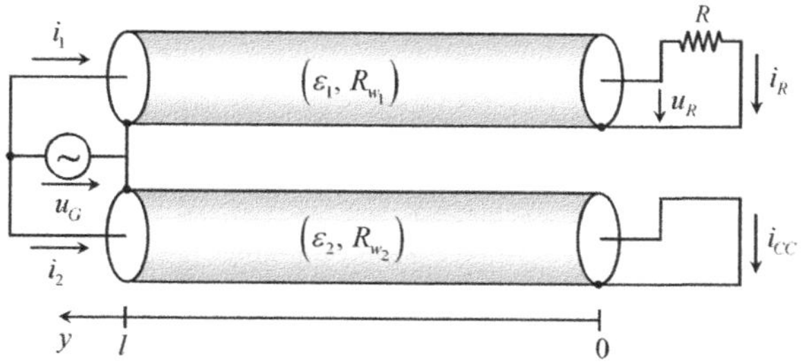

Figure 9.37 A voltage generator drives two lossless coaxial cables of length l. While one of the cables is matched, the other is short-circuited

Answers

Q_1 $v_k = c/\sqrt{\varepsilon_r} \rightarrow v_1 = 1.5 \times 10^8\,\text{m/s};\; v_2 = 1.0 \times 10^8\,\text{m/s}.$
$\lambda_1 = 1.5\,\text{m};\; \lambda_2 = 1.0\,\text{m}.$

Q_2

$$C_k = \frac{1}{v_k L} \rightarrow C_1 = 111\,\text{pF/m};\; C_2 = 250\,\text{pF/m}.$$

$$R_{w_k} = \sqrt{\frac{L}{C_k}} \rightarrow R_{w_1} = 60\,\Omega; \quad R_{w_2} = 40\,\Omega$$

Q_3 $R = R_{w_1} = 60\,\Omega.\; \overline{\Gamma}_1 = 0;\; \overline{\Gamma}_2 = -1.$
$\overline{U}_G = \sqrt{2}\,48\,e^{j0}\,\text{V};\; \overline{U}_R = \sqrt{2}\,48\,e^{j\pi}\,\text{V}.\; \overline{I}_1 = \sqrt{2}\,0.8\,e^{j0}\,\text{A};\; \overline{I}_R = \sqrt{2}\,0.8\,e^{j\pi}\,\text{A}.$
$\overline{I}_2 = 0;\; \overline{I}_{cc} = \sqrt{2}\,1.2\,e^{j\pi/2}\,\text{A}.$

Problem 9.9.9

Consider an aerial ($\mu = \mu_0$, $\varepsilon = \varepsilon_0$) lossless transmission line, of length $l = 1\,\text{m}$, subjected to a time-harmonic regime. The generator voltage is $u_G(t) = \sqrt{2}\,(U_G)_{rms}\cos(\omega t)$, with $(U_G)_{rms} = 100\,\text{V}$. The resistors included in the load circuit are characterized by $R = 1\,\text{k}\Omega$. As shown in Figure 9.38, two standing wave patterns were recorded, one with the switch S open, the other with the switch closed.

Q_1 Why can you say that the line is matched when S is closed? Determine R_w and v.

Q_2 Consider the situation with S open. By analyzing the standing wave pattern, determine the wavelength λ, the operating frequency f and the phase constant β. Evaluate $\overline{\Gamma}$ and SWR, and obtain $(U_{rms})_{\min}$.
Find the complex amplitudes $\overline{U}_G, \overline{I}_G, \overline{U}_L$ and $\overline{I}_L$.
Determine the standing wave pattern concerning the evolution of $I_{rms}(y)$.

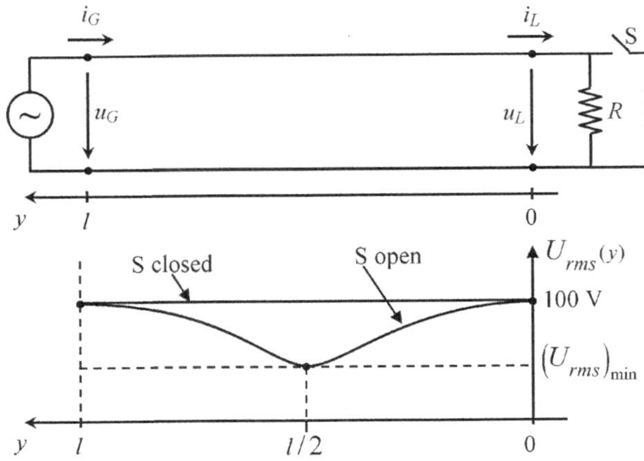

Figure 9.38 A transmission line is terminated by a load configuration that depends on the status of the switch. The corresponding voltage standing wave patterns are shown for S closed and open

Answers

Q_1 With S closed ($R_L = R/2 = 500\,\Omega$) the line is matched because $U_{rms}(y) =$ constant. $R_w = R_L = 500\,\Omega$; $v = 1/\sqrt{\mu_0\varepsilon_0} = 3 \times 10^8\,\text{m/s}$.

Q_2 From $\lambda/4 = l/2$, you get $\lambda = 2\,\text{m}$. $f = v/\lambda = 150\,\text{MHz}$. $\beta = 2\pi/\lambda = \pi\,\text{rad/m}$.

$$\overline{\Gamma} = \Gamma = \frac{R - R_w}{R + R_w} = \frac{1}{3}; \text{ SWR} = \frac{1+\Gamma}{1-\Gamma} = 2; \ (U_{rms})_{min} = \frac{(U_{rms})_{max}}{\text{SWR}} = 50\,\text{V}$$

$$\overline{U}_G = \sqrt{2}\,100\,\text{V}; \ \overline{U}_L = -\sqrt{2}\,100\,\text{V}; \ \overline{I}_G = \sqrt{2}\,100\,\text{mA}; \ \overline{I}_L = -\sqrt{2}\,100\,\text{mA}.$$

The standing wave pattern concerning the evolution of $I_{rms}(y)$ is characterized by two minima $(I_{rms})_{min} = 100\,\text{mA}$ at the line ends, $y = 0$ and $y = l$, and by a maximum $(I_{rms})_{max} = 200\,\text{mA}$ occurring at $y = l/2$.

Appendix A

Formulas from Vector Analysis

$$\mathbf{A}\cdot\mathbf{B} = \mathbf{B}\cdot\mathbf{A}$$

$$\mathbf{A}\times\mathbf{B} = -\mathbf{B}\times\mathbf{A}$$

$$(\mathbf{A}\times\mathbf{B})\cdot\mathbf{C} = \mathbf{A}\cdot(\mathbf{B}\times\mathbf{C})$$

$$\mathbf{A}\times(\mathbf{B}\times\mathbf{C}) = (\mathbf{A}\cdot\mathbf{C})\,\mathbf{B} - (\mathbf{A}\cdot\mathbf{B})\,\mathbf{C}$$

$$\frac{d}{d\xi}(\mathbf{A}\cdot\mathbf{B}) = \mathbf{A}\cdot\frac{d\mathbf{B}}{d\xi} + \frac{d\mathbf{A}}{d\xi}\cdot\mathbf{B}$$

$$\frac{d}{d\xi}(\mathbf{A}\times\mathbf{B}) = \mathbf{A}\times\frac{d\mathbf{B}}{d\xi} + \frac{d\mathbf{A}}{d\xi}\times\mathbf{B}$$

Differential Operators

$$\text{curl grad } V = 0$$

$$\text{div curl } \mathbf{A} = 0$$

$$\text{lap } V = \text{div grad } V$$

$$\text{lap } \mathbf{A} = (\text{lap } A_x)\ \vec{e}_x + \left(\text{lap } A_y\right)\ \vec{e}_y + (\text{lap } A_z)\ \vec{e}_z$$

$$\text{curl}\,(\text{curl } \mathbf{A}) = \text{grad div } \mathbf{A} - \text{lap } \mathbf{A}$$

$$\text{div}\,(V\mathbf{A}) = V \text{ div } \mathbf{A} + \text{grad } V\cdot\mathbf{A}$$

$$\text{curl}\,(V\mathbf{A}) = V \text{ rot } \mathbf{A} + \text{grad } V\times\mathbf{A}$$

$$\text{div}\,(\mathbf{A}\times\mathbf{B}) = (\text{curl } \mathbf{A})\cdot\mathbf{B} - \mathbf{A}\cdot(\text{curl } \mathbf{B})$$

Electromagnetic Foundations of Electrical Engineering J. A. Brandão Faria

Rectangular coordinates (x, y, z):

$$\text{grad}\, V = \frac{\partial V}{\partial x}\vec{e}_x + \frac{\partial V}{\partial y}\vec{e}_y + \frac{\partial V}{\partial z}\vec{e}_z$$

$$\text{lap}\, V = \frac{\partial^2 V}{\partial x^2} + \frac{\partial^2 V}{\partial y^2} + \frac{\partial^2 V}{\partial z^2}$$

$$\text{div}\, \mathbf{A} = \frac{\partial A_x}{\partial x} + \frac{\partial A_y}{\partial y} + \frac{\partial A_z}{\partial z}$$

$$\text{curl}\, \mathbf{A} = \left(\frac{\partial A_z}{\partial y} - \frac{\partial A_y}{\partial z}\right)\vec{e}_x + \left(\frac{\partial A_x}{\partial z} - \frac{\partial A_z}{\partial x}\right)\vec{e}_y + \left(\frac{\partial A_y}{\partial x} - \frac{\partial A_x}{\partial y}\right)\vec{e}_z$$

Cylindrical coordinates (r, ϕ, z):

$$\text{grad}\, V = \frac{\partial V}{\partial r}\vec{e}_r + \frac{1}{r}\frac{\partial V}{\partial \phi}\vec{e}_\phi + \frac{\partial V}{\partial z}\vec{e}_z$$

$$\text{lap}\ V = \frac{1}{r}\frac{\partial}{\partial r}\left(r\frac{\partial V}{\partial r}\right) + \frac{1}{r^2}\frac{\partial^2 V}{\partial \phi^2} + \frac{\partial^2 V}{\partial z^2}$$

$$\text{div}\ \mathbf{A} = \frac{1}{r}\left(\frac{\partial (rA_r)}{\partial r} + \frac{\partial A_\phi}{\partial \phi} + r\frac{\partial A_z}{\partial z}\right)$$

$$\text{curl}\ \mathbf{A} = \left(\frac{1}{r}\frac{\partial A_z}{\partial \phi} - \frac{\partial A_\phi}{\partial z}\right)\vec{e}_r + \left(\frac{\partial A_r}{\partial z} - \frac{\partial A_z}{\partial r}\right)\vec{e}_\phi + \frac{1}{r}\left(\frac{\partial (rA_\phi)}{\partial r} - \frac{\partial A_r}{\partial \phi}\right)\vec{e}_z$$

Spherical coordinates (r, θ, ϕ):

$$\text{grad}\ V = \frac{\partial V}{\partial r}\vec{e}_r + \frac{1}{r}\frac{\partial V}{\partial \theta}\vec{e}_\theta + \frac{1}{r\sin\theta}\frac{\partial V}{\partial \phi}\vec{e}_\phi$$

$$\text{lap}\ V = \frac{1}{r^2}\frac{\partial}{\partial r}\left(r^2\frac{\partial V}{\partial r}\right) + \frac{1}{r^2\sin\theta}\frac{\partial}{\partial \theta}\left(\sin\theta\frac{\partial V}{\partial \theta}\right) + \frac{1}{(r\sin\theta)^2}\frac{\partial^2 V}{\partial \phi^2}$$

$$\text{div}\ \mathbf{A} = \frac{1}{r^2}\frac{\partial (r^2 A_r)}{\partial r} + \frac{1}{r\sin\theta}\left(\frac{\partial (A_\theta \sin\theta)}{\partial \theta} + \frac{\partial A_\phi}{\partial \phi}\right)$$

$$\text{curl}\ \mathbf{A} = \frac{1}{r\sin\theta}\left(\frac{\partial (A_\phi \sin\theta)}{\partial \theta} - \frac{\partial A_\theta}{\partial \phi}\right)\vec{e}_r + \frac{1}{r}\left(\frac{1}{\sin\theta}\frac{\partial A_r}{\partial \phi} - \frac{\partial (rA_\phi)}{\partial r}\right)\vec{e}_\theta + \frac{1}{r}\left(\frac{\partial (rA_\theta)}{\partial r} - \frac{\partial A_r}{\partial \theta}\right)\vec{e}_\phi$$

Curl Theorem (Stokes Theorem)

$$\int_{S_s} \text{curl}\ \mathbf{A}\cdot\mathbf{n}_S\ dS = \oint_{\mathbf{S}} \mathbf{A}\cdot d\mathbf{s}$$

Divergence Theorem (Gauss Theorem)

$$\int_V \text{div}\ \mathbf{A}\ dV = \int_{S_V} \mathbf{A}\cdot\mathbf{n}_o\ dS$$

Appendix B

Lorentz Transformation

Movement along the z axis:

$$\begin{cases} x = x' \\ y = y' \\ z = \dfrac{1}{\beta}(z' + vt') \\ t = \dfrac{1}{\beta}\left(t' + \dfrac{vz'}{c^2}\right) \end{cases}$$

$$\beta = \sqrt{1 - \left(\frac{v}{c}\right)^2}$$

$$c = \frac{1}{\sqrt{\mu_0 \varepsilon_0}}$$

Electromagnetic Foundations of Electrical Engineering J. A. Brandão Faria

Appendix C

Elements of Complex Algebra

$$\bar{z} = a + jb = z\, e^{j\theta} = \sqrt{a^2 + b^2}\, e^{j\,\arctan(b/a)}$$

$$\bar{z}_1 + \bar{z}_2 = (a_1 + a_2) + j(b_1 + b_2)$$

$$\bar{z}_1 \bar{z}_2 = z_1 z_2\, e^{j(\theta_1 + \theta_2)}$$

$$\bar{z}^* = (a - jb) = z\, e^{-j\theta}$$

$$\bar{z}\bar{z}^* = z^2$$

$$\frac{1}{\bar{z}} = \frac{e^{-j\theta}}{z}$$

$$\sqrt{\bar{z}} = \pm\sqrt{z}\, e^{j(\theta/2)}$$

$$\bar{z}^n = z^n\, e^{jn\theta}$$

$$\cos\theta = \frac{e^{+j\theta} + e^{-j\theta}}{2}$$

$$\sin\theta = \frac{e^{+j\theta} - e^{-j\theta}}{2j}$$

$$\cos(j\theta) = \cosh\theta$$

$$\sin(j\theta) = j\sinh\theta$$

Electromagnetic Foundations of Electrical Engineering J. A. Brandão Faria

Appendix D

Elements of Fourier Analysis

Periodic signals of period T:

$$s(t) = s_{av} + \sum_{k=1}^{\infty} S_k \cos(\omega_k t - \phi_k); \;\; \omega_k = 2\pi k/T$$

$$s_{av} = \frac{1}{T}\int_0^T s(t)\,dt; \;\; S_k = \sqrt{A_k^2 + B_k^2}; \;\; \phi_k = \arctan\left(B_k/A_k\right)$$

$$A_k = \frac{2}{T}\int_0^T s(t)\cos(\omega_k t)\,dt; \;\; B_k = \frac{2}{T}\int_0^T s(t)\sin(\omega_k t)\,dt$$

Non-periodic signals:

$$s(t) = \int_0^{\infty} S(\omega)\cos\left(\omega t - \phi(\omega)\right)d\omega$$

$$S(\omega) = \sqrt{A^2(\omega) + B^2(\omega)}; \;\; \phi(\omega) = \arctan\left(B(\omega)/A(\omega)\right)$$

$$A(\omega) = \frac{1}{\pi}\int_{-\infty}^{\infty} s(t)\cos(\omega t)\,dt; \;\; B(\omega) = \frac{1}{\pi}\int_{-\infty}^{\infty} s(t)\sin(\omega t)\,dt$$

Electromagnetic Foundations of Electrical Engineering J. A. Brandão Faria

Bibliography

Publications Involving the Author Concerning the Subject Matter of This Book

Books

J. Brandão Faria, *Multiconductor Transmission-Line Structures*, John Wiley & Sons, Inc. (USA), 1993.

J. Brandão Faria, *Óptica*, Editorial Presença (Portugal), 1995.

Journal Publications

J. Brandão Faria, 'Abnormal coupling between a two-wire line and a single conductor', *ETZ Archiv*, vol. 10, no. 1, pp. 19–23, 1988.

J. Brandão Faria, 'Capacitor modeling at high frequencies: a pedagogical example of the application of distributed circuit analysis for introductory electromagnetics or microwave courses', *IEEE Transactions on Education*, vol. 35, no. 3, pp. 214–216, 1992.

J. Brandão Faria, 'On the time-domain transmission-line equations', *Microwave and Optical Technology Letters*, vol. 22, no. 3, pp. 194–197, 1999.

J. Brandão Faria, 'A new magnetic displacement sensor and linear actuator device', *Journal of Applied Physics*, vol. 87, no. 9, pp. 7076–7078, 2000.

J. Brandão Faria, 'On the segmentation method used for analyzing nonuniform transmission lines: application to the exponential line', *European Transactions on Electrical Power*, vol. 12, no. 5, pp. 361–368, 2002.

J. Brandão Faria, 'On the transmission matrix of 2n-port reciprocal networks', *Microwave and Optical Technology Letters*, vol. 33, no. 3, pp. 151–154, 2002.

J. Brandão Faria, 'Conservation laws for time-domain signals propagating along multiconductor transmission-line structures', *IEE Proceedings – Microwaves, Antennas and Propagation*, vol. 150, no. 5, pp. 375–378, 2003.

J. Brandão Faria, 'The polarization ellipsoid revisited', *IEEE EMC Society Newsletter*, no. 198, pp. 38–39, 2003.

J. Brandão Faria, 'A pedagogical example of the application of transmission-line theory to study conductor grounding effects', *IEEE Transactions on Education*, vol. 47, no. 1, pp. 141–145, 2004.

J. Brandão Faria, 'On the partial-capacitance scheme for multiple conductor systems', *Microwave and Optical Technology Letters*, vol. 47, no. 4, pp. 346–349, 2005.

J. Brandão Faria, 'Electromagnetic field approach to the modeling of disk-capacitor devices', *Microwave and Optical Technology Letters*, vol. 48, no. 8, pp. 1467–1472, 2006.

J. Brandão Faria, 'Permittivity and conductivity evaluation of cylindrically shaped homogeneous bodies: basic principles and quadrupole system measurement method', *International Review of Electrical Engineering*, vol. 2, no. 5, pp. 648–654, 2007.

J. Brandão Faria, 'Application of a harmonic expansion method approach to the computation of L and C matrices for open-boundary inhomogeneous multiconductor transmission-line structures with strong proximity effects present', *Electrical Engineering—Archiv für Elektrotechnik*, vol. 90, no. 5, pp. 313–321, 2008.

J. Brandão Faria, 'The role of Poynting's vector in polyphase power calculations', *European Transactions on Electrical Power*, pp. 1–6. Published online 29 February 2008, www.interscience.wiley.com/journal/etep. DOI 10.1002/etep.248.

J. Brandão Faria and M. Almeida, 'Accurate calculation of magnetic-field intensity due to overhead power lines with or without mitigation loops with or without capacitor compensation', *IEEE Transactions on Power Delivery*, vol. 22, no. 2, pp. 951–959, 2007.

J. Brandão Faria and J. Borges da Silva, 'Wave propagation in polyphase transmission lines: a general solution to include cases where ordinary modal theory fails', *IEEE Transactions on Power Delivery*, vol. 1, no. 2, pp. 182–189, 1986.

J. Brandão Faria and M. Guerreiro Neves, 'Accurate evaluation of indoor triplex cable capacitances taking conductor proximity effects into account', *IEEE Transactions on Power Delivery*, vol. 21, no. 3, pp. 1238–1244, 2006.

Further Reading

Classics

C. Balanis, *Advanced Engineering Electromagnetics*, John Wiley & Sons, Inc., 1989.

A. Frühling, *Cours d'Electricité*, Dunod, 1966.

W. Jackson, *Classical Electrodynamics*, John Wiley & Sons, Inc., 1975.

B. Jancewicz, *Multivectors and Clifford Algebra in Electrodynamics*, World Scientific, 1988.

K. Krauss and K. Carver, *Electromagnetics*, McGraw-Hill, 1973.

K. Kupfmüller, *Electricité Théorique et Appliquée*, Dunod, 1959.

J. Maxwell, *A Treatise on Electricity and Magnetism*, Dover, 1954.

M. Plonus, *Applied Electromagnetics*, McGraw-Hill, 1978.

S. Ramo, J. Whinnery and T. Duzer, *Fields and Waves in Communication Electronics*, John Wiley & Sons, Inc., 1965.

K. Simonyi, *Foundations of Electrical Engineering*, Pergamon Press, 1963.

A. Sommerfeld, *Electrodynamics*, Academic Press, 1950.

J. Stratton, *Electromagnetic Theory*, McGraw-Hill, 1941.

Recent Titles

W. Hayt and J. Buck, *Engineering Electromagnetics*, McGraw-Hill, 2005.

D. Misra, *Practical Electromagnetics*, John Wiley & Sons, Inc., 2007.

C. Paul, *Electromagnetics for Engineers*, John Wiley & Sons, Inc., 2004.

Z. Popovic, *Introductory Electromagnetics*, Prentice Hall, 1999.

E. Rothwell and M. Cloud, *Electromagnetics*, CRC Press, 2001.

F. Ulaby, *Fundamentals of Applied Electromagnetics*, Prentice Hall, 2006.

S. Wentworth, *Fundamentals of Electromagnetics with Engineering Applications*, John Wiley & Sons, Inc., 2005.

Index

Printed and bound by CPI Group (UK) Ltd, Croydon, CR0 4YY
07/10/2024
14569933-0002